202-331-7775

ASIA'S CLEAN REVOLUTION
INDUSTRY, GROWTH AND THE ENVIRONMENT

Contributing Editors: David P. Angel and Michael T. Rock

❝ In this new century and millennium, the people of the world need a vision of possibility and promise; of prosperity and wellbeing for all. Asia has offered the world that vision in the past. The great ancient Asian societies gave the world science, music, poetry, elemental foundations of mathematics. And in the recent past Asia has offered us the Green Revolution and the Quality Revolution, both aimed at addressing some of society's most pressing needs. Now it is time for a Clean Revolution. In collaboration with the United States and with the support of programmes such as those offered by the US–Asia Environmental Partnership, Asia again can lead the world, offering today the vision of a clean and healthy future that is so urgently needed. This book, while lucidly presenting the challenges, also offers a compelling compendium of the programmes that work. I hope leaders the world over will absorb its message and chart the course for a Clean Revolution now. ❞

Kathleen McGinty, Chair,
US Council on Environmental Quality (1995–98)

❝ The authors address Asia's burgeoning environmental concerns by going to the roots of environmental performance: the policies of trade, technology, investment and governance. By seeking to improve the environmental content of these roots and focusing on the 'intensity' of materials, energy, pollution and waste in economic growth, the book recognises the critical importance of policy creativity and industrial efficiency—prior to the emplacement of a whole new generation of industrial capacity. ❞

The Honorable Don Ritter, ScD, US Congressman and now President of the National Environmental Policy Institute, Washington, DC

❝ Global shifts in industrial regimes are creating new growth opportunities and creating new environmental and social inequities at the same time. Coming to grips with the complexities of economic development, environmental quality and social equity requires a broad, research-based foundation, and this book provides the needed grounding for policy-makers, scholars, activists and business strategists. ❞

Kurt Fischer, Co-founder and US Co-ordinator,
The Greening of Industry Network

❝ *Asia's Clean Revolution* puts forward a superb diagnosis of the environmental and economic challenges presented by rapid urbanisation and industrialisation in Asia. The book should be on the reading list of policy-makers in Asia, the United States, and around the world. ❞

James Gustave Speth, Dean of the Yale School of Forestry and Environmental Studies; former head of the United Nations Development Programme

---

❝ The Asian crisis will soon be over and in the coming years Asia will be back on a path to development. Past development has brought resource degradation and pollution in Asia. And this has occurred while industry is still in its infancy in the region. The future development path will be based on more intensive agricultural development, a broader base of industrial development and accompanied urban growth. We must not pursue this development path 'business-as-usual', but must follow the road toward clean shared growth in Asia. This book shows the path worth taking in meeting the sustainable development challenges of the future. ❞

Emil Salim, Chairman, National Economic Council, Republic of Indonesia

---

❝ *Asia's Clean Revolution* makes a most profound point: that reconciling economic and environmental goals will be possible only through a transformation in technology and industrial organisation—a shift unprecedented in scope and pace, to new technologies and systems that will dramatically reduce environmental impact per unit of prosperity. In other words, nothing short of a 'clean revolution' will do.

This book constitutes a superb diagnosis and prescription for a more sustainable growth path in the context of the rapidly evolving Asia region. The various ideas and proposals represent an important, transformative vision for the changing global circumstance—as well as a positive, powerful initiative for the environment. ❞

Owen Cylke, US–Asia Environmental Partnership and Winrock International, USA, and Somporn Kamolsiripichaiporn, Chulalongkorn University, Thailand

---

❝ As East Asia recovers from the devastating financial crisis of 1997–98, it will be crucial for the region to avoid the temptation of short-sighted environmental neglect in order to retrieve industrial competitiveness, e.g. by co-ordinating environmental policies. Countries in the region urgently need to more effectively address the fast-growing problem of industrial pollution. While the more developed countries in the region have had some success in developing competent and effective environmental authorities, lower-income countries remain keen to attract 'dirty' low-tech industries. This timely volume draws important lessons from recent experience in the region to suggest appropriate policies for achieving sustainable development on the basis of late industrialisation. Most crucially, new firms and industries will need special help and incentives to effectively integrate appropriate environmental considerations into accelerated technological development policies. ❞

K.S. Jomo, Professor of Economics, University of Malaya

ASIA'S CLEAN REVOLUTION

INDUSTRY, GROWTH AND THE ENVIRONMENT

CONTRIBUTING EDITORS
DAVID P. ANGEL and MICHAEL T. ROCK

Greenleaf
PUBLISHING
2000

© 2000 Greenleaf Publishing Limited

Published by Greenleaf Publishing Limited
Aizlewood's Mill
Nursery Street
Sheffield S3 8GG
UK

Typeset by Greenleaf Publishing.
Printed and bound, using acid-free paper from managed forests, by
Creative Print & Design (Wales), Ebbw Vale.

All rights reserved. No part of this publication may be reproduced,
stored in a retrieval system, or transmitted, in any form or by any
means, electronic, mechanical, photocopying, recording or otherwise,
without the prior permission in writing of the publishers.

British Library Cataloguing in Publication Data:

  Asia's clean revolution : industry, growth and the
    environment
    1. Industries - Asia - Environmental aspects   2. Economic
    development - Environmental aspects - Asia
    I. Angel, David P.   II. Rock, Michael T.
    333.7'095

  ISBN 1874719330

CONTENTS

Acknowledgements   7

Preface   8
    Owen Cylke and Somporn Kamolsiripichaiporn

1. Toward clean shared growth in Asia   11
    David P. Angel, Michael T. Rock and Tubagus Feridhanusetyawan

PART 1: *Framing the Issues*   38
    David P. Angel and Michael T. Rock

2. Technology and environmental performance: leveraging growth and sustainability   41
    George R. Heaton, Jr, and Budy Resosudarmo

3. Globalisation and the environment in Asia: linkages, impacts and policy implications   63
    Daniel Esty, Mari Pangestu and Hadi Soesastro

4. Public policies to promote cleaner shared industrial growth in East Asia   88
    Michael T. Rock, Ooi Giok Ling and Victor Kimm

5. Industrialising cities and the environment in Pacific Asia: toward a policy framework and agenda for action   104
    Michael Douglass and Ooi Giok Ling

6. Civil society and the future of environmental governance in Asia   128
    Lyuba Zarsky and Simon S.C. Tay

PART 2: *Case Studies in Innovation*   155
    David P. Angel and Michael T. Rock

7. **Putting pressure on polluters: Indonesia's PROPER programme**
A CASE STUDY FOR THE HARVARD INSTITUTE FOR INTERNATIONAL
DEVELOPMENT 1997 ASIA ENVIRONMENTAL ECONOMICS POLICY SEMINAR   157
    Shakeb Afsah and Jeffrey R. Vincent

8. Water pollution abatement in Malaysia   173
    Jeffrey R. Vincent and Rozali Mohamed Ali with Khalid Abdul Rahim

9. Toward more sustainable development:
the environment and industrial policy in Taiwan   194
    Michael T. Rock

10. Measuring up: toward a common framework for
tracking corporate environmental performance   209
    Daryl Ditz and Janet Ranganathan

Postscript   246
    Melito Salazar and Warren Evans

Bibliography   251

List of Abbreviations   267

Authors' Biographies   270

Index   274

ACKNOWLEDGEMENTS

We gratefully acknowledge support provided by the US–Asia Environmental Partnership for the work underlying this book.

Chapters 7 and 8 were previously available as case studies of the Harvard Institute for International Development (HIID) and are reproduced here with the permission of the HIID.

Chapter 9 was previously published as 'Toward Sustainable Development: The Environment and Industrial Policy in Taiwan', *Development Policy Review* 14 (1996): 255-72 and is reproduced here with the permission of Basil Blackwell Publishers.

Chapter 10 was previously published as a research report of the World Resources Institute (WRI)—D. Ditz and J. Ranganathan, *Measuring Up: A Common Framework for Tracking Corporate Environmental Performance* (Washington, DC: World Resources Institute, 1997)—and is reproduced here with the permission of the WRI.

PREFACE

Owen Cylke and Somporn Kamolsiripichaiporn

The world's economic future will be determined in significant part by what happens in rapidly industrialising countries—particularly in Asia—where capital, consumption, manufacturing and technology will again become a growth engine for the global economy. Significantly, the world's environmental future is equally dependent on what happens in Asia where economic, population and urban growth and environmental stress are converging most forcefully.

Recognising that resolution of the tension between growth and the environment early in the third millennium may be the test by which posterity will most stringently grade our generation, the US–Asia Environmental Partnership organised a set of international conversations in 1998 to develop a better understanding of the Asian context and the rapidly changing global circumstance. Based on those conversations, and in close collaboration with the Greening of Industry Network, the Partnership felt confident in 1999 to step back from its project and programme work to look at frameworks for policy intervention, commissioning a series of papers both by Asian and by American authors. This book is the product.

In considering the issues and rather remarkable ideas put forward in the various chapters of this book, it may be useful to step back from the Asia Crisis to recall that from 1965–90 the 23 countries of East Asia grew faster than all other regions of the world. These same countries were also unusually successful at sharing the fruits of growth. As a result of rapid, shared growth, living standards improved dramatically over a 30-year period. Indeed, there is good reason for the period to have been characterised as an 'economic miracle'. Asia will probably return to this kind of growth regime, maybe not to 8%–10% growth, but more likely 7%–8% than 4%–5%.

All the economies in Asia today are urban-organised and industrial-led. Indeed, the major development forces at work in the region will only accentuate these trends over the next 20 years. Think about it. Some 80% of the industrial infrastructure in just 20 years will consist of plant that is not on the ground today. Where 50% of global manufacturing already takes place outside the OECD (Organisation for Economic

Co-operation and Development) economies, perhaps as much as 75% will take place in the industrialising economies of Asia and Latin America by 2020.

Indeed, Asia may account for as much as 60% of world income by 2020. Perhaps 400 million will be added to the region's population over the next 20 years, and maybe 200 million will move into the consuming class and into cities. It is the *momentum* of these forces, not just the numbers, that suggests that the world's environmental future will be determined in significant part by what happens in Asia and in other rapidly industrialising and urbanising regions. The subject of this book, then, is truly important.

Context and perspective is particularly important, and here the book is especially useful. The OECD economies are not developing economies, and North America is not Asia. The drivers of a cleaner environment may emerge from different sources than they did in the OECD, from concerns about scarcity in addition to public health, suggesting a policy priority for efficiency over pollution. The chapters in this book are filled with this kind of insight, arising as they do from intercontinental co-authorship. Similarly, this book calls our attention to the reality that today is not tomorrow, and that now is not then. Simple, perhaps, but recall that Rachel Carson wrote *Silent Spring* in 1962 and that the *Brundtland Report* was issued in 1987 (WCED 1987). The Greening of Industry Network was launched in Asia in 1998, and the authors of the chapters in this book are debuting their work just today—at the dawn of a new millennium.

Against this backdrop, *Asia's Clean Revolution* makes a most profound point—that reconciling economic and environmental goals will be possible only through a *transformation* in technology and industrial organisation—a shift unprecedented in scope and pace, to new technologies and systems that will dramatically reduce environmental impact per unit of prosperity. In other words, nothing short of a 'clean revolution' will do. The book also succeeds in laying out the contours of the challenge and the full range of policy approaches necessary to reconcile economic and environmental goals. Quite possibly, political leaders in Asia will face no greater challenge in the decades ahead—a challenge that will require continuing effort at the highest levels of governance, and including international co-operation on a scale seldom seen. The chapters in this book suggest a template for the required effort.

The opportunity for transformation, of course, does not present itself everywhere with equal promise. The OECD economies operate a largely ageing industrial infrastructure that depends on control, retrofit and remediation strategies to deal with pollution. Asia, on the other hand, appears to have most of the ingredients for a 'clean revolution': increasing public awareness and concern for the environment, improving environmental regulation and enforcement, rapid industrial growth (from a limited base), and increasing pressure for a cleaner environment from the community on the one hand and from the international marketplace on the other.

The book also identifies the new forces that will drive development and sustainability in the 21st century. Globalisation is obviously one of them. It is linking economies and cultures and empowering individuals and groups as never before. It is opening immense new opportunities, and at the same time generating a potent backlash. And this combination of forces is compelling more and more governments and companies to recognise that sustainable development, industrial transformation, a cleaner environment, increased corporate responsibility, greater citizen participation and concern for those left out of the growth phenomenon must move from a *choice* to an *imperative*. Certainly, there is no one code, no one set of guidelines, no one size that fits all; but this book reflects the growing recognition that there are principles to shape and work by, best practices to consider and adopt and new partners with whom to co-operate and innovate.

Asia's Clean Revolution presents a template for the desired transformation. The core ideas are sound, illustratively:

1. Understanding the environmental challenge in Asia to include energy, water, materials and waste in addition to pollution
2. The importance of bringing economic growth and the environment into strategic alignment
3. Factoring considerations of environmental quality into new investment
4. Encouraging public policy to move beyond the environmental consequences of economic activity to the very operations of that activity itself
5. Refocusing public policy on environmental behaviour and drivers
6. Building on and strengthening the pro-environmental pressures emerging from the global marketplace and local communities

The policy agenda is equally sound:

1. Getting the signals straight by restructuring the macro framework and establishing sustainability goals
2. Rationalising environmental policy by strengthening the regulatory baseline while at the same time introducing performance-based policies and leveraging those pro-environmental pressures emerging from the community and international marketplace
3. Policy integration by expanding the range of governmental agencies promoting a cleaner environment and sustainability, such as ministries of industry
4. Improving governance, at the international, regional, national, local and corporate levels
5. Leveraging the information revolution by making measurement matter

In reading the chapters in this book, we discerned almost 40 action categories under each of the five policy rubrics, and perhaps 80 specific initiatives that governments, corporations and communities can take.

This book constitutes a superb diagnosis and prescription for a more sustainable growth path in the context of the rapidly evolving Asia region. The various ideas and proposals represent an important, transformative vision for the changing global circumstance—as well as a positive, powerful initiative for the environment.

TOWARD CLEAN SHARED GROWTH IN ASIA

David P. Angel, Michael T. Rock and Tubagus Feridhanusetyawan

The environmental problems of developing Asia are now well documented. The combination of rapid urban–industrial growth and *de facto* 'grow now and clean up later' environmental strategies have resulted in low energy efficiency within industry, natural resource depletion, materials-intensive production, polluted rivers and groundwater supplies and unhealthy air in many Asian cities. According to the Asian Development Bank (ADB 1997), average levels of air particulates in Asia over the period 1991–95 were approximately fivefold higher than in OECD (Organisation for Economic Co-operation and Development) countries and twice the world average (see Table 1.1). Measures of water pollution, such as biochemical oxygen demand (BOD) levels and levels of suspended solids, were also substantially above world averages. Prior to the current crisis,

Table 1.1 **Environmental conditions in Asia, average 1991–95**

Source: Lohani 1998

	Asia	Africa	Latin America	OECD	World
Air pollution					
Particulates (mg/m^3)	248	29	40	49	126
SO_2 (mg/m^3)	0.023	0.015	0.014	0.068	0.059
Water pollution					
Suspended solids (mg/l)	638	224	97	20	151
BOD levels (mg/l)	4.8	4.3	1.6	3.2	3.5

energy demand in Asia was doubling every 12 years and demand for electricity was growing two to three times faster than gross domestic product (GDP), resulting in major increases in greenhouse gas emissions.

Important efforts to address many of these environmental problems are currently under way within the region. Over the past ten years, greater resources have been committed to pollution control and to the remediation of existing pollution. Investments have also been made in urban infrastructure, particularly in water supply and sanitation systems, and, to a lesser extent, in mass transit systems. In addition, environmental regulatory systems are being strengthened in a majority of countries within the region. There is also a variety of initiatives under way in many East Asian countries which use market-based instruments, information disclosure, public participation, clean-technology diffusion and other innovative policy approaches. Although actual results vary widely from country to country, evidence suggests that enhancements in regulatory activity are yielding important incremental improvements in the environmental performance of industrial firms within the region (O'Connor 1994; Rock 1996b; USAEP 1997; Vincent 1993; World Bank 1999).

What these regulatory initiatives have not done is change the basic structural relationship between urban–industrial growth and the environment, and the attendant trajectory of increased energy and materials use, pollution and resource depletion. Indeed, environmental policy as we know it today—in Asia and in the rest of the world—is not directed towards such a transformative goal. Born of the concerns and expectations of OECD economies in the 1960s and 1970s (such as improving local and national environmental quality, minimising health and safety risks and reducing pollution from a mature urban–industrial capital stock), environmental policy is only beginning to come to terms with the global sustainability challenge. To be sure, there have been a multitude of important innovations in environmental regulatory approaches and policy instruments, as well as increased efforts to tailor regulatory approaches to local economic, social and political conditions. But the core focus of environmental policy for industry continues to be that of reducing negative environmental 'outputs' (such as pollution and waste) and on improving local environmental 'outcomes' (such as air and water quality). What the growing body of scientific evidence on deforestation, climate change and resource depletion has taught us, however, is that the sustainability challenge goes beyond pollution and declining local environmental quality to human-induced degradation of the biosphere. Although current environmental policy yields incremental improvements in energy and materials efficiency (as a means of reducing pollution and improving local environmental quality), these plant-level improvements are typically overridden by the scale effects of energy—and materials—intensive economic development.

Nowhere in the world is the challenge of 'changing course'—of shifting to patterns of economic development that are less intensive in use of energy and materials and in production of pollution and waste—more urgent than in the rapidly industrialising economies of Asia. Most of Asia is in the midst, not at the end, of an urban–industrial-led development transition unparalleled in its scale and intensity. Gross national product (GNP) per capita in East Asia has grown at an average annual rate of 5.5% over the past

30 years, or more than twice the rate of the OECD economies (World Bank 1998a). This has in many ways been a development triumph, reducing poverty and improving life expectancy within the region. But the average GNP per capita of Indonesia, Malaysia, the Philippines and Thailand (the so-called second-tier newly industrialised countries [NICs]) in 1996 was still a modest $2,392, or approximately one-tenth of that in the high-income OECD economies. On average, roughly half of the workforce in these four countries in 1990 was still employed in agriculture. Six out of ten people live outside of urban areas.

Future increases in per capita income within the region will almost certainly entail continuing massive shifts in economic structure, from agriculture into industry and from rural areas into cities. What this is likely to result in is the largest increase in urban population in human history. By one estimate, the urban population in the East Asian NICs, including China, will increase from about 550 million in 1995 to almost 1.2 billion in 2025 (WRI 1997). Asia's share of global output, which was roughly 10% in 1950 and 30% in 1995, is expected to reach 55%–60% by 2025 (Radelet and Sachs 1997: 46). The rate of growth may be in doubt; the direction of change is not.

In the absence of new policy interventions, the likely impacts of such large-scale industrial and urban growth in Asia over the next 30 years are also reasonably predictable. Even with substantial improvements in environmental regulation and a significant shift toward cleaner technology within the region, the ADB (1997) predicts declining environmental quality under a 'business-as-usual' scenario in the lower-income countries of South-East Asia, such as Indonesia and the Philippines, and South Asia. This conclusion is also shared by projections for air quality from the World Bank (1997d). Recent work suggests a similar finding for energy consumption. For example, Carmichael and Rowland (1998) project that current pollution prevention programmes, if widely implemented in Asia, have the potential to yield a 30% improvement in energy efficiency of economic activity by 2020. But, even if such efficiency improvements are achieved, energy usage and attendant greenhouse gas emissions will still double over this time-period (Carmichael and Rowland 1998). Asia is predicted to overtake the OECD economies as the largest source of greenhouse gas emissions worldwide sometime between 2015 and 2020.

It is this shadow of the future—the large-scale increase in urban and industrial activity in Asia forecast for the next three decades—that demands a new policy response. Given the very likely continuing major shift from agriculture to industry, and from rural areas into cities, the critical challenge in Asia is to reduce substantially the energy, materials, pollution and waste intensity of urban–industrial activity in ways that support continued improvement in socioeconomic welfare. This is the challenge of clean shared growth in Asia. Our goal in this book is to lay out a policy framework to address this challenge, that is, to turn the trajectory of future urban and industrial activity in Asia toward patterns of development that are less intensive in terms of use of energy and materials and production of pollution and waste in a dynamic of continuous improvement and superior performance.

Growth in East Asia over the past three decades has been tightly linked to the increasing globalisation of the world economy. Trade and export-oriented industrialisation have been at the leading edge of the development model pursued within the region. Foreign direct investment by multinational corporations as well as portfolio investment from both Asia and the OECD have been important drivers of economic growth (Dua and Esty 1997). More generally, East Asia has benefited from intensified international flows of information and technology. Development in the region is increasingly influenced by cultural processes operating on an international scale, ranging from norms of consumption to concepts of governance and business management. In this context, efforts to develop a policy framework for clean shared growth inevitably confront the structure of the contemporary global political economy and the attendant opportunities for, and limits to, policy intervention. Specifically, what are the implications of economic globalisation for the strategies that might be pursued in East Asia and elsewhere to promote improvements in socioeconomic welfare and the environment?

Held *et al.* (1999) identify three broad schools of thought regarding globalisation. The first school, labelled 'hyperglobalist', recognises in globalisation the emergence of a new economic age in which the historic role of nation-states is superseded by a new order of global governance, economy and civil society. In its neoliberal form, the hyperglobalist thesis identifies increasing trade and economic integration worldwide as a key driver of improvement in socioeconomic welfare. For neo-Marxists, these same processes of economic integration lead to inequality and environmental degradation rather than clean shared growth (Greider 1997). The second school of thought, labelled by Held *et al.* (1999) the 'sceptics', argues that the claims of a new global economy are grossly exaggerated and that the level of global economic integration today is actually less than that observed during the late 19th century. Although there has been a trend toward increasing international trade and investment over the past 50 years, this has been associated predominantly with the growth of regional trading blocs (North America, Asia–Pacific and Europe) rather than with a truly global economy (Boyer and Drache 1996; Hirst and Thompson 1996). By this account, the power of nation-states is changing but is not necessarily diminished, often working indirectly through the influence of powerful governments on regional and international organisations, such as the North American Free Trade Agreement (NAFTA) and the International Monetary Fund (IMF).

The third school of thought is labelled 'transformationalist' (Held *et al.* 1999), and it is this set of ideas that comes closest to our own analysis of the dynamics of the global economy today. Contrary to the views of sceptics, the transformationalist thesis recognises in globalisation a powerful set of economic, cultural and social forces that are indeed restructuring economies and societies around the world, and the attendant opportunities for policy intervention in support of improved socioeconomic welfare and reduced environmental degradation. At the core of the globalisation dynamic is the intensification and geographical extension of economic, social and cultural linkages (Dicken 1998). The medium of these linkages ranges from information flows to international capital investments, media images to organisational partnerships, but contrary to the views of hyperglobalists the outcome of these processes of interconnection and interlinkage are essentially contingent, neither necessarily positive nor necessarily negative with respect to socioeconomic welfare and the environment in different parts of the world. Specific outcomes are open to the influence of public and private policy, of governance systems and of public participation at a wide variety of scales, from that of the individual community to national government policies and the actions of regional and international organisations such as Asia–Pacific Economic Co-operation (APEC), Association of South-East Asian Nations (ASEAN), and the World Trade Organisation (WTO). By this account, globalisation constitutes a range of powerful forces that can potentially be directed to particular societal goals through the development of appropriate policies and governance structures. The challenge is to identify the forms of intervention at different geographical scales that support improvements in socio-economic welfare and the environment and to promote the successful articulation of policies pursued with communities, regions, national governments and regional and international organisations.

It is on this basis that we approach the challenge of clean shared growth in Asia. The policy framework we propose draws on four core themes. First, addressing the challenge of clean shared growth requires harnessing the powerful forces of economic and social change at work within the global economy today. Intensified flows of investment, technology and information, as well as the increasing interconnectedness of producers and consumers, manufacturers and suppliers, and firms and investors, present significant policy opportunities for turning the trajectory of urban–industrial development in rapidly industrialising and urbanising Asia. The greater availability and ease of access of information worldwide, for example, provides a significant opportunity for impacting the environmental performance of industry. But the likely impact of information availability depends in part on the development of low-cost, standardised, transparent and verifiable systems of environmental performance measurement at the scale of the factory, industry and industrial sector. Similarly, the likely impacts of intensified flows of technology worldwide on economic development and the environment depend in part on the ability of firms and regions to use, adapt and enhance such technologies effectively.

Second, the sustainability challenge facing rapidly industrialising Asia today is qualitatively different from the environmental concerns that underlay the emergence of mature environmental regulatory systems within OECD economies during the 1960s and 1970s. The context within which regulatory systems must operate diverges in important ways even among Asian economies. Understanding this context is critical to successful policy intervention. Accordingly, policy development must begin by identifying the economic, political and social conditions in Asia and the world that constitute the structure of the sustainability challenge and the attendant opportunities for change. Foremost among these conditions is the anticipated future expansion in urban and industrial activity and the need to shape this trajectory of future economic development. In addition, the policy response must reflect the profoundly globalised character of many economic and social processes in Asia today, from investment and trade, to sources of information and technology, and locations of end-user markets.

Third, given that future increases in per capita income in much of Asia will almost certainly entail major shifts from agriculture into industry, and from rural areas into cities, the key policy focus must be on substantially reducing the energy, materials, pollution and waste intensity of urban–industrial activity. Reductions in energy, materials, pollution and waste intensity must continuously offset ongoing expansion in urban and industrial activity (Rock et al. 1999a). This probably requires improvements that go well beyond what is required to meet baseline local environmental and health needs and beyond the improvements achieved through existing environmental regulatory approaches. And, because the sustainability challenge in Asia includes such global concerns as greenhouse gas emissions and resource depletion, it is imperative that the policy approach address the actual use of energy and materials (i.e. intensity of use) and not just ameliorate the pollution and waste by-products of urban–industrial activity. Framed in these terms, the policy focus necessarily moves beyond controlling pollution to influencing basic processes of investment, technology change and market development, or what we call the 'denominator' of economic activity.

Fourth, we argue that the focus on intensity, and on the urban–industrial process itself, expands the range of drivers and points of entry that can be harnessed to the goal of improved environmental performance. The array of possible drivers of improved environmental performance is wide, from community pressure to market demand, supplier relations, international agreements and environmental regulation. Effective environmental regulation will be crucial to success. But, within the policy domain, shaping the energy and materials intensity of urban–industrial activity is as much an issue of technology, trade, urban and industrial policy as it is of environmental policy *per se*. The openness of East Asian economies to trade, investment and technology, and the increasing globalisation of markets and information flows, suggest that these economic processes are potentially powerful drivers of improved environmental performance. Public policy has a key role to play in bringing these drivers to bear on the investment and technology decisions of private industry, in fostering a clear economic and environmental performance orientation among firms that promote a dynamic of continuous improvement, and by fostering systems of civic ordering (e.g. public–private partnerships) and private ordering (e.g. management cultures) that support these goals.

The context for clean shared growth in Asia today

How can public policy best address the challenge of clean shared growth in developing Asia today? The most common response to this question has been to look to the environmental regulatory systems of the United States and other OECD economies as potential models for policy intervention. There is certainly much to be gained from the OECD experience, as evidenced by the progress made over the past three decades in reducing industrial pollution and improving environmental quality within these economies. But there are at least two reasons to suppose that a broader frame of reference will be valuable. First, regulatory approaches within the OECD are themselves undergoing a process of reassessment and change (OECD 1997). This re-evaluation is linked to a growing interest in the role that civil society, information, the corporate sector, technology innovation and markets can play as drivers of improved environmental performance (see e.g. Davies and Mazurek 1996; Heaton and Banks 1997). Second, the environmental problems and concerns, and the economic, social and political context within which environmental policy was developed within the OECD economies in the 1960s and 1970s are very different from those of developing Asia today (a partial list of these differences would include levels of foreign direct investment, external trade, scale of urban areas, rate of industrial growth and effectiveness of legal systems). To take but one example, whereas the OECD economies had a relatively mature urban–industrial capital stock and infrastructure, developing Asia is in the midst of very rapid urban–industrial growth. The economic and political context has profound implications for the policy response.

Our analysis begins with the character of the development process in East Asia today, and the global economic, political and technological context within which that development is taking place. Our comments are necessarily broad, for in some instances there is as much variation within the developing economies of East Asia as there is between East Asia and the OECD. Our key focus is on the developing economies of South-East Asia, but, as we show below, the challenge and the opportunity of clean shared growth in Asia entails a set of structural characteristics that have broad significance within the region.

Industrial-led development

First, industrialisation and technology catch-up are central elements of the development model that brought rapid shared growth to the economies of East Asia.[1] Typically, development involved a shift from agriculture into labour-intensive and resource-intensive industries, in the first instance, and then into more knowledge-intensive and technology-intensive industries, such as electronics, as well as producer services. As shown in Table 1.2, the structure of output varies considerably across Asian economies with the contribution of agriculture to GDP decreasing over the period 1980–96 for many low-income economies in South-East and South Asia. Two consequences follow for our analysis. First, because much of Asia seeks to emulate the model of development pursued by the first-tier Asian NICs, rapid industrialisation will undoubtedly remain at the core of the sustainability challenge within the region. Second, because the development model pursued by the Asian NICs depends on technology catch-up and the cultivation of local innovation capability, influencing the pattern of technology investment and change is a key opportunity for policy intervention.

Past and future growth

As shown in Table 1.3, many of the rapidly industrialising economies of East Asia achieved remarkable rates of growth in GDP per capita over the period 1965–96. But, even with this growth, most low-income countries in Asia, including the second-tier East Asian NICs (Indonesia, Malaysia, the Philippines and Thailand), are still in the early stages of a

1 This is not to denigrate the importance of intensification of agriculture or of massive investments in basic education, healthcare, infrastructure and family planning to the success of the East Asian shared growth model.

18 ASIA'S CLEAN REVOLUTION

	Agriculture		Industry		Manufacturing*		Services	
	1980	1996	1980	1996	1980	1996	1980	1996
South Asia	38	28	25	28	17	19	37	44
East Asia	28	20	44	44	32	33	28	36
China	30	21	49	48	41	38	21	31
Indonesia	38	28	26	29	18	20	36	43
Korea	15	6	40	43	29	26	45	51
Malaysia	22	13	38	46	21	34	40	41
Philippines	25	21	39	32	26	23	36	47
Singapore	1	0	38	36	29	26	61	64
Thailand	23	11	29	40	22	29	48	50

* Industry includes manufacturing which is also reported as a separate category.

Table 1.2 **Structure of economic output (% of GDP) for selected countries, 1980 and 1996**
Source: World Bank 1998d

Table 1.3 **Growth in GNP per capita, 1965–96: average annual growth (%)**
Source: World Bank 1998d

South Asia	2.2
East Asia	5.5
China	6.7
Indonesia	4.6
Korea	7.3
Malaysia	4.1
Philippines	0.9
Singapore	6.3
Thailand	5.0

profound urban–industrial development transition. Most of the industrial stock that will be in place 25 years from now is not on the ground today.[2] What this means in practice is that the first three decades of the 21st century will likely witness the most prodigious expansion of urban–industrial activity in Asia in the history of the world. This is both a threat to sustainability and an opportunity to shape, at an early stage, the energy, materials, pollution and waste intensity of new urban–industrial investment. If actions are taken now there is a once-in-a-country's-lifetime opportunity to achieve a more sustainable growth trajectory. This opportunity is significantly different from that faced by the OECD countries when they launched their environmental programmes in the 1970s. Then the problem was not how to make the new urban–industrial capital stock cleaner but rather how to retrofit a large existing capital stock with end-of-pipe controls to reduce emissions after they were produced.

Private industry

The development process in Asia today is driven overwhelmingly by private capital. As of the mid-1990s, public investment (largely development assistance) constituted only 10% of new capital flows in East Asia. Of the dominant private investment, approximately 50% was foreign direct investment by international business. The implication is clear. Achieving clean shared growth depends on greening the new investment and technology choices of private business. Public policy must focus on promoting conditions under which such greening will take place. The prospects for such policy intervention may never have been better. Many leading companies are undergoing an unprecedented reassessment of their own role in a sustainability transition (Fischer and Schot 1993; Graedel and Allenby 1995; IHDP 1999; Roome 1998; Socolow et al. 1994). Sometimes this takes the form of the greening of supply chains and the identification of win–win opportunities in economic and environmental policies, including those affecting technological change and openness to the world economy (Heaton and Banks 1997; Wheeler and Martin 1992). Sometimes it takes the form of new international voluntary environmental management standards and private law models of environmental regulation, such as the ISO 14000 series, or of industry codes of conduct (as in the chemical industry's 'Responsible Care' programme) (Roht-Arriaza 1995). And sometimes it takes the form of corporate disclosure and accountability (as in the rapid growth of corporate environmental reporting and green accounting).

Latecomer status

As O'Connor (1994) has noted, the latecomer status of Asian NICs within the global economy is an important context for policy response. This is particularly the case with

2 In a pre-crisis study of Indonesia, the World Bank (1994a) projected that 85% of capital stock that would be in place by 2020 is not in place today. Even with the current recession and slower and delayed growth, the significance of new investment remains. At an annual growth rate of manufacturing output of 7.25% (just half the growth rate maintained during the 1990s prior to the crisis), manufacturing output doubles every ten years.

respect to technology, where Asian NICs have access to an array of environmentally advanced technologies developed within OECD economies. There is now considerable evidence that newer plant and equipment developed mostly within the OECD economies tends to be or can be made cleaner than existing plant and equipment (Arora and Cason 1995; Christensen et al. 1995; Nelson 1994; Wheeler and Martin 1992). This means that it is now, or soon will be, technically and economically possible for manufacturers in the NICs to import, adopt, adapt, modify and innovate on an industrial capital stock that will tend to be cleaner simply because it is newer.[3] The policy challenge is to promote the selection and use of such clean technology in new urban and industrial investment. Since the majority of technology design and development remains centred with OECD economies, technology policy in the OECD will be crucial to clean shared growth in Asia and indeed around the world. To repeat, the critical technology is not that of end-of-pipe pollution control, but new and improved product and process technologies that are designed to achieve higher efficiencies in energy and materials use.

Urban-based growth

Industrialisation in East Asia has been closely allied with highly concentrated urban growth. The Bangkok region, for example, accounts for almost one-half of Thailand's GDP and a little more than 75% of manufacturing value added (World Bank 1994b: 8). Four cities on Java (Jakarta, Suraya, Bandung and Semarang) account for 36% of Java's and 27% of Indonesia's industrial output (World Bank 1994b: 75). As a consequence, much of East Asia's industrial pollution is concentrated in urban areas. This means that efforts to address industrial pollution need to be framed in the context of addressing allied urban environmental problems. It also means that urban governance, land-use control and urban infrastructure investments are likely to be important elements of the policy response. As shown in Figure 1.1, many Asian countries are predicted to experience very large increases in urban population over the first three decades of this century. The total urban population of the East Asian NICs, including China, is expected to approximately double by the year 2025 to almost 1.2 billion people (WRI 1997).

Globalisation

The shared growth miracles of Asia occurred in tandem with rising openness to trade and investment within the world economy. As shown in Table 1.4, trade as a percentage of GDP increased substantially in East Asia between 1970 and 1996. In broad terms, open economies increase the interdependence of Asia and the global economy at large, whether this be in terms of market demand, investment and technology supply or the global impacts of economic change (such as climate change). Open economies are

---

3 We do not, however, have good data on the extent to which manufacturers are availing themselves of cleaner technology within South-East Asia. As most existing policy presumes a 20%–30% improvement in energy and materials efficiency simply through the use of newer, cleaner technology, this becomes a critical policy issue.

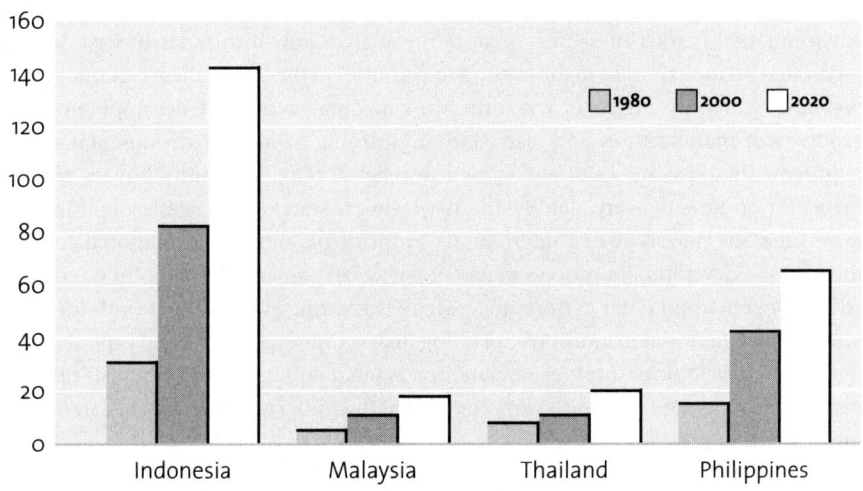

Figure 1.1 **Urban population (in millions)**
Source: WRI 1997

Table 1.4 **Trade as percentage of GDP, 1970 and 1996**
Source: World Bank 1998d

	1970	1996
South Asia	14	30
East Asia	17	58
China	5	40
Indonesia	8	27
Korea	37	69
Malaysia	80	183
Philippines	43	94
Singapore	232	356
Thailand	34	83

already exposing manufacturers in the NICs to an increasingly wide range of pro-environmental market pressures. Because the shared growth miracles in East Asia are predicated on the export of manufactures to countries in the OECD, these external market pressures will only increase over time. In our view, successful developing-country exporters of manufactures will learn and are learning to meet environmental market requirements the same way they learned to meet developed-country buyers' requirements for on-time delivery, quality and packaging requirements (Keesing 1988). At the same time, the importance of international investment, and of multinational corporations, to the development process in Asia today creates an additional point of entry for addressing environmental performance. There is growing evidence that both local and international financial institutions operating in developing countries and capital (stock) markets in developing countries operate in ways that either reward or punish firms for their environmental behaviours. Environmental due diligence in lending is an increasingly common practice in Asia.

Governance and civil society

Despite strong pressures toward global convergence, there remain substantial differences between South-East Asian societies and countries such as the United States in many dimensions of governance and civil society, ranging from the strength of legal systems, to traditions of strong public policy planning (as in the industrial and technological planning pursued by the first-tier NICs), to the role of regional governance systems (such as ASEAN and APEC), to the presence of non-governmental organisations (NGOs) and transparency in financial transactions. It is also within these areas of governance structure that there is the greatest variation among countries within South-East Asia. Public policy must recognise and respond to these differences and to the specific social and political context of individual countries.

Having said this, there are several general trends that are likely to be of importance for policy intervention. As in other parts of the world, the growing availability of information and the growth of NGOs is creating a new force within governance structures of Asia. To varying degrees in each of the NICs, citizens, communities and organised groups in civil society are placing increased pressure on governments and private sectors to improve environmental quality. In some places, such as Indonesia, public-sector environmental agencies are taking advantage of public and community pressure to devise low-cost enforcement strategies that take advantage of the concern of firms for their (environmental) reputations. In our assessment, achieving clean shared growth will depend in part on the ability of Asian countries to harness these processes of private and civic ordering.

Economic crisis

Jointly, the above seven elements (industrial-led development, future growth, private capital, latecomer advantages, concentrated urbanisation, globalisation and governance) define the critical context within which a policy response must be developed to meet the challenge of clean shared growth in Asia. But no analysis of context would be complete without reference to the economic and fiscal crisis that dominated the region at the end of the 1990s.

To begin with, it is important to note that the effects of the crisis and the prospects for a return to past patterns of rapid growth vary widely within the region. Thailand, Indonesia and South Korea were hardest hit and experienced a substantial real contraction in output. Per capita income in Indonesia, for example, fell from approximately US$1,100 per capita in 1996 to US$460 per capita in 1998. Malaysia and the Philippines were also hard hit. The economies of Singapore, Taiwan and China have not experienced the freefall in output visible elsewhere (World Bank 1998a).

The immediate environmental effects of the economic crisis included cutbacks in regulatory activity, delayed investments in new technology and reductions in overall levels of economic activity and attendant pollution (Afsah 1998; World Bank 1998a). The crisis will probably also affect the sectoral composition of economies, especially in the lower-income countries of South-East Asia. Most manufacturing industries with a high import content have suffered from deteriorating exchange rates and a slowdown in export opportunities. Industries with high local content, such as natural resource-based industries, will probably increase in importance in the short term.

It is also possible that the economic crisis will trigger more broadly ranging changes in policy practices and patterns of development. Attention has focused in this regard on two particular themes, namely, disclosure/transparency and commitments to an open economy. In the aftermath of the crisis, many observers voiced support for a partial retreat from the growing openness of trade and investment within the region (Lim 1998). Certainly, the volatility of capital flows, especially debt flows, in East Asia intensified the crisis. The World Bank (1998a) reports that in the space of one year net capital flows reversed by more than US$100 billion, but as yet there has been little evidence of a broad withdrawal from open economies. Indeed, for most Asian NICs, export growth remains a key strategy for recovery from the crisis, albeit dependent on economic recovery in Japan and continued economic growth in North America and Europe. In addition, foreign direct investment exhibits less short-term volatility than other forms of international private capital investment.

We conclude that, despite the profound and traumatic effects of the crisis, the economic and political context for policy response outlined above remains valid. Although the timing and pace of future economic growth is uncertain, the key opportunity remains that of influencing the energy, materials, pollution and waste intensity of new urban–industrial investment. Reductions in intensity will continue to depend on both environmental regulation and increasingly on a performance-based dynamic of continuous improvement within firms. What is perhaps of greatest uncertainty is the strength and balance of different drivers of environmental performance, such as market demand, community pressure and supply-chain management, within East Asia.

The issue of intensity

Having established the context for policy response, we now examine in more detail the nature of the environmental challenge facing East Asia in coming decades. Environmental problems within the region are in part the result of the sheer pace of growth of the last 25 years of the 20th century, but these problems also reflect an initial emphasis in many Asian economies on pollution-intensive and energy-intensive manufacturing and resource processing industries and the limited attention paid to pollution control and pollution abatement. One consequence of this has been a high energy intensity of GDP. And, although the ratio of commercial energy use to GDP is high relative to OECD standards, commercial energy use per capita is low and expected to rise very rapidly over coming decades. For the four second-tier NICs (Indonesia, Malaysia, the Philippines and Thailand), commercial energy use per capita measured in kilogram oil equivalents was 820 kg in 1996, as compared with 5,123 kg in the higher-income OECD economies (World Bank 1998d). When combined with expected increases in population, the consequence is a very large increase in energy use (see Fig. 1.2) and greenhouse gas emissions (see Fig. 1.3) predicted for developing Asia over the next 20 years.

Reducing pollution from industry and improving local air and water quality have been important environmental regulatory priorities within the region, especially since the early 1990s. There is now a wide variety of efforts under way to strengthen environmental regulatory systems and to promote pollution prevention and clean production (USAEP 1997; World Bank 1999). Given these ongoing activities, two critical questions need to be answered. The first is whether current policy approaches, if widely adopted and enforced, are sufficient to reduce industrial pollution and improve local environmental quality within developing Asia. The second is whether a policy approach focusing on regulation

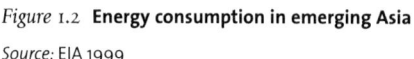

Figure 1.2 **Energy consumption in emerging Asia**
Source: EIA 1999

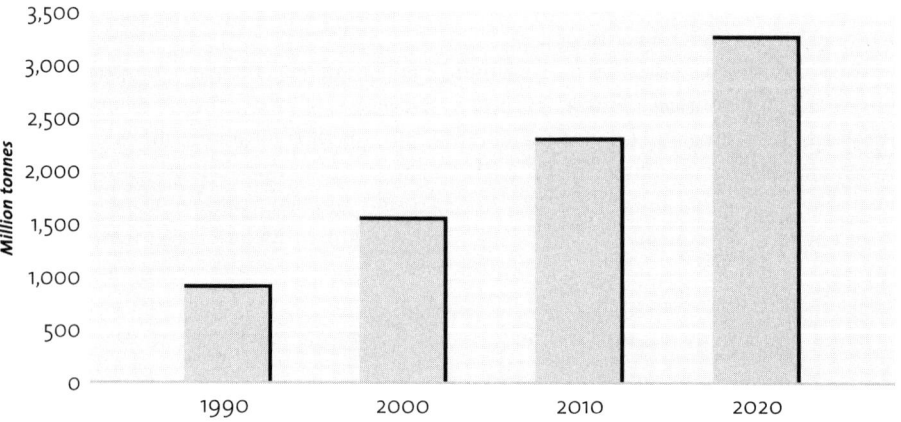

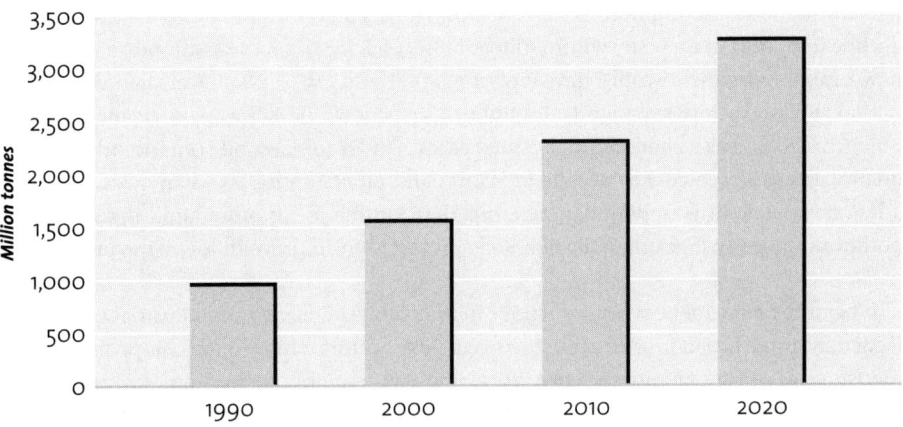

Figure 1.3 **Carbon dioxide emissions in developing Asia**
Source: EIA 1999

of industrial pollution and improving local environmental quality will simultaneously resolve broader environmental concerns, such as resource depletion, escalating fossil-fuel consumption and climate change. Our main conclusion with respect to the first question is that projected outcomes depend critically on the energy, materials, pollution and waste intensity of new investment in East Asia (about which there is a paucity of good data). With regard to the second question, we conclude that current policy approaches—focused as they are on controlling pollution—are insufficient to the challenge, hence our call for a new policy approach to build on and complement existing activity.

Future levels of industrial pollution and environmental quality within east Asian economies are sensitive to two critical modelling assumptions, namely, the rate of growth of industrial output and, most especially, the energy, materials, pollution and waste intensity of industrial activity. Prior to the economic crisis of the late 1990s, most researchers predicted continued rapid growth in industrial output and a reduction in some measures of pollution intensity of industrial activity within the first-tier and second-tier Asian NICs. Actual levels of pollution and environmental quality depend on the relative strength of these two often countervailing trends within individual countries. The analysis is complicated in that growth and pollution intensity are likely to be interdependent variables.

By way of illustration, one influential World Bank (1994a) study of Indonesia projected disastrous increases in pollution in that country by the year 2020, amounting to a tenfold increase in water pollutants, a fifteenfold increase in emissions of suspended particulates into the air, and a nineteenfold increase in emissions of bioaccumulative metals, such as mercury and lead. The pollution intensity of industrial activity with respect to particulates, BOD and toxics (but not bioaccumulative metals) was predicted to decline modestly in Indonesia over this 25-year period as a result of shifts into industrial sectors that were less polluting. But these declines were overridden by a predicted thirteenfold increase in industrial output by 2020. A second World Bank study predicted a more rapid decline in pollution intensity within industrial sectors, based on the assumption that all new capital investment would have emissions that were 25%–50% lower than existing capital stock (World Bank 1997d). Pollution intensity would fall as new, cleaner technology increased as a share of total capital stock. Under this second scenario urban air quality was still predicted to decline in Jakarta and other South-East Asian cities, but at a less severe rate. It is only under scenarios that assume much more rapid declines in pollution intensity that urban air quality is projected to improve in lower-income East Asian NICs.

What is the basis for possible reductions in the pollution intensity of industrial output? Generally, three broad processes are involved. First, sectoral shifts in the composition of industrial activity away from resource processing into assembly and technology-intensive industries typically reduces the overall pollution intensity of industrial output (though not of toxics and bioaccumulative metals). Second, if new capital investment is cleaner than existing capital stock, then the process of new investment and growth reduces the average pollution intensity of industrial output. Third, strengthened regulatory requirements and enforcement, public pressure, market demand and other drivers of improved environmental performance support more effective pollution control of existing industrial activity (e.g. retrofitting improved end-of-pipe control equipment) and, more importantly, support investment in best available technologies that meet world-class environmental standards, as well as more general pollution-prevention activities, such as encouraging recycling and re-use, and planning decisions that minimise demand for energy and materials.[4]

If the experience of the OECD and first-tier East Asian NICs is a guide, then we should expect that with rising incomes in East Asia all three of these processes will contribute to reducing the pollution intensity of industrial activity, offsetting the effects of increased industrial output. Hettige et al. (1997) modelled these effects for water pollution by using data on emissions from a set of developed and newly industrialising countries for the period 1977–89. They found that pollution intensity of industrial activity falls sharply with income up to income levels of US$6,000 per capita and is stable thereafter. These types of finding, along with growing evidence that East Asian NICs are moving to strengthen environmental regulation systems, led some observers to conclude that the region is on the road to reduced industrial pollution and improved environmental quality.

Three points of caution need to be voiced in this regard. First, the anticipated improvement in environmental quality is largely a feature of the first-tier NICs and the medium-income economies of South-East Asia. Thus, even with reductions in pollution intensity of industrial output, air and water quality are expected to continue to decline in lower-income economies, such as Indonesia, as growth in industrial output exceeds the rate of improvement in reduction of pollution intensity. It is important to recall in this regard that all the second-tier NICs had per capita incomes in 1996 of less than

---

4   Such policies go beyond the firm *per se*, as for example in urban planning activities that minimise transportation flows associated with industry.

US$4,500 and that per capita incomes in China and much of South Asia were less than US$1,000 per capita (see Fig. 1.4). In much of Asia environmental conditions will get worse before they get better (Lohani 1998). Second, as we have seen, the projections by the World Bank and others of lower net industrial pollution in certain East Asian NICs depend critically on assumptions that new capital investment will be 25%–50% cleaner than existing capital stock. In our assessment, this assumption requires careful empirical evaluation (e.g. by documenting the extent to which firms investing in Asia today are choosing, maintaining and extending cleaner technologies).

Third, our core concern is with the environmental challenges of rapid urban–industrial growth that are *not* fully addressed through controlling industrial pollution at the plant level. The most visible of such challenges are the ever-expanding use of energy and materials, resource depletion and global climate change. In the area of energy, for example, even with a projected 30% improvement in energy efficiency, energy usage is expected to double by the year 2020. It is on this basis that we propose a policy approach that focuses directly on the energy, materials, pollution and waste intensity of urban–industrial activity, as opposed to addressing these issues indirectly and partially through pollution control.

The drivers of change

The focus on intensity necessarily brings us to the industrial process itself, to shaping the basic processes of investment, technology change and market development within the

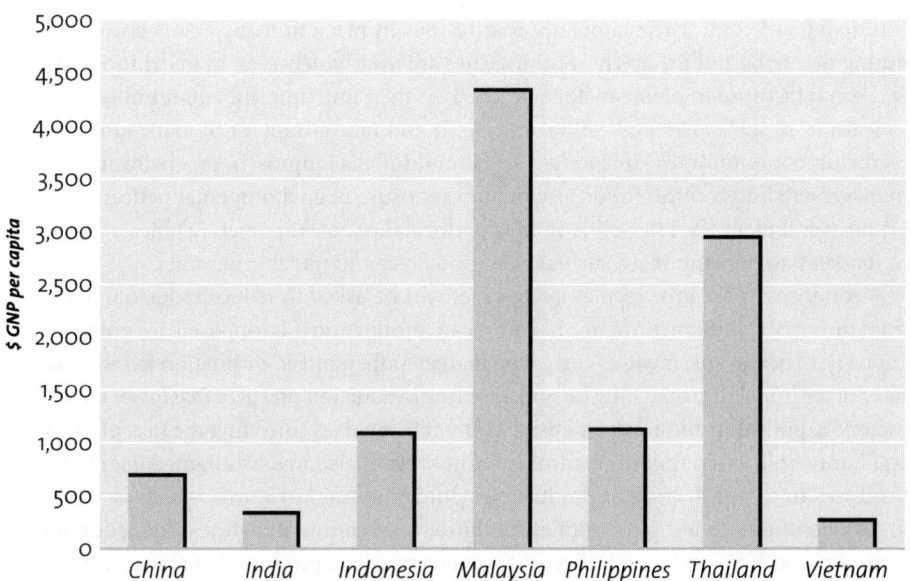

Figure 1.4 **GNP per capita (US$), selected Asian countries, 1996**

Source: World Bank 1998d

industrial economy. What determines these investment, technology and market decisions as they relate to environmental performance? The traditional answer to this question is environmental regulatory policy. But environmental regulatory policy is only one means of shaping investment and technology decisions; and it has been most successful in bringing firms and industries into compliance with particular performance standards through end-of-pipe pollution control. Environmental regulatory policy has been less successful in moving firms beyond compliance and engendering a self-sustaining dynamic of continuous improvement. By moving beyond pollution control to the industrial process itself, we are able to call on a range of additional policy drivers, from industrial and urban policy to various forms of private and public governance.

Accordingly, the policy framework we propose supports multiple strategies focusing on a variety of performance drivers. In developing this framework, we recall the elements of the development context laid out earlier in this chapter, including the key challenge of driving down the energy, materials, pollution and waste intensity of future urban–industrial activity. We begin with the role that environmental policy, including environmental regulation, might play in meeting this challenge.

Environmental policy

Our analysis suggests that environmental regulatory systems in Asia will be called on to support three types of goal. First, regulatory systems have a responsibility to protect public health. Most likely this will be achieved through the establishment of clear and consistent ambient environmental standards that are in turn linked to discharge limits for individual facilities. Once ambient and discharge standards are set, regulators must ensure compliance. Some countries, particularly in the first-tier NICs, have gone a long way towards doing just this. In others there remains a need to strengthen the legal authority and the institutional and technical capability of regulatory agencies. It is unlikely to be the case that Asian economies will adopt fully the kinds of traditional command-and-control regulatory approaches put in place in many OECD economies during the 1970s and 1980s. These approaches are now widely seen as being too costly and too difficult to implement (Russell 1990). At the same time, the opportunity exists to learn from the experiences within the OECD and take advantage of many emerging regulatory opportunities, including: (1) a multimedia approach to environmental management and control; (2) greater public disclosure of environmental performance; (3) firm- and plant-level flexibility in how performance goals are met; (4) increased use of market-based instruments, such as pollution taxes and tradable permits.

Second, environmental regulatory agencies will be asked to reduce industrial pollution through pollution prevention and clean production (as opposed to pollution control). Crucially this involves correcting market, policy and co-ordination failures that discourage firms from searching for and adopting production practices that lower energy, materials, pollution and waste intensity. Partly this involves correcting the bias of much regulatory policy to pollution-control solutions, but it also involves harnessing market processes to the goal of clean production. Unlike pollution control, which is almost always derivative of regulatory policy, pollution prevention and clean production are responsive both to regulatory policy and other competitive pressures in the marketplace, such as materials costs and opportunities for win–win economic savings. Here the key advance will be in reducing the often high transactions costs and learning costs associated with clean production alternatives.

Clearly, initiatives pursued under the first two goals described above (protecting public health and supporting clean production) will contribute towards the goal of clean shared growth. However, resultant improvements in energy and materials efficiency, in particular, are unlikely to be sufficient to the scale of improvement that is required to offset continued urban–industrial growth. Accordingly, we anticipate that the challenge of clean shared growth will place new demands on environmental regulatory systems. It is to this third set of expectations, linked to substantial improvements in energy, materials, pollution and waste intensity, that we now turn.

Extending environmental policy

In the context of clean shared growth in developing Asia, environmental policy must make three critical contributions. First, it must articulate clear environmental and developmental goals. One result of the regulatory re-invention process within OECD economies has been recognition of the importance of clear and consistent goals that communicate priorities and directions to all segments of society, and set benchmarks for measuring progress (NAPA 1995; Steinzor 1998). Such a goal-driven and outcome-driven policy framework fits well with approaches taken to past development successes—in agriculture, primary education, export-led industrialisation and others—by East Asian NICs. But it must also reflect emerging claims for greater public participation and transparency in the policy-making process.

Second, environmental regulation—and regulators—must support the integration of policy-making and policy implementation across multiple organisational domains, from industry to trade and technology, from public policy to corporate management, from local environmental quality to global and regional environmental concerns and from national policy to regional and international regulation. Traditional environmental policy has been pursued largely as an issue of environmental regulation and largely within environmental ministries or departments. The need now is to introduce environmental goals into line ministries and then to integrate environmental regulatory policy with industry, trade and technology policy. Such integration must also co-ordinate policy-making at multiple geographical scales, from the local to the national, regional (e.g. ASEAN) and international (e.g. WTO).

Third, environmental policy must support a performance orientation on the part of firms and industries. Crucial to this will be improvements in the quality and quantity of information available on the environmental performance of firms. It should by now be clear that information is a powerful policy tool. Information will increasingly become a driver of change both for firms, for whom it is the foundation of performance-based management, and for government and society at large. Fortunately, there is substantial evidence that information development and disclosure are promoting improved environmental performance of industrial firms in East Asia. Indonesia has a colour-coded environmental rating and disclosure programme for major water polluters that has improved environmental performance (see Chapter 7). China's sustainable cities index programme, which annually rates, ranks and publicly reports on the environmental performance of its major cities, appears to be influencing the location of industrial activity, the rate of growth of urban infrastructure and plant-level investments in pollution control.

To date, information on environmental performance has largely been used as a tool of environmental regulation, that is, to assess whether firms are meeting externally imposed regulatory standards. Increasingly, this information will be used in at least four other ways:

1. As a tool of strategic management by firms that are seeking to optimise their own environmental performance, whether it be with respect to energy efficiency or life-cycle impacts of products

2. As an element of financial and risk management by banks, investors, insurance companies and the financial community at large, in the same way that standardised accounting information is currently used

3. As part of a marketing and corporate management strategy—increasingly, purchasing decisions inside and outside of the firm will be linked to information on environmental performance, whether this be through market regulations, the ISO 14000 series, green labelling or supply-chain management

4. As a tool by which communities, citizens and organised groups in civil society hold governments (their own as well as others), as well as the private sector, accountable for performance

Public policy must foster and harness these trends toward performance-based management in the public and private sectors.

Industrial and technology policy

One of the conclusions of our analysis is that meeting the challenge of clean shared growth is as much an issue of industrial and technology policy as it is of environmental policy *per se*. This is because successful efforts to reduce the energy, materials and pollution intensities of industrial production will depend critically on the development and deployment of new technologies and on the capability of firms to use existing plant and equipment efficiently, to know how to improve on it and innovate with it and how to manage efficiently the process of technical change and technology acquisition. Unless firms can do these things and do them well, there may be significant limits to their ability to ratchet up into skills-intensive industrial growth (as Korea and Taiwan have done in different ways [Kim 1997; Wade 1990]) and to respond to pressures from regulatory agencies, communities and markets that would push them in a direction that lowers energy, material, pollution and waste intensities.

If we are right, three important implications follow. First, policies that promote firm-level technical learning and capabilities acquisition that are necessary for the second-tier NICs to recover from the current economic crisis and gain competitiveness in skills-intensive manufactures are likely to be equally good for the reduction of energy, material, pollution and waste intensity. They should make it easier for firms to engage in better housekeeping practices and minor process innovations that prevent pollution. They should make it possible for firms to 'stretch' existing plant and equipment by substantially modifying it to reduce energy and materials use. They should also make it easier for firms to evaluate imported plant, equipment and technology. Second, because intensity reduction is or will be a relatively new activity for industrial firms in the NICs and Asia more generally, industrial firms in the NICs and in Asia are likely to need industry-specific and technology-specific information (and specialised technical training) on how to do this. This is the just the kind of information and specialised training that institutions that are part of the national technology infrastructure (such as industrial technology institutes or standards agencies) are good at providing. They should be encouraged to provide such information and training to overcome information failures and the high transactions costs associated with reducing energy, materials, pollution and waste intensities. This is most likely to be true for small and medium-sized enterprises (SMEs), and governments in the NICs would be advised to consider expanding existing multinational corporation (MNC) and SME linkage programmes designed to do this to include environmental considerations (Battat *et al.* 1996). It may also make sense to consider developing such programmes for large domestic firms and their suppliers.

Not surprisingly, doing all of this and doing it well also requires governments to invest in national technical capability-building by supporting education, particularly in engineering (and environmental engineering) and by investing in institutions that test materials, inspect and certify quality standards (including environmental quality standards such as the ISO 14000 series), calibrate measuring instruments and provide difficult-to-obtain information (including in the area of clean technologies). As the experiences of Korea and Taiwan demonstrate, large investments in literacy, in education and in engineering training make it easier for firms to acquire technical capabilities (Tan and Batra 1995).

For the short and medium term, most technology and capital equipment will continue to be sourced from OECD economies. Accordingly, the kinds of investment and technology transfer, adoption and use policies described here have to extend across international boundaries and from OECD economies to developing Asia. Within this international domain, trade and investment policy emerges as an additional crucial policy lever for shaping technology choices and more generally shaping the environmental policy of industry.

Trade and investment policy

As discussed earlier in this chapter, one of the defining contexts for clean shared growth in Asia is the degree to which development has occurred in parallel with increasing economic integration on a global scale, and especially in terms of levels of international trade and investment, as well as the prominent role of MNCs in the development process. This suggests that international multilateral regulation of investments and market processes, as well as private-law models of international business regulation (such as the ISO 14000 series) can be used as policy tools to promote reductions in energy, materials, pollution and waste intensity of new industrial investment in developing Asia.

To date, there has been only mixed success in building in environmental concerns into multilateral agreements. Policy initiatives toward harmonisation of environmental standards within multilateral trade agreements and introduction of environment riders within the multilateral agreement on investment have achieved only modest gains, partly because of concerns on the part of developing countries that environmental standards will be a form of indirect protectionism (e.g. see Esty and Gentry 1997). More progress has been made with explicit international environmental agreements, such as the Montreal Protocol and the Framework Convention on Climate Change. Despite the limited achievements to date, multilateral agreements remain an important domain for policy intervention.

The recent explosion in private capital flows—both direct foreign investment by multinational corporations and of portfolio flows—means that states can no longer flout what Matthews (1997: 57) calls the *de facto* rules being set by markets. This can be seen most clearly in Asia's current financial crisis. With respect to the environment, we are witnessing the rapid emergence of what Roht-Arriaza (1995: 486-99) refers to as a global private-law model of environmental regulation. This private-law model relies, at least in part, on private, producer-based systems of international environmental standards. Sometimes this takes the form of MNCs imposing their home-country practices on their subsidiaries located elsewhere (Brown *et al.* 1993) or on their suppliers. Sometimes this takes the form of private business organisations, such as the World Business Council for Sustainable Development (WBCSD), the Coalition for Environmentally Responsible Economies (CERES), the Global Environmental Management Initiative (GEMI), the International Chamber of Commerce (ICC)[5] or the Chemical Manufacturers' Association (CMA),[6] creating their own standards for measuring and comparing the environmental impacts of their activities (Roht-Arriaza 1995: 497-99). Sometimes, as with the ISO 14000 series, it takes the form of development of a wide-ranging set of private-law international environmental standards governing everything from the development of acceptable private-sector environmental management systems to environmental auditing, product life-cycle analysis and environmental labelling (Roht-Arriaza 1995).

These global links are of especial importance to the externally oriented export economies of East Asia. To date, most attention has focused on developing a regulatory structure to match the global economy, whether through WTO, the General Agreement on Tariffs and Trade (GATT), APEC or other international and regional organisations. It is possible that these emerging international regulatory bodies will be important drivers of improved environmental performance, but it is *more* likely that the market processes of globalisation described above will be of importance. Having said that, it is far from clear that private-law regulation will develop as a powerful driver of superior environmental

---

5 In 1991 the ICC created the *Business Charter for Sustainable Development* in response to recommendations in the 1987 Brundtland Commission report on the environment (Roht-Arriaza 1995: 498).

6 In North America, the Chemical Manufacturers' Association (CMA) developed its own 'Responsible Care' programme in 1991. The programme binds CMA members to a set of principles regarding safe chemical development, use and transfer as well as to a code of management practices (Roht-Arriaza 1995: 498).

performance in Asia. Our recommendation in this regard is for the urgent development of effective systems of performance measurement. A good place to begin is with the development of a standardised, low-cost, scaleable approach to measuring the energy, materials, pollution and waste intensity of production and of technology choices. The measurement systems should include information on greenhouse gas emissions. We must then investigate how such information can be used most effectively by firms, consumers, suppliers, municipalities and the like. As with price and quality, the ability to measure environmental performance is crucial to harnessing market processes and private management systems to the goal of clean shared growth.

Urban policy

Currently, in developing Asia rapid industrialisation and rapid urbanisation are tightly intertwined. The majority of industrial activity occurs within urban areas, and the environmental impacts of industrial activity are amplified by concentrated urban form. Precisely because cities are the locus for so much of the industrial transformation under way in developing Asia today, they are also the focal point for civic engagement with the development process. Such civic engagement at the urban scale will probably be critical to achieving clean shared growth, and urban policy can make an important contribution by enhancing governance capabilities within cities, by promoting effective models of public–private partnership and by enhancing the quality and quantity of information available on the environmental performance of industry within cities.

Why is enhanced urban governance and civic engagement likely to be of importance to clean shared growth? First, there is some evidence to suggest that successful policy development and implementation benefits from the opportunity to take account of place-specific conditions accorded by devolution of decision-making to the local level. This is especially the case where policy decisions require difficult trade-offs among multiple potentially competing priorities, such as improving environmental conditions, reducing unemployment and improving economic welfare. Second, collaborative governance and civic engagement promote broad 'ownership' of the regulatory and policy initiative. Such broad ownership occurs when all participants (community groups, business, government, unions and others) have a stake in the success of a policy approach and recognise that systemic failure is a worse option than seeking compromise and common ground. Such ownership is likely to be of particular importance as developing Asia continues to undergo wrenching changes not just in economic structure but also in social and political organisation.

Beyond this critical issue of promoting enhanced governance capability at the urban scale, urban policy can support clean shared growth by enhancing urban management, most especially in the areas of location and land-use planning and in the provision of urban infrastructure. In much of developing Asia, urbanisation is taking the form of massive 'mega' cities with population concentrations of upward of ten million people. Such urban concentration increases the stress on dissipative systems within the natural environment and creates extreme localised environmental problems (e.g. with water supply and waste disposal). Location and land-use policy has a major role to play by structuring where industries locate and how people move from home to work.

Cities also have to do a better job at managing urban infrastructure. Available evidence suggests that infrastructure expansion is not keeping up with rapid population and economic growth. This is partly because of the fact that city governments often lack the capacity to tax the new income created within cities as a means of financing infrastructure investments. But in some countries it is also a result of the lack of competition among public service providers, lack of managerial autonomy and outcome accountability and poor relations between service providers and end-users (World Bank 1994b).

Governance and civil society

Although cities will be the locus for civic engagement and greater public pressure on governments and the policy process, this urban-scale activity will be but one part of the continuing development of civil society in Asia in coming decades. Indeed, one of the great unknowns of the development transformation that will unfold in Asia over the next 30 years is the way in which civil society will develop and the attendant forms of governance that will emerge. This dimension of the development context also will likely show a high degree of variation among Asian societies.

Most analyses of civil society focus on the growth of NGOs and the role that information access now plays in supporting increasingly powerful international networks of NGOs. Rapid reductions in the costs of communicating information, over fax machines, telephones and the Internet, now mean that NGOs in one country can and are easily reached by NGOs in any other country. The impact of this on private and public policy-making within and between countries has been amply demonstrated by Matthews (1997). As she sees it, NGOs now deliver more development assistance than the entire United Nations system (excluding the World Bank and the IMF), and they are increasingly successful in pushing even the largest and strongest governments around. Thus when the United States and Mexico attempted to hammer out a trade agreement behind closed doors, NGOs in both countries forced both governments to negotiate more openly and to pay greater heed to health and safety, pollution, consumer protection and labour practices than either had intended (Matthews 1997: 54).

What the future holds in this regard is far from clear. Under these conditions of uncertainty, our recommendation is for a proactive process of participatory engagement with community groups, NGOs and others. In short, engagement carries a higher probability of benefit, and lower risk, than exclusion. Although we cannot as yet systematically demonstrate that civic engagement will succeed as a driver of clean shared growth, we do know that exclusion from the policy process undermines effectiveness and legitimacy of outcome. As with the case of private business, the bedrock of such civic engagement is information. Here we cite the success of the PROPER environmental rating and disclosure programme in Indonesia to secure improvements in environmental performance (see Chapter 7).

Summary and conclusions

There is no road map to cleaner shared growth in developing Asia, but we can already recognise (1) the scale of the challenge, (2) the important points of strategic entry and (3) the likely driving forces of change. *We submit that the critical opportunity lies in reducing the energy, materials, pollution and waste intensity of new urban and industrial investment.* Success in this endeavour requires harnessing 'new' drivers of environmental performance, namely information, globalisation and technology. This would entail acceptance of clean shared growth as a strategic goal, focusing on the development of clean technologies, industries and urban infrastructure as sources of economic advantage and environmental improvement. This commitment must be communicated as a clear 'market signal' through which to influence the future technology and investment decisions of private firms, capital markets, urban municipalities and consumers.

Leveraging of private industry and capital as the principal agents for developing clean products and production processes would also be required. This can only be achieved by strengthening the internal and external drivers of the technology and investment decision. Partly this will involve the establishment and enforcement of environmental standards. The new policy opportunity, however, involves promoting the use of information, performance measurement, market pressure and supplier linkages as a driver of superior performance. Globalisation has substantially enhanced the likely effects of such public and market drivers

Public policy has three roles: that of (1) environmental regulation, (2) of fostering information and market pressures and (3) the use of economic and technology policy, land-use control, licensing and other instruments to promote clean investment and technology decisions directly.

What would a policy agenda look like that supports these goals? The policy matrix would begin with a portfolio of environmental regulatory institutions, policy instruments and capacities for implementation and enforcement. Effective environmental regulatory systems are the bedrock of clean shared growth. Such regulatory systems begin with clear performance expectations for firms and industries; these performance expectations must be consistently enforced and backed by appropriate responses to non-compliance, ranging from penalties to technical assistance. The regulatory expectations must in turn be supported by effective policy instruments, and here healthy debate continues around the appropriate mix of command-and-control, market-based instruments and information-based policies, such as performance disclosure. These policy instruments must be tailored to the range of firms and industries present within an economy, including SMEs, and foster innovation in new policy approaches, such as sector-based and place-based environmental management. The environmental regulatory system must progressively be built around concepts of pollution prevention and clean production, going beyond end-of-pipe pollution control. It must also address issues of resource pricing and infrastructure financing and development. Also, the environmental regulatory system must encourage a wide portfolio of private initiatives, ranging from the ISO 14000 series and supply-chain management to environmental due diligence in investment.

To varying degrees, the developing economies of South-East Asia have begun to put in place the elements of such an environmental regulatory system. The first-tier NICs (Hong Kong, Singapore, South Korea and Taiwan) have the core elements of a command-and-control environmental regulatory system in place. Here the important priority is to move beyond end-of-pipe pollution control toward the use of market-based instruments, pollution prevention and clean production and to strengthen non-regulatory drivers that support the achievement by firms of superior performance that goes beyond compliance with baseline environmental standards. These developing regulatory systems must increasingly search for more efficient ways to yield continuous improvement in environmental performance, especially among SMEs. Among the second-tier NICs (Indonesia, Malaysia, the Philippines and Thailand) and China there is an urgent need as a next step to strengthen the basic institutions of environmental regulation and to ensure the development of effective capacities for implementation and enforcement.

In some economies (Thailand and the Philippines), environmental agencies operate without landmark environmental legislation that empowers them to set ambient and emissions standards, monitor performance and enforce compliance (see Rock *et al.* 1999b). In others (Indonesia and Thailand), regulatory agencies have no authority to monitor, inspect or enforce facility-specific emissions standards. In virtually all of these economies, regulatory agencies lack both sufficient technical capacity and sufficient resources to manage national environmental protection programmes effectively. Even as these basic capacities of environmental regulation are being strengthened, the second-tier Asian NICs must draw on the experience of industrial economies around the world and move aggressively toward the adoption of market-based instruments, pollution prevention, clean production and superior environmental performance.

If environmental regulation is the bedrock of clean shared growth, much of the new policy opportunity—and the urgent policy imperative—involves going beyond compliance in a dynamic of continuous improvement leading to superior environmental performance. It is only with such superior performance that the scale effects of rapid urban–industrial growth will be offset by parallel reductions in the energy, materials, pollution and waste intensity of economic activity. Here, as a first step towards such a transformative development dynamic, we highlight four broad policy initiatives. In various forms, these initiatives are appropriate to all of the Asian NICs and are supplemental to environmental regulatory policy *per se*.

The first initiative involves strengthening non-regulatory drivers of environmental performance. The range of *potential* non-regulatory drivers is broad, including market demand, community pressure, cost reduction, market development and others. The key to harnessing these drivers, we argue, is the development and disclosure of transparent, low-cost, scaleable and standardised information on the environmental performance of production lines, enterprises, firms, industrial sectors, urban areas and national economies. Various experiments with such performance metrics are under way in Asia, such as the PROPER programme in Indonesia (see Chapter 7) and China's urban environmental indicators programme (Rock *et al.* 1999b). An important policy priority is the development and widespread adoption of effective systems of environmental performance measurement and disclosure.

The second initiative involves the identification and implementation of development goals for clean shared growth. Here we would highlight four dimensions of the goal-setting process: it must be directed to the next 20 years of development transformation in the region (2020); it must be framed in terms of energy, materials, pollution and waste intensity of urban–economic activity and in terms of basic economic processes of investment and technology change; it must be forged in partnership with business and other constituencies; and it must facilitate the identification and co-ordination of environmental performance goals across a range of relevant policy domains, from industry to technology and trade.

The third initiative focuses specifically on enhancing the capability of firms and industries in Asia to develop, use, adapt, adopt and improve on product and process technologies and associated manufacturing practices in a dynamic of continuous improvement. In the absence of such *in situ* industrial and technological capabilities, it will be extremely difficult for Asian economies to achieve the kind of economic transformation needed to improve substantially both environmental performance and socio-economic welfare. There is no one model for how such technological and managerial capacity-building might be achieved. Success in South Korea was heavily dependent on large firms (Kim 1997); in Taiwan it involved close collaboration between industry and government-sponsored R&D centres (Rock 1996b).

The fourth priority initiative is that of enhancing institutional capabilities for clean shared growth at both the local and the national scale and in terms of Asia's participation in regional and international environmental agreements. Increasingly, many policy decisions about clean shared growth are being devolved down to local communities and urban areas; it is important that the devolution of responsibility also be accompanied by the development of institutional capability and resources for policy implementation. At the same time, considerable attention needs to be given to the interplay among different levels of governance, from the local to the national to the international, and to the horizontal interplay among different policy domains.

We conclude with one final comment. It is important to recognise how little we know about many specific aspects of clean shared growth. For example, we lack systematic data on the characteristics of current industrial investment and technology flows in Asia. How clean, or dirty, is current investment? How effective are current policy initiatives, such as information clearinghouses and clean production round tables? To what extent do current production technologies present win–win opportunities in Asia, yielding economic and environmental benefits? There is an urgent need for systematic research and evaluation to parallel the kinds of policy innovation outlined in this book.

PART 1
Framing the Issues

David P. Angel and Michael T. Rock

IN CHAPTER 1 WE INTRODUCED THE BROAD ARCHITECTURE OF A POLICY APPROACH to the challenge of promoting clean shared growth among the newly industrialising economies of East Asia. We defined clean shared growth as a process of sustainable economic development that yields improvements in environmental quality (locally and globally) and continued enhancement of socioeconomic welfare, including reductions in poverty. Our analysis suggests that meeting the challenge requires a transformative approach to policy that yields more than incremental improvements in the environmental performance of industry, or what we have labelled a 'clean revolution' in urban–industrial economies. Others have envisioned the process of change in terms of a 'sustainability transition' in the relation of economic system to the environment (NRC 1999). In the context of the rapidly industrialising economies of East Asia, we identify reductions in the energy, materials, pollution and waste intensity of economic activity as a critical goal. The critical opportunity lies in influencing the technology choices and management of new urban and industrial investment.

The range of processes involved in such a development dynamic is broad indeed, ranging from issues of governance to processes of technology change. Similarly, many different groups are necessarily involved, from firms to local governments, community organisations to multilateral banks. The next five chapters examine in more detail five of the core process dynamics that impact on the challenge of clean shared growth: investment and technology change, economic globalisation, public policy, urbanisation, and governance and civil society. In each case, the authors examine the character of the development process in East Asia and the opportunities available to promote clean shared growth in Asia and the world.

In Chapter 2, George R. Heaton, Jr, and Budy Resosudarmo argue that technology has a pivotal role to play in reducing the energy, materials, pollution and waste intensity of economic activity. Three policy imperatives are identified: namely, the accelerated diffusion of best practice, incremental innovation in environmental technology and the harnessing of emerging technologies. Fundamentally, this would entail a departure from the past practice of OECD economies, where environmental problems have been addressed over the past 30 years primarily through end-of-pipe pollution control.

In Chapter 3, Daniel Esty, Mari Pangestu and Hadi Soesastro examine the linkages between international economic integration and the environment in East Asia. These authors identify trade agreements and multilateral agreements on investment, both at a regional level (such as the Asia–Pacific Economic Co-operation [APEC]) and internationally (such as the General Agreement on Tariffs and Trade [GATT]), as critical venues for shaping the environmental performance of economic activity. They conclude that progress has been limited in introducing environmental priorities into these international governance structures, and that 'the gap between what has been done thus far and what is realistically needed is widening'.

Chapter 4, written by Michael T. Rock, Ooi Giok Ling and Victor Kimm, considers the role of public policy in promoting cleaner shared growth in East Asia. These authors argue that getting policies right in three areas—in environmental policy, in trade and resource pricing policies and in industrial, investment promotion and technology policies—is critical to the success of cost-effective reductions in the energy, materials and pollution intensity of economic activity. They suggest a portfolio of short-term and medium-term policy initiatives that might be pursued by countries at different levels of economic development in East Asia, from Taiwan to Vietnam.

In Chapter 5, Michael Douglass and Ooi Giok Ling examine the interconnection between urbanisation, industrial growth and the environment in East Asia. Many judge rapid urbanisation to be one of the greatest economic and environmental challenges facing Asia today. The authors identify three broad opportunities for addressing this challenge: namely, capacity-building at the local and urban scale, the promotion of collaborative governance and the development of inter-city networks as venues for exchanging ideas and information.

Several of these ideas are also discussed by Lyuba Zarsky and Simon S.C. Tay in Chapter 6, where the authors examine civil society and the future of environmental governance in Asia. Here the importance of collaborative governance and stakeholder involvement is taken one step further to suggest that the prospects for ecological sustainability in Asia and elsewhere in the world rest largely on effective environmental governance. Several models of governance are examined. Even within East Asia itself the range of political and social contexts suggests that no one single model of governance is likely to succeed in all places. Sensitivity to economic, social and cultural context will be a critical facet of effective policy response within the emerging globalised economy.

2
TECHNOLOGY AND ENVIRONMENTAL PERFORMANCE
Leveraging growth and sustainability

George R. Heaton, Jr, and Budy Resosudarmo

Whatever one's view of global environmental prospects—fast worsening or slowly improving—there can be little disagreement that environmental quality—in Asia, the United States or Europe—still falls vastly short of what most people want. Though faced with this imperative, the principal decision-makers who control its realisation—that is, firms and policy-makers—too often parse the problem as a choice between economic growth today and long-term environmental sustainability. Technology offers a way out of this dilemma. Key to economic growth and environmentally friendly products, processes and systems, technological change is a neutral motive force that can be channelled toward whatever goals society chooses. There is no reason that new technologies, motivated by public policy and private gain, cannot effectively co-optimise along environmental and economic dimensions.

Technologies that improve environmental quality are hardly new. Indeed, a strong paradigm of what environmental technology is and how to elicit it developed within the nations of the Organisation for Economic Co-operation and Development (OECD) about 30 years ago. Essentially single-purpose, environmental technologies were developed to satisfy regulation-mandated pollution limitations and clean-up after the fact. Rarely did they emphasise *ex ante* design changes that could avoid environmental insult. A global industry with sales of more than US$400 billion operates on the basis of this paradigm, with its most buoyant markets today in industrialising countries where environmental policy is taking root.

From the benefit of experience, the pollution-abatement and clean-up paradigm can be judged a useful, but limited, first-generation approach. For the longer term, however, its tendency to prolong the lives of inherently dirty or resource-intensive technologies makes it a doubtful platform from which to launch an environmentally sustainable society—or to promote economic growth. Happily, a different approach—clean, shared growth—is beginning to be envisioned. Dematerialisation of industry, intelligent process controls, eco-friendly products, alternative agriculture and industrial firms whose environmental and economic agendas coincide are some of its elements. To achieve this transformation of technology, a parallel transformation must be forged in the realm of public policy: environmental policies reinvented and merged with polices toward industrial technology and investment, cast as a catalyst for technological change.

The essential question in this chapter is: What are the prospects and possibilities for an environmental technology transformation in Asia? Because no data speaks directly to the point, its answers are only suggestive, not definitive. It is clear, however, that the potential and the stakes are enormous. Asia's population, economic growth and increasing environmental footprint make it perhaps the most important pivot on which the world's environmental future will turn. And, while Asian environmentalism will strengthen, it is by no means certain what form it will take. In neither Asia nor the OECD countries has the character of environmental technology yet focused on long-term investments that can prevent pollution, cut resource intensity and design for the environment. The argument herein maintains that Asia's historic high growth and high capital investment pattern, coupled with its technical capabilities, economic and institutional structures, remain poised to realise this potential: a fundamentally different approach from the technology retrofit that has characterised environmental policy in the OECD. To define and achieve it will require important changes in perspective and practice, both in Asia and among the OECD nations, who are inevitable partners in the endeavour.

The discussion begins with a review of technology's critical role in achieving environmental quality and economic growth. It then dissects the process of technological change and proposes three imperatives that employ different aspects of it as forces for environmental improvement. After reviewing pivotal features of the Asian situation, it proposes a more specific package of short-term and long-term environmental technology policies.

Technology: the critical variable

One way of looking at the relationship between human activity and the environment is to conceptualise it as a mathematical identity. Seen thus, aggregate environmental impact becomes a function of the total number of people, the amount of their economic activity per capita and the intensity of environmental insult that their patterns of production and consumption imply. Expressed formulaically in terms of pollution, population and gross national product (GNP), the relationship is usually stated:[1]

$$\text{pollution} = \frac{\text{pollution}}{\text{GNP}} \times \frac{\text{GNP}}{\text{population}} \times \text{population}$$

[1] The authors are indebted to J. Gustave Speth for discussion of this identity and to Marian Chertow for work tracing its origin and evolution. Both are now at the Yale University School of Forestry and Environmental Affairs, New Haven, CT, USA.

Naturally, this identity is only an abbreviated shorthand, and, naturally, it is subject to dispute. The very word 'pollution', for example, needs to be expanded into the concept of the total 'environmental footprint' of a society's activity, including aspects as diverse as its patterns of resource usage and its destruction of species in natural ecosystems. Similarly, the connection between GNP per capita and the amount and intensity of pollution is much more complex than the equation can express;[2] so too is the dynamic between population growth and economic growth.

Despite its limitations, the equation still focuses attention on elements that can make a real difference to environmental quality. By reducing it and using slightly different language, these factors come into relief:

environmental footprint = pollution intensity × economic growth

When one further considers what variables underlie both economic growth and pollution intensity, technology quickly emerges as key. For advanced industrial economies such as the United States, it is well established that technology is the main source of productivity improvements, without which economic growth would stagnate (see the literature cited in Council on Competitiveness 1988). In the development model seen in the emergent economies in Asia, assimilation of externally generated technology has been at the core of first-stage growth; creation of new technology and technological adaptation provide the basis of its continuation (on the Korean experience, see Kim 1997).

The impact of technology on the environment emerges as less unidirectional, but no less critical. Early on, environmentalists tended to blame technology for environmental degradation. In a literal sense, the criticism hit the mark: many post-WWII technologies—synthetic pesticides for example—appeared to be damaging ecosystems more seriously and faster than traditional practices. But, more subtly, it was soon recognised that the real culprit was poor, ill-informed or misdirected design choices made by humans rather than anything inherent in technology. Applied correctly, technology is equally the enabler of environmental progress: today, for example, the revolution in agricultural biotechnology holds the key to eco-friendly pest control.

Although technology is the variable that underlies improvements both in economic wellbeing and environmental sustainability—and can accomplish the two goals simultaneously—it is important to realise that it is a dependent variable, with no preordained course. In other words, technology is endogenous, with its form and direction dependent on the signals its creators receive from the cultures, markets and institutions in which they work. Studies tracking technological change in firms have consistently demonstrated this: innovations motivated by market needs, as opposed to laboratory discoveries, are far more successful. Similarly, the alacrity with which firms can produce new technology in response to environmental demands—either from consumers or from regulators—is beyond question: the quick rise of substitutes for phosphate-based detergents in the early 1970s, for polychlorinated biphenyls (PCBs) in the 1980s and

---

2 The well-established literature on dematerialisation in highly industrialised societies, the 'reverse Kuznets curve' and particularly Shakeb Afsah's recent work all speak to this issue.

chlorofluorocarbons (CFCs) in the 1990s are all examples that make the point (Heaton 1990).

Technology is thus the variable in the environmental equation that is uniquely flexible, applicable in both the short term and the long term and is relatively value-neutral. Population trends, in contrast, obviously offer only a long-term alternative—and present significant moral dilemmas. And, even if reduced economic growth were to represent a pollution-limiting strategy, it would hardly be palatable. In order to consider what possibilities technology offers as a lever for both economic and environmental improvement, its pathways, its prospects in Asia and the policies to influence it are examined further below.

Pathways of technological change and investment

Technology is the application of science to useful purposes. Though science and technology are often linked, it is a mistake to dwell on this connection, particularly in the Asian environmental context. Science—new knowledge and understanding—is not one of the major factors limiting progress toward environmental improvement in Asia. Deployment of better technology certainly is. Beyond this, new science is not always necessary to the creation of new technology, in Asia or elsewhere. Indeed, technological innovation—which includes both the 'hardware' of machines and physical processes as well as the 'software' of its management and use—often takes place in the absence of scientific discovery. The steam engine, for example, predated theoretical understanding of why it worked; and the organisation of mass production—an essential technology of modern life, with profound environmental implications—drew little, if at all, on science.[3]

Technological innovation is the first commercial application of a new technology. In virtually every society today, technological innovation occurs overwhelmingly in private firms—not in universities, government labs, the military or research institutes. Although these latter institutions are certainly critical in establishing and diffusing the knowledge base on which technological innovation depends, firms are ubiquitously the main generators and deliverers of new technology to society. This statement holds as true for the 'environment industry' as it does for more traditional sectors.

Most innovations are not radical leaps forward; in fact, the preponderance are incremental improvements to existing products and processes. If radical innovations typically require systematic and extensive research and development (R&D) over a long time-period, incremental efforts proceed at a more modest, continuous pace. Nor does being innovative require PhDs or other advanced degrees. But what does seem to distinguish innovative from non-innovative firms is the former's attentiveness to the signals sent by the market and society, their flexibility and the open, non-hierarchical management style that they combine with technical acumen.

---

3   The above said, it is true that the connection between science and technology is argued to be closer today than before—an effect termed 'telescoping' of research-intensive technologies (e.g. biotechnology and computers).

Innovative firms generate technically successful and economically viable new products, new processes and new systems. They may sell them or use them internally. In addition, all firms, innovative or not, are the consumers of innovations. This process, by which innovations spread to subsequent users, is termed 'diffusion'. In successful diffusion, the recipient must discover and capture the value of technology generated elsewhere. This is a fundamentally different process from innovation. The most successful diffusion efforts rest on good information about available technology, adequate capital to acquire it and the technical resources to adapt it in a new circumstance. In fact, most technologies need adaptive reworking sooner or later; and when adaptation becomes extensive enough it is virtually indistinguishable from incremental innovation.

'Technology transfer' is another way of describing diffusion, especially when it occurs through sales or turnkey transfers of technologies in licences of intellectual property. Particularly in developing countries, technology transfer has been perceived as the main mechanism to acquire technologies developed externally. It tends to be visualised as a quicker alternative to innovation, particularly when a country's technical capabilities are insufficient. Although this formulation has its appeal, it can also be dangerous to the extent that it gives the misimpression that the societal capability for overall technological development is less than an integrated system involving all stages of the innovation process. Moreover, pursuit of technology transfer often overlooks the need for capabilities and relationships that can accomplish the hard work of adapting, integrating and renewing technology in a new context. Indeed, to the extent that the lure of technology transfer diverts attention from building supplier–user co-operation over the long term, it can be a dysfunctional concept.

In all countries, technological change is closely associated with capital investment. Investment in new plant and equipment both creates demand for technological innovation and affords the locus for its deployment. In Asia, the investment pathway to technological change is particularly critical for two reasons: the combination of extraordinarily high rates of capital turnover, and low rates of indigenous R&D.[4] From an environmental point of view, this situation holds both potential and danger: the opportunity to move environmentally friendly technologies into place faster than elsewhere, via new investment, and the danger of over-reliance on externally generated solutions to local environmental problems.

The last essential point about technological change is that it is driven overwhelmingly by demand. Just as firms that rely on 'technology-push' strategies of laboratory-based invention are routinely less successful than those that innovate to suit their customers' needs, so too are national strategies that emphasise increasing the supply of environmental technologies likely to be less successful than those that augment demand for environmental quality. But the concept of 'demand' is by no means equivalent to that of 'the market'. Indeed, it is clear that free markets will come nowhere close to producing the amount of environmental quality any society wants and that governments must therefore augment demand through regulation, information strategies, financial incentives and other means. But, from the vantage point of the innovating firm, the source of

4   Of course, rates of R&D vary significantly across the Asian region (see discussion below), though high rates of investment have been universal.

demand is irrelevant as long as it presents viable commercial opportunities for new technology to fill.

For an environmental technology transformation: three imperatives

Diffuse best practice

The need for an 'environmental transformation of technology' has been apparent—and advocated—for some time (Heaton et al. 1991). One part of the argument rests on a critique of technologies in current use—consumer products, industrial processes, infrastructure systems—from the viewpoint of long-term environmental sustainability. Here, little doubt surrounds the premise that the technologies now employed vastly undershoot the environmental performance that would be possible by adopting others, already developed. With air and water pollution levels now worse than in any other region, Asia amply illustrates this proposition (see Chapter 1, Table 1.1 on page 11). Clearly, technical problems are not the factor that is impeding the fuel economy of US automobiles from doubling (for discussion of current research programmes with this goal, see Roos et al. 1997). Similarly, readily available building products—from sun-sensitive 'smart' window coatings, to long-life bulbs, to standard insulation—could halve US residential energy consumption (Kelly 1990); and substitution of electric-arc steel-making for the basic oxygen process could move scrap inputs in steel from 30% to 100% (Heaton et al. 1991). That these improvements do not occur has little to do with technical capability. The same point can be made about every economy and most economic units: almost none operates at the state of the art.

Diffusion is thus the first imperative in an environmental technology transformation. Policies to promote faster, wider diffusion of today's better but unused environmental practices need not focus on the traditional supply-side features of technology policy—R&D, education of scientists and engineers, etc. Rather, they must increase demand for improved technologies—through deliberate use of regulation or other industrial standards. They need to decrease the cost of such technologies relative to their established competitors—through mechanisms as diverse as taxation and commodity pricing. Last, they need to facilitate the channels of information and acquisition for the entities that are their consumers.

Incremental innovation: rethinking environmental technology

Worldwide sales of the 'environmental industry' now amount to over US$450 billion yearly (USDC 1998). Although there is not yet a good, internationally accepted definition or database for the industry,[5] the technologies that the industry sells typically have

5 Work on this issue has been episodic, but continues at the OECD and the US Department of Commerce (see especially USOTA 1994).

environmental improvement as their sole purpose. These are commonly categorised into four sub-groups: pollution control, damage remediation and restoration, pollution monitoring and assessment, and damage avoidance (NSTC 1994). Although data is not available to show the breakdown of expenditures among these four categories, it is safe to say that such data would rank them in the order they are listed above, with pollution control the largest and pollution avoidance much the smallest.

Another way to view the industry is through the functions it provides, which include services, sales of equipment and the delivery of environmental resources (water, clean energy, recovered materials). Here, the data is better, as shown by Table 2.1, which contrasts revenues of the industry in the United States, Europe and Japan.

The data in Table 2.1 supports a number of important points about the character of environmental technology within the OECD.[6] Most tellingly, one sees that only a trivial amount—no more than 0.5%—is prevention-oriented. In all three venues, water treatment works and utilities absorb approximately 30% of the total. Beyond this, the pattern of expenditure derives from the regulatory regimes that are ubiquitously based on pollution control standards for individual media: air, water, waste. Naturally, there are some differences in emphasis. The United States, for example, spends a considerably greater percentage on air pollution compared with the other regions, and its consulting, engineering and remediation services outpace those of the others. Japan focuses on solid-waste management much more heavily than do the other regions, and Europe and Japan both emphasise resource recovery more than does the United States.

Historical data such as that in Table 2.1 does not necessarily indicate future directions. Although data on patterns of environmental industry R&D would go far toward showing the trajectory of environmental technology, this are not yet available. A few facts are known, however. First, the industry's R&D as a whole is very low in comparison with other technology-intensive sectors, and it is highly concentrated in a few areas (USDC 1998). In addition, throughout the 1990s, the stock market performance of the US environmental industry has significantly underperformed industry averages, and the amount of venture capital investment has fallen by a factor of ten (USDC 1998). Particularly in the so-called 'valley of death' between the generation of a new idea and its commercialisation, financing for new environmental technology has been in extremely short supply (NSTC 1994).

The above notwithstanding, the trajectory of environmental technology development today should not be viewed in entirely negative terms. On the contrary, particularly among the major 'polluting' sectors—chemicals, paper and pulp, resource extraction, large-scale manufacturing, electronics, etc.—there appears to have been an important change in mentality and in internal environmental practice: acceptance of environmental sustainability as a core mission of the company, whether in domestic or foreign investment (Smart 1992). Evidence of what this means for the process of technological innovation—though almost entirely anecdotal—is nevertheless telling. Major firms in the United States, for example, report the transformation of their R&D processes as they

---

6 It should be noted that, although the table covers only one year and two sites, data does exist over time for the United States, Japan, Europe and Asia. Asian data is presented below.

	US		Europe		Japan	
	$ billion	(%)	$ billion	(%)	$ billion	(%)
EQUIPMENT						
Water and chemicals	16	9.3	10.5	7.9	5.6	6.4
Air pollution control	15.4	9.0	7.3	5.5	3.3	3.8
Instruments and information	1.8	1.0	1.6	1.2	1.0	1.1
Waste management	10.7	6.2	9.1	6.8	8.6	9.9
Process and prevention	0.9	0.5	0.5	0.4	0.5	0.6
SERVICES						
Solid waste management	32.7	19.0	29.5	22.1	29.6	34.0
Hazardous waste management	5.9	3.4	5.2	3.9	3.8	4.4
Consulting and engineering	14.2	8.3	8.4	6.3	1.1	1.3
Remediation	8.3	4.8	3.7	2.8	1.1	1.3
Analysis	1.2	0.7	1.0	0.7	0.5	0.6
Water treatment works	24.6	14.3	21.8	16.3	9.6	11.0
RESOURCES						
Water utilities	27.0	15.7	19.7	14.8	12.2	14.0
Resource recovery	11.6	6.8	13.6	10.2	9.2	10.6
Environmental energy	1.4	0.8	1.5	1.1	1.0	1.1
TOTAL	**171.8**		**133.4**		**87.1**	

NB *US$ are used throughout; percentages do not always sum to 100 as a result of rounding errors*

Table 2.1 **Revenues of the environmental industry: US, Europe and Japan, 1996**
Source: USDC 1998

move toward the integration of environmental goals with overall firm strategy (Heaton *et al.* 1992). Policy experiments deliberately designed to encourage technological innovation have proliferated in the United States and Europe, with significant success (Anex 1999; Norberg-Bohm 1997).

There is also some evidence of an emergent new approach to the design of environmental technology. For example, a major survey of the environmental industry by the US

Department of Commerce has shown that the market for 'tacked-on' pollution-control equipment and waste-management services appears to be in long-term decline. Taking stock of this situation, the industry has come to believe, according to this survey, that its best prospects lie in a three-part 'reinvention' in which it would:

- Sell value, not only technical 'fix-its'
- Deliver total resource productivity rather than environmental control
- Integrate environmental management with customers' overall business strategy

Realising this approach would represent a major rethinking of the environmental technology paradigm of the past 30 years.

Still, the compelling conclusion from trends to date is that the environmental business in the OECD nations has been structured largely as a technology retrofit exercise with a particular, limited purpose: compliance with regulation. The corollary is its lack of consonance with the larger dynamic of technological change in the firms that are its clients. If these trends were unfortunate in the OECD context—prolonging the life of old technology rather than transforming it—they may prove disastrous if applied to Asia. Whether the economies of Asia revert to high growth or not, the imperative will still be to integrate environmental and growth objectives. If the dominant paradigm of pollution-control technology from the OECD nations is transferred, this kind of co-optimisation will not be possible, and environmental technology will continue to be placed outside of the mainstream of industrial development.

Radical change: harness emerging technology

The third imperative for an environmental transformation of technology is to harness the potential of emerging technological revolutions—information technology, biotechnology and new materials—in ways that promote environmental sustainability. In each case, their potential is enormous: intelligent manufacturing systems with zero waste; pest-resistant crops that eliminate the need for chemical pesticides; materials designed for total recyclability (Heaton et al. 1992). Throughout the OECD countries, an increasing percentage of private investment is moving toward these 'high technologies', and public R&D programmes are increasingly focused on their promotion (Heaton 1997a). The argument is only just beginning to be made that, without the kinds of radical innovations these technologies hold, progress toward an environmentally sustainable society will be marginal (Anex 1999; Norberg-Bohm 1997).

Efforts to tap the potential of these technologies for environmental improvement have only just begun. Japan has probably gone farthest in terms of funding, with the inauguration some five years ago of the Research Institute for Innovative Technology for the Earth (RITE), the world's largest environmental technology research facility (Heaton 1997a). The European Community has recently made the environment a major feature of its technology promotion 'plans' (Heaton 1997a). In the United States, the Clinton–Gore Administration proposed a large-scale Environmental Technology Initiative (ETI) early in its first term, but this is now largely moribund (Heaton and Banks 1997).

Without doubt, virtually all of the R&D and the radically new innovations arising from these fields will continue to be located in the OECD countries. However, their eventual diffusion into the Asian context is a matter of equal importance. Thus, the technology and environmental policies throughout Asian countries could benefit enormously over the long term from technical capabilities that focus on the application of these emerging technologies in the Asian context.

Pivotal underpinnings of technological change in the Asian situation

It bears repeating that the current environmental and technological situation in Asia presents a crossroads of pivotal significance, to both Asia and the world. If Asian patterns of social and economic organisation have differed from those that characterised the development trajectory within OECD countries, the technologies employed to date have not. By and large, the 'Asian miracle' has been built on the same resource-intensive, highly polluting, unsustainable technologies commonly employed in the West. If these technologies must change, where will the locus of change lie and on what principles will it be designed? More than one answer is plausible.

One possibility is that Asia will follow the OECD lead, in both the design of environmental policies and in the deployment of technologies to implement them. By and large, this would mean end-of-pipe, clean-up technologies and reliance on technology transfer into Asia from OECD vendors. The likelihood of this scenario is based on history and convenience: the inertia that leads to a continuation of past patterns. The obvious downsides are that it is unlikely to ameliorate current environmental conditions to an acceptable extent, to reverse the tendency toward degradation or to augment Asian technological independence. Another possibility is that Asia can chart its own course. Alternative environmental policies tailored to the Asian situation can be pioneered, and new, clean technologies can be developed internally.

If the choice between the two pathways as so stated appears stark, it is unlikely to be so in reality. Technologies and policies will continue to be imported into Asia from the OECD nations, but they will also continue to be adapted and modified to suit local situations, and new technological and policy approaches, drawn from a now vast store of accreted capability, will be developed in Asia. The following discussion outlines some of the pivotal underpinnings of the trajectory of environmentally oriented technological change in Asia. These include: aptitude for change, technical capability, regional investment patterns and the direction of environmental investment.

Aptitude for change

One of the benefits of fast economic growth in Asia is almost a tautology: the aptitude for growth and change. In fact, however, societies exhibit different capabilities and

2. TECHNOLOGY AND ENVIRONMENTAL PERFORMANCE *Heaton and Resosudarmo* 51

	1980	1985	1990	1992	1994	1996	1997
Indonesia	20.9	23.1	28.3	25.8	28.3	31.1	n/a
Malaysia	30.4	27.6	33.6	35.1	38.7	41.5	42.8
Thailand	29.1	28.2	41.1	40.0	41.2	41.7	35.0
Taiwan	33.9	19.1	23.1	24.9	24.1	21.2	23.4
Singapore	46.3	42.5	39.5	36.4	32.2	34.1	36.1
Japan	32.2	28.2	32.3	30.7	28.7	29.6	31.3

Table 2.2 **Ratio of gross domestic investment as a percentage of GDP in Asian countries**
Sources: APEC 1998

proclivities for change, and these are, to some extent, learned characteristics. In many Asian societies, the 1980s and 1990s forced public institutions, private firms and individuals to become adept—through experience—at the skills and activities that confer success in a rapidly changing economy. Yearly GNP growth has averaged more than 5% from the mid-1960s to the mid-1990s; export production has been continuously recalibrated to suit new market needs; foreign capital and technology have revolutionised the structure of industry. Indeed, it is ironic that Japan, the region's first mover, may now be the country most resistant to change.[7]

Another factor that reveals the aptitude for change is the relationship between gross domestic product (GDP) and gross domestic investment. As Table 2.2 indicates, two basic tendencies can be seen: a high and increasing rate of domestic investment among the less-developed countries (i.e. Indonesia, Malaysia and Thailand) and a high but decreasing rate of domestic investment among the more developed countries (i.e. Japan, Taiwan and Singapore).

The domestic investment data in Table 2.2 extrapolates to what is already well known: the extraordinarily fast rate of capital turnover all Asian societies were experiencing before the recent economic crisis. Since capital investment is one of the primary vehicles for technological change (typically, diffusion), these numbers also offer a surrogate measure for the habitual willingness with which these societies accept new technologies.

Going beneath the quantity of investment, it needs also to be emphasised that Asian economies have experienced dramatic structural shifts, away from traditional agricultural pursuits toward industry and services. The combined scenarios of high capital investment and sectoral shifts have produced the well-known increases in the 'toxic

---

7   Indeed, the core technology policy debate in Japan today revolves around exactly this issue: has Japan become so conservative and rich that its capability for growth and change has atrophied?

intensity of production' at the root of environmental crisis. On the other hand, Asia's habituation to technological, economic and structural change as the usual case may also give it a unique aptitude—surpassing that of any other region—to accept the changes necessary to turn new investment in an environmentally friendly direction.

Technical capacity understated by conventional measures

Conventional measurements of national scientific and technological capabilities tend to focus on a few well-known inputs. For the Asian region, however, such data is both an imperfect and a misleading characterisation of the capabilities that can be applied to the process of technological change, particularly in the case of environmental technology.

As Table 2.3 shows, Asian R&D scenarios present a three-tiered pattern of countries over time: highly R&D-intensive societies, such as Korea and Japan, which are now essentially on a par with the United States and Europe in the level of investment; societies with increasing R&D intensity, such as in Singapore and Taiwan, bringing them to a moderate level by worldwide standards; and countries with a continuing low level of R&D, such as Indonesia.

A well-recognised limitation of R&D data is that it only measures organised R&D. To the extent that technological change relies on less formal activity than R&D and focuses on incremental innovation and adaptation rather than radical technical breakthroughs, measurements of national R&D may understate a country's true capability.

The most important input to the process of technological change is certainly technically skilled people, whether or not they are engaged in R&D. Numbers of recipients of technical degrees within a given population are typically used to tally this input. As Table 2.4 indicates, the proportion of recipients of masters' and doctoral degrees in Japan, Singapore, Korea and Taiwan is similar to or exceeds that in the United States, thus giving these countries exceptional human capabilities in technology development.

Another important educational phenomenon for Asia is the number of students who receive degrees abroad in science and engineering. By the early 1990s more than 125,000 Asians were receiving such degrees in the United States. This figure, which tripled during

Table 2.3 **Investment in R&D as a percentage of GDP, selected countries**
Sources: BPPT 1993; NSF 1993, 1997, 1998

	Japan	Singapore	Korea	Taiwan	Indonesia	US
1980	2.0	0.3	0.6	0.7	0.3	2.3
1985	2.6	0.7	1.4	1.0	0.3	2.7
1990	2.9	0.9	1.9	1.7	0.3	2.6
1995*	2.8	n/a	2.8	1.8	0.3	2.5

* *Estimate*

	Japan		Singapore		Korea		Taiwan		US	
	M	PhD	M	PhD	M	PhD	M	PhD	M	PhD
1975	121	41	8	3	n/a	n/a	n/a	1	173	19
1980	131	54	19	8	58	5	56	2	162	17
1990	209	88	54	22	145	22	184	15	150	17
1994	292	91	n/a	n/a	n/a	37	n/a	28	169	18

Table 2.4 **Recipients of masters' (M) and doctoral (PhD) degrees per million people for selected countries**

Sources: BPPT 1993; NSF 1993, 1997, 1998

that decade, represented about two-thirds of all US science and engineering degrees given to foreign students and about one-quarter of all such degrees. Although the vast majority of these degrees went to citizens of China, Japan, Taiwan, India and Korea, other countries in the region were not unrepresented (NSF 1996).

What is less appreciated is the augmentation of technical ability that will result from dramatic increases in Asian students going to Japan. Recruitment of Asians to Japan through the provision of generous scholarships has become a Japanese priority, with a goal of 100,000 such students by the year 2000. Most of these will be drawn from China, Indonesia, Thailand, Pakistan and Bangladesh. Even now, foreign students, largely from Asia, receive about 40% of Japanese degrees in science and engineering (NSF 1996). Besides serving as an indicator of internal technical capability, the tendency by Asians to study abroad reflects an increasing integration into the world economy and an accompanying reduction of disparities in the level of science and technology, at least among Asian élites.

Another measure of the degree to which Asian individuals and firms can compete at the highest levels is offered by Table 2.5, which tracks US patents granted to Asian inventors. Although Japan vastly outpaced all other countries (accounting in 1997 for about half as many patents as those held by Americans), Korea and Taiwan were also significant presences in the late 1990s, and the remaining countries listed—Indonesia, Malaysia and Singapore—have all made dramatic increases from a low base.

The quantity and distribution of Asian technical capability that this various data shows suggests two important possibilities for future environmental technology development. First, aggregate technical capability in Asia is likely to be adequate to support much more indigenous development of environmental technology, if public policy and other factors push in this direction. Second, the tiered structure of Asian R&D suggests that a new pattern of intra-regional specialisation may be emerging. Breaking with past reliance on the United States and Europe as the engine of new technology, the high-R&D Asian societies could well become independent sources of new environmental technology for those that are less R&D-intensive. This possibility is supported by the data below on regional investment.

	Japan	Singapore	Korea	Taiwan	Indonesia	Malaysia	US
1971	4,006	4	2	0	2	0	55,467
1981	8,387	4	17	80	1	1	38,019
1991	21,027	15	404	904	2	12	47,569
1997	24,314	111	1,828	2,490	12	26	57,876

Table 2.5 **US patents granted to inventors from Asian countries**
Sources: NSF 1993, 1996, 1998; USDC 1998

Increasingly regional technology and investment

One of the trends that emerged in Asia during the 1980s was a movement toward stronger regionalism, in which flows of technology and investment inside the region derived increasingly from other countries within the region. This trend is highlighted in Table 2.6, which tracks the origin of foreign direct investment (FDI).

The data in the table highlights the rise of newly industrialising economies (NIEs) as a source of capital in Asia. This is particularly true for Malaysia, the Philippines, Thailand and Indonesia, where in 1990 capital from the NIEs accounts for a third to half of all FDI. The relative decline of the United States as a source of FDI is also apparent; only in Korea was it still the prime investor in 1990. In many countries, Japan became an increasing investment presence during the 1980s. When Japan's and the NIEs' FDI contributions are aggregated, the reliance of Asia on itself—not the United States and Europe—becomes all the clearer.

The implications increasing regionalism may hold for environmental policy are provocative. If Japan and other Asian sources increase their dominance in capital investment, then it stands to reason that the paradigm of their environmental policies—and technologies—will provide an increasingly important model as well. If their approach essentially replicates that taken earlier in the United States and Europe, then the same pattern of environmental technology development can be expected throughout Asia. At the same time, however, newly important Asian sources of capital could also be the lever to a transformed approach in the region.

The trajectory of environmental technology

The lack of good studies tracking patterns of investment and R&D for environmental technology—either in the OECD countries or in Asia—represents a serious drawback to determining the trajectory technology development will take. To the extent that the past determines the future, it presents a troubling indication. Table 2.7 arrays expenditure patterns for environmental technologies in Asia in 1996 as compared with those in OECD countries. Making aggregations across categories, one sees that over 50% of the total environmental investment in Asia is allocated to the combination of water utilities, water

2. TECHNOLOGY AND ENVIRONMENTAL PERFORMANCE *Heaton and Resosudarmo*   55

		US	Japan	NICs*
	1986	3.3	11.1	23.7
MALAYSIA	1988	12.6	27.9	35.3
	1990	6.2	31.8	39.9
	1986	28.7	28.5	10.2
PHILIPPINES	1988	12.6	27.9	35.3
	1990	6.2	31.8	39.9
	1986	19.5	36.0	9.2
TAIWAN	1988	12.7	40.7	12.2
	1990	25.9	39.7	11.9
	1986	7	43.2	15.7
THAILAND	1988	10.8	49	27.4
	1990	7.7	19.2	62.2
	1986	16	40.6	10.5
INDONESIA	1988	16.6	5.8	34.7
	1990	1.7	25.6	29.7
	1986	35.4	38.9	4.5
KOREA	1988	22.2	54.3	1.2
	1990	39.5	29.3	2.6

* Newly industrialised countries: Taiwan, Singapore, Hong Kong and Korea

Table 2.6 **Origins of FDI in selected Asian countries (%)**

Source: NRIISAS 1995

treatment and water equipment and chemicals. This is a considerably larger share than in Japan, Europe or the United States, where the same aggregation of categories accounts for 30%–40% of the total. In most other categories, environmental technology investment patterns in Asia do not depart strongly from the OECD model. Process and prevention technology, for example, is less than 1% of the total in every region. This data suggests that the focus of environmental investment in Asia to date has been even more strongly on abatement and clean-up—above all for water—than was the case in the OECD nations.

	US	Europe	Japan	Asia
EQUIPMENT				
Water and chemicals	9.3	7.9	6.4	14.1
Air pollution control	9.0	5.5	3.8	4.7
Instruments and information	1.0	1.2	1.1	1.0
Waste management	6.2	6.8	9.9	6.8
Process and prevention	0.5	0.4	0.6	0.5
SERVICES				
Solid waste management	19.0	22.1	34.0	17.8
Hazardous waste management	3.4	3.9	4.4	2.6
Consulting and engineering	8.3	6.3	1.3	4.2
Remediation	4.8	2.8	1.3	2.1
Analysis	0.7	0.7	0.6	0.5
Water treatment works	14.3	16.3	11.0	14.1
RESOURCES				
Water utilities	15.7	14.8	14.0	23.6
Resource recovery	6.8	10.2	10.6	5.8
Environmental energy	0.8	1.1	1.1	2.1
TOTAL	100	100	100	100

NB *Percentages do not always sum to 100 as a result of rounding errors*

Table 2.7 **Comparative expenditures for environmental technology (as a percentage of total expenditure)**

Source: USDC 1998

These patterns of investment point again to the watershed Asian countries are facing today. Clearly, Asian investments in environmental technology to date have been calculated to ameliorate the lack of basic environmental infrastructures; clearly, they have followed the pattern employed in OECD countries; and, clearly, OECD-based firms have been the suppliers in this technological trajectory. If a future paradigm of indigenous, clean technology is to be developed in Asia, it must discard the pattern that this data implies and be based on a new set of environmental policies and prescriptions. These are discussed below.

Policies for technology transformation in Asia

If it is easily plausible to envision the force of environmentalism strengthening throughout Asia over the coming decades, it is nevertheless far from clear what form its implementation will take. As conceived in the OECD countries, environmental policy thus far has largely focused on mandating desired environmental conditions, with only subsidiary attention given to the technologies and management practices through which they will be achieved. The consequence has been primary reliance on clean-up, end-of-pipe, remediative technology.

If Asia is to chart a new way for itself, it must develop new policies. In particular, it must develop policies that focus as much on technology as on environmental desiderata. The goal of such policies should be to recast how the environmental impacts of technological choices are considered and coped with by investors, engineers, managers and all those who affect the development and deployment of technology. An agenda for such an environmental technology policy, encompassing an appropriate conceptual framework, macro-policies, short-term technology diffusion, medium-term green design and long-term harnessing of radical technological change, is set out below.

Avoiding an inappropriate paradigm

Much of the paradigm for environmental policy—and its implementing technologies—originated in the United States during the 1960s. At that point, the US economy was at the apogee of a particular production paradigm and industrial structure. Much the same may be said for Europe and Japan: all were manufacturing-based, wedded to large-scale mass production and dependent on industrial facilities that were often old. The industries at the bedrock of these economies—automobiles, chemicals, steel, consumer durables—were resource-intensive, polluting, economically conservative and technologically rigid. The technological revolutions that dominate today's economy—information, biotechnology, miniaturisation—were then only nascent, and the intense global interconnectedness that now characterises economies and corporations alike had only begun to unfold.

These conditions dictated the design of environmental policy. To a large extent, OECD environmental policy through the 1970s and 1980s was almost entirely a domestic affair: countries establishing their own standards based on local conditions, capabilities and politics. A 'command-and-control' regulatory process set the agenda, relying on a tradition of 'rational' government decision-making that emphasised rigorous scientific and economic analysis, large administrative and data resources, and adversarialism, whether intellectual or legal. Last, the implementation of regulation relied overwhelmingly on technological retrofit: the adoption of known or close-to-available technologies that could be integrated into the industrial status quo.

In Asia, today's circumstances push in precisely opposite directions. First, Asia's environmental problems are as much the world's as its own. Global warming illustrates this from a purely physical point of view, and the interdependent flow of capital, products and technology makes it an economic reality. The impetus for change will thus arise both from external and from internal pressures. Second, the Asian economies are likely to continue to be much more hospitable to structural change—and much more dependent on new investment—than was the case within the OECD countries. Designed-in avoidance of environmental damage can thus assume higher priority. Third, the technical capabilities of Asian societies are not heavily weighted toward R&D and policy analysis, nor are their administrative resources so ample. This suggests that Asian environmental polices cannot rely on the formal, legalistic and technical policy mechanisms common in the OECD countries. Last, the traditional policy and power dynamic in Asia—in which industrial interests and ministries play a major role—may well afford a more natural pathway for policies that harmonise environmental and economic goals than was ever the case in countries where these functions have traditionally been divided or at odds.

Structuring the macroclimate

Because environmental policies are generally superimposed on a pre-existing policy framework, they encounter a common difficulty: entrenched economic, institutional and policy structures that are at best orthogonal to, and at worst openly in conflict with, their purposes. Such policies form a macroclimate that conditions and channels technology development. Without changes in them, the goal of technological transformation for a sustainable economy will be difficult to achieve.

Many of the most profound hurdles to the commercialisation of environmentally superior technologies reside in commodity and resource pricing. Energy prices, though certainly the largest single factor, are far from the only problem. Agricultural subsidies, underpricing of virgin resources, and economic disadvantages to recycling all play a part. Even among explicitly environmental policies, pricing schemes to make pollution, resource use and waste expensive are vastly under-utilised. At the other end of the policy scale, lack of an environmental focus in the education of managers and engineers effectively encourages the ignoring of environmental factors in technology design and investment choices. The skills of environmental analysis and green design—which are teachable—clearly offer a key to long-term environmental improvement.

A full discussion of the macroclimate for technological change is far beyond the scope of this chapter. The main point of listing some of them here is to emphasise that technological change is as much a function of endemic social features as it is of scientific and technical knowledge, and that these features need to be considered as part of the overall innovation and environmental policy system.

Policies for the immediate term: promote incremental innovation and diffusion of the state of the art

Technological innovation is both a long-term and a disruptive process. To the extent that it poses radically improved alternatives to the technological status quo, it will eventually drive the technologies that represent the status quo out of use. In the short term, however, the technologies in current use can usually be substantially improved—if they are challenged.[8] Evidence from many industries in OECD countries makes the case that the source and nature of the challenge is less important than its force—environmental regulatory pressures, for example, may motivate as effectively as international competition. (Studies that have surveyed the connection between environmental regulation and technological change in various industries are discussed in Heaton 1997b.)

What this suggests for an environmental technology policy in Asia is the need for a first-tier strategy to promote environmentally oriented incremental innovations and to diffuse state-of-the-art technologies not yet widely employed. Since the potential supply of such technologies is already ample throughout OECD countries, the essence of the strategy in Asia should be demand-side enhancement. Four main elements should have prominence: regulation, information dissemination, enhancing demand through the multinational corporation (MNC) supplier chain and technical assistance.

Regulatory commands offer probably the most effective policy mechanism for diffusing currently available technology. From the point of view of fostering innovation, however, regulation tends to entrench the status quo by legal fiat, which can create substantial barriers to environmental innovators. The complex relationship between regulation and technological change has been discussed elsewhere (Heaton and Banks 1997). The essential point here is that regulation needs to target both diffusion and innovation. In the former case, it should be directed at upgrading or eliminating existing environmentally offensive technologies. For new technologies, regulation should create a lenient and flexible climate, thus giving innovators a chance to experiment with new approaches.

Environmental databanks offer another common attempt to increase technology diffusion, through dissemination of information about currently available technologies. Although such information has its place, it is not nearly as effective in augmenting the demand for superior technologies as is information about company environmental performance. In the United States, for example, the Toxic Release Inventory (TRI) has enjoyed considerable success as a means of demand enhancement. Some policies in Asia—notably Indonesia's system of coding corporate environmental performance—operate on a similar principle (Ditz and Ranganathan 1997). Such efforts need to proliferate and expand.

Technical assistance is frequently seen as a supply-side means of increasing environmental capacity. However, experience with programmes that focus on instructing companies about how to implement clean production suggests that this approach is a weak motivator. On the other hand, when technical assistance combines environmental efficiency with improvements in product quality and manufacturing cost, its recipients can become highly motivated to undertake a package of improvements. So restructured, technical assistance can become an effective demand enhancement.

A last, and dramatically under-utilised, mechanism to increase the demand for environmentally superior technology resides within the MNC community. Although MNCs' relationships to their supplier chain affiliates have often been seen as a supply-side

8 This point is discussed at some length in Utterback (1994). One example is vacuum tubes, which made major improvements in response to the transistor challenge.

vehicle for the diffusion of improved environmental technology, much less appreciated—and potentially more powerful—is the degree to which they can enhance environmental demand. Given their economic clout, MNCs could improve environmental demand through mechanisms much like those used by government procurement agencies; that is, promising to purchase goods and services whose environmental characteristics exceed prevailing norms.

Medium-term policies: green design of new technology

All technologies proceed through a design phase, when the possibilities for how to configure them are weighed in terms of functionality and other benefits, as well as against cost and other drawbacks, such as pollution or waste. The design phase is inherently creative, though by no means research-intensive; indeed, good designs may result without any research. It is also a time when problems can be designed out before they arise. This is the essence of pollution prevention.

One way of looking at the cause of environmental problems in OECD nations is to see such problems as design failures: an absence of environmental consciousness early enough to avoid damage. The same may be said of Asia today, only more so. If the pollution intensity of Asian economies, particularly in the manufacturing sector, is actually getting worse, then, clearly, the root failure is in the way products, processes, facilities and systems are being designed.

The pathway to changing the environmentally destructive design paradigm begins with asking who controls it and what motivates them. Three thrusts suggest themselves. First, the design of technologies embodied in MNCs' new investments deserve special attention. MNCs routinely assert that their investments in developing countries meet 'the same' environmental standards as those in the home country. This rubric probably needs to be changed. In fact, it is plausible to maintain that developing-country technologies need to be cleaner, given the faster trajectory of pollution intensity they appear to be on. As importantly, transplanting 'the same' technologies as those back home allows for little design on-site and thus frustrates exactly the capability that needs augmentation.

Most Asian governments are already highly attuned to interactions with MNCs. Pushing the MNCs' new investments toward better environmental standards should not represent an excursion into new territory. Nor should it pit one country against the other if it is done publicly. Indeed, the public pro-environmental positions taken by many MNCs can be leveraged in the effort, as well as the involvement of the non-governmental organisation (NGO) community. There is also an important, thus far under-utilised, role for OECD governments and multilateral organisations in this arena.

Second, the industrial and environmental policies in the Asian countries need to join hands—perhaps even be merged—particularly with respect to domestic investment. US policy structures in particular show the dangers of separating technology, industrial and environmental policy, with promotional programmes for new technology that under-emphasised environmental needs, and regulatory programmes that seemed anti-industrial. The Japanese approach may offer a more harmonious mechanism, folding environmental enforcement into an industrial ministry, the Ministry of International Trade and

Industry (MITI), which has long been the largest influence on patterns of domestic investment.

Last, because design is as much an ethos or mentality as it is a technical enterprise, the sensitivity of designers to the environment must be enhanced. This is basically an educational effort, but one that needs to be undertaken as much by the community of practising engineers and managers as through traditional venues for education and training.

Long-term policies: harnessing radical technological change

Although emergent revolutions in technology—biotechnology, miniaturisation, information systems, new materials—hold immense applications for environmental improvement, public policy in most countries is hardly coming to terms with them. In the United States, for example, the vast majority of publicly funded R&D categorised as 'environmental' is in fact environmental science, with almost none devoted to technology development.[9] Some other countries may be considering the long-term technological possibilities more systematically. The Netherlands, for example, has a forecasting process—'backcasting'—that attempts to infer current actions from future technological prospects; and Japan has mounted the world's largest environmental technology research programme in the form of RITE (Heaton 1997a).

Although R&D programmes such as RITE are certainly not a universal approach, this is not to say that other countries should not mount systematic attempts to harness the environmental potential of emergent technology. In fact, technology policies in Asian countries—notably Korea and Singapore—have already demonstrated success in an analogous area, that of industrial technology (Kim 1997). The thrust of such policies historically has not been so much to do research or technology development as to create the capabilities and underpinnings—technical, legal, institutional, managerial—that allow each society to internalise and exploit technologies on the horizon. Because the environmental technology policies of Asian countries must largely be implemented as initiatives to reduce the environmental intensity of new investment rather than to be a retrofit of the old, the integration of radically new technological possibilities could comprise an important aspect. Its first step should simply be to establish analytical, information and planning capabilities that will take account of technological trends.

Conclusions

Despite the acclaim innovators receive, there is much to be said for being a 'second mover': avoiding arduous and expensive R&D, bypassing inevitable mistakes of first

9 In fact, the dearth of environmental technology development was a main rationale behind the Clinton–Gore Environmental Technology Initiative in 1993, which has largely disappeared (see Heaton and Banks 1997).

iterations, and crafting technologies appropriate to particular circumstances. Policy innovations are no less subject to this dynamic. Applying this precept to environmental policy, the Asian countries are well poised to benefit from others' history. After 30 years' experience, the virtues and drawbacks of OECD policies have come into relief, and a fertile climate of policy 're-invention' has emerged. The pivotal role of new technology in achieving environmental improvement and economic growth is among its most important realisations.

To create an environmentally sustainable technological trajectory for OECD economies, the immense inertia of mature capital and technology, institutions, policies and slow economic growth must all be overcome. Asian economies—habituated to change—present a radically different opportunity. If an increasing intensity of environmental insult is to be reversed, the design and implementation of new technologies will be key. This imperative does not mean ignoring the need for immediate abatement and remediation of current pollution with known techniques, but it does mean relegating the 'end-of-pipe' clean-up strategy that has so dominated environmental policy and technology thus far to much lower prominence.

Asian societies can thus chart a new course for environmental policy. Indeed, it may even be counterproductive to speak of environmental policy singly, without reference to the overall 'innovation systems' in which new technology is developed, acquired, designed and implemented. Because the innovation system is focused simultaneously on change in the short, medium and long term, so too must an environmental technology policy function in all these time-frames. As technological innovation is a social phenomenon, the societal levers to influence its rate and direction are many: R&D funding, education, regulation, industrial licensing and resource pricing offer only a partial list. These and others must be enlisted to make technology a lever for growth and sustainability.

GLOBALISATION AND THE ENVIRONMENT IN ASIA
Linkages, impacts and policy implications

Daniel Esty, Mari Pangestu and Hadi Soesastro

The inescapable linkage between economic aspirations and environmental goals has emerged in full force in Asia since the 1990s. A focus on 'limits to growth' (Meadows 1972) has given way to a recognition that economic development can promote environmental progress (WCED 1987) and that poverty is a source of serious environmental degradation. In practice, however, the goal of achieving 'sustainable development' remains elusive. Given expected increases in the scale and pace of short-term urban–industrial growth outlined in Chapter 1, the key to achieving more sustainable development in Asia lies in promoting policies that reduce the materials, energy and pollution intensities of the region's economies. The purpose of this chapter is to examine the ways in which globalisation, in general, and trade and investment liberalisation, in particular, can be used to promote more sustainable outcomes in Asia.

Globalisation is, of course, a multi-dimensional trend. At its core is an economic transformation linking capital and product markets across national borders. The emergence of a set of inherently global-scale pollution and resource-management issues (e.g. climate change, the thinning of the ozone layer, depletion of marine fisheries) is another dimension of globalisation. Globalisation also encompasses other linkages, including communications (e.g. the Internet), culture (e.g. Pokémon) and media (e.g. the universality of CNN) that have begun to create a global 'civil society'.

This chapter examines the linkages between economic integration and environmental protection in Asia. It seeks to identify points of leverage within the globalisation process that can be used to improve environmental sustainability throughout the region and promote dynamic and cleaner development of Asia. The chapter is divided into three sections. In the first section the forms and trends of globalisation in the Asia–Pacific region are discussed. The nature of the economic integration process that has taken place in the region and the crucial environmental issues that have emerged are emphasised. The following section identifies the points of leverage within the globalisation process that can be used to ensure that continued economic integration across Asia will not lead to unsustainable development and increased environmental degradation. The final section offers a set of policy implications and a series of conclusions.

Forms and trends of globalisation and environment in the Asia–Pacific

Globalisation takes many forms. In this chapter, which focuses on economic interdependence, the term 'globalisation' refers to the linkages between economies, through trade, investment, production and financial markets. Recognition of the ecological interdependence of nations and the transboundary elements of many environmental harms is another concrete form of globalisation. A broader understanding of globalisation must also encompass the globalisation of ideas, values and culture made possible by the rapid and inexpensive dissemination of information through improved communications, computers and information management technologies. This form of globalisation is critically important because it drives changing views on issues such as the environment in the Asia–Pacific region. However, given scope and space limitations, this last aspect is not discussed in this chapter.

Globalisation: trade, production and investment

Increased economic integration and interdependence

International linkages, especially through trade, have played an important role in East Asian expansion. In recent years, other international linkages that involve international factor mobility have become important, including foreign direct investment (FDI), portfolio investment, services, and strategic alliances between corporations. Growth, trade and FDI in East Asia have reinforced each other to sustain the continuous economic expansion that de-linked the region from global recession in the early 1990s. There has also been increased economic interdependence within the region through these linkages.

For most of the Asia–Pacific economies, nearly half of all exports are destined for the Asia–Pacific region, and especially Japan. The Philippines and South Korea are exceptions to this rule, with around 40% of their markets located in the Asia–Pacific. Although this percentage is lower than that of intra-European trade, which is above 60%, and closer to the levels of intra-North American trade, it is occurring without the advantage of a formal trade agreement.

More sophisticated measures of regional interdependence, such as gravity coefficients, indicate that there is greater intra-regional bias in the Asia–Pacific than found either in

Europe or in North America.[1] The bias declined considerably compared with the postwar period as a result of the liberalisation of the global economy, the development of the region's economies with greater acceptances of products from outside of the region and a greater US role in East Asia. However, since the mid-1980s the bias has increased, indicating that the rapid expansion of East Asian markets has led to enhanced integration of East Asia, beyond the growth of intra-regional trade. The coefficient also remains higher than that for other regions (Petri 1995).

The export-base-motivated FDI from Japan and the East Asian newly industrialising economies (NIEs) to South-East Asia and China reflects an ongoing process of specialisation among countries in the region and helps explain the high levels of intra-regional trade and its composition. Around 75% of intra-regional trade is in intermediate and capital goods, suggesting that most of the trade flows are complementary. There is also a high degree of intra-industry trade (World Bank 1998a).

Petri (1995) suggests that the acceleration of international linkages in East Asia was reinforced by various factors. First, technological improvements and associated cost reductions in transportation and communications infrastructure co-ordinated cross-border production and trade. Businesses in the region also gained experience and information, and built networks to facilitate these linkages.

These developments were reinforced by a second set of favourable policy changes as East Asian economies liberalised trade and investment regimes across the board, or at least developed regional growth areas or special economic zones. The liberalisation of markets and foreign investment regimes promoted increased trade and investment linkages that have in turn produced market-driven economic integration within the Asia–Pacific region (Watabe and Yamaguchi 1996).

Unilateral liberalisation by the East Asian economies initially began in response to falling commodity prices and world recession. This process was also driven by the increased competition for markets and investment that resulted from the liberalisation of former socialist and communist nations after the end of the Cold War. Throughout Asia, the rise of 'subregional economic zones', highly economically integrated and geographically contiguous areas separated by political boundaries, occurred as a result of such market forces and private-sector linkages (Chia and Lee 1993).

The various regional agreements and the General Agreement on Tariffs and Trade (GATT) Uruguay Round commitments (for the member Asia–Pacific economies) have not yet played a major role in opening up the Asia–Pacific markets. They have had more of a complementary and facilitating role, mainly because most of these commitments began to take effect only in 1995, whereas the process of unilateral liberalisation began in the mid-1980s. These agreements, particularly the commitments resulting from the Uruguay Round negotiations, have reinforced unilateral commitments in the Asia–Pacific region

1   The gravity coefficient has been widely used in the analysis of intra-regional trade bias and is calculated by dividing the bilateral trade share of a partner by the partner's share in world trade. A coefficient greater than one indicates that the partner is more important in the bilateral relationship than in global trade: that is, it enjoys a positive bilateral bias. East Asian gravity coefficients are typically larger than one and much higher than those found either in Europe or North America.

and, in some cases, have made them more binding. Asia–Pacific nations have actively participated in multilateral efforts to cut tariffs and boost trade, and, even though some countries fell short in their commitments, especially on binding tariffs, the region played a role in helping to bring the Uruguay Round of negotiations to a successful close (Bergsten 1994; Funabashi 1995). It is interesting to note that the Asian crisis has not led to a major retreat from trade liberalisation on behalf of any country other than Malaysia (even in Kuala Lumpur the backtracking has largely been focused on investment liberalisation rather than on freer trade).

Regionalism

In addition to participation in the Uruguay Round, the development of regional agreements in the Asia–Pacific region has also accelerated. This acceleration has been, in part, a response to perceived competition from other regional agreements such as the European Union and the potential for a Greater Europe encompassing Western Europe and the North American Free Trade Agreement (NAFTA). In 1992, the Association of South-East Asian Nations (ASEAN) Free Trade Agreement (AFTA) was signed, creating a target of limiting tariffs on intra-ASEAN goods to below 5% within 15 years. Quantitative restrictions on products experiencing tariff reduction were also to be reduced. Subsequently, the deadline for 'free trade' was moved up ten years (to 2003), and discussions took place on widening the agreement to include services and investment. The free trade agreement in ASEAN is more of the 'traditional' type since it focuses mainly on trade liberalisation and is now essentially limited to tariff reduction. The ASEAN Free Investment Area does not yet involve investment liberalisation and is more concerned with the joint promotion of investment. It is unlike NAFTA and the more recent free trade agreements that are comprehensive in scope, incorporating trade, investment, environment and competition policy.

The centrepiece of Asia–Pacific regional trade and investment liberalisation is the Asia–Pacific Economic Co-operation (APEC) forum, launched in 1989 (for a detailed history of APEC, see Funabashi 1995). The APEC forum requires no formal negotiations or agreements and relies on voluntary unilateral initiatives proposed by its members. APEC is based on the concept of 'open regionalism' in which liberalisation is offered mainly on a most-favoured-nation (MFN) basis. This structure is very much in line with the market-driven integration that was already taking place in the region. APEC was meant to provide the framework and peer pressure necessary to ensure the continuation of liberalisation. At the same time, given its unique membership of developing and developed nations, APEC recognises the need to include measures that facilitate trade and investment and that balance liberalisation with economic and technical co-operation.

The APEC process took off in 1993 with high-level political commitment given at the first informal APEC meeting in Seattle, WA, USA. APEC has since moved ahead by:

- Setting specific goals for free trade and investment in the region by 2010 for developed nations, and by 2020 for developing nations

- Defining a set of principles to underlie its three-pronged liberalisation, facilitation and economic and technical framework

- Reinforcing multilateral initiatives such as the Information Technology Agreement negotiations that took place at the World Trade Organisation (WTO) in 1996 (Dua and Esty 1997)
- Developing various agreements on facilitation
- Creating a programme of individual and collective action in a wide range of areas, including the environment

In 1995, APEC's 21 economies accounted for 45% of global exports (as opposed to 34% in 1965) (Dua and Esty 1997).[2]

These various processes at the multilateral and regional levels have functioned to reinforce the unilateral process of liberalisation in the region. The fact that many Asia–Pacific economies were already opening up eased the acceptance of these multilateral and regional agreements. Indeed, unlikely champions of free trade and investment arose, including ex-President Suharto of Indonesia. Without Indonesia's support in 1992, AFTA would have been stillborn; moreover, Indonesia's subsequent role as the host of APEC in 1994 played a key role in setting the goals for the organisation.

Although some effort has been put into regional (and global) trade agreements, globalisation and integration in the Asia–Pacific region has centred on market-driven trade and investment flows (Lawrence 1996). Appreciation of the currencies and rising costs in the more advanced Asia–Pacific economics, namely Japan and the Asia–Pacific NIEs (Korea, Taiwan, Hong Kong and Singapore), led to a relocation of more labour-intensive industries and processes. Such a relocation of production was made possible by liberalisation in the home and host economies as well as by technological advances. Thus, the globalisation in product, service and capital markets that has occurred in the Asia–Pacific region, particularly since the mid-1980s, resulted largely from voluntary and non-binding processes of liberalisation and fulfilling global commitments rather than regional agreements or co-operation.

Regional co-operation has been important to the Asian economies as a reinforcing mechanism as well as a confidence booster, encouraging continued unilateral reforms and strengthened multilateral initiatives. This support can counter the danger of increased protectionism and the potential for a slowdown, or even reversals, in the process of liberalisation. Beyond building confidence to push ahead, regional co-operation can be important in building capacity by increasing exchange of information and policy experiences and fostering technical and economic co-operation on pertinent issues such as the environment. This could facilitate the implementation of improved and more sustainable policies and better prepare members and their policy-makers to support global and multilateral initiatives. APEC is in fact in a position to shape the course of Asia–Pacific development because of the range of its membership, the economic and technical co-operation at the heart of the organisation and the balanced approach of

2 The majority of these nations are Asian. The member economies include: Australia, Brunei, Canada, Chile, China, Chinese Taipei, Hong Kong, Indonesia, Japan, Malaysia, Mexico, New Zealand, Papua New Guinea, the Philippines, Singapore, South Korea, Thailand, and the United States. In 1998 three more members, Peru, Russia and Vietnam, were added.

liberalisation, facilitation and economic and technical co-operation that it so strongly emphasises. But doubt remains over how successful the APEC process will be, and the track record of the past several years is not encouraging. Thus, regional co-operation may not ultimately be able to lead the drive for sustainable development, but it can still play an important complementary role in support of multilateral and unilateral initiatives.

Globalisation and environmental harms

Transboundary environmental problems

Asia's environmental problems can be placed into three geographic categories: local/national, regional, and global (Dua and Esty 1997).[3] The impacts of local/national environmental problems are confined within the borders of the nation that creates the harm. Within the Asia–Pacific region, the pollution of internal water resources, the disposal of domestic waste and the release of particulates and lead into the air can all be considered local/national environmental problems. The existing pattern of economic growth in the Asia–Pacific region has created or exacerbated a wide range of local/national environmental problems (Angel *et al.* 1999). The expansion of exports made possible by increased integration and globalisation of the Asia–Pacific economies is linked to many of these problems. For instance, along with the expansion of textile exports by Thailand and Indonesia has come an increase in the pollution of rivers by the textile industry. It has been estimated that in some West Javan rivers the textiles industry is responsible for 70% of total pollution loadings (Intal 1996).

Regional and global environmental harms originate within the borders of one or more nations, but the impacts of these harms are transboundary in nature. These impacts traverse national borders and affect other nations within the Asia–Pacific region (regional harms) and across the world (global harms). Transboundary environmental problems are on the rise in Asia. Transboundary air pollution, ranging from the release of sulphur dioxide by China (which has caused acid rain problems in South Korea and Japan) to the recent forest fires in Indonesia (which have created serious haze problems for neighbouring nations), can be traced to economic changes across the region including both a scale-up of industry and an increase in consumption of polluting goods (such as cars). Examples of specific regional harms include the depletion of Pacific fisheries, which may have detrimental impacts on a number of Asia–Pacific nations, as well as problems stemming from the release of sulphur dioxide (Dua and Esty 1997). Global harms include greenhouse gas emissions from nations within the Asia–Pacific region that contribute to climate change across the planet.

3   It is, of course, important to note that not all harms can be limited to only one of these categories. For instance, some environmental problems, such as deforestation, may have local effects (soil erosion, land degradation and water pollution) and global effects (reduction in the world's supply of carbon sinks which can mitigate greenhouse gas emissions) (Dua and Esty 1997: note 18).

Such regional and global impacts demonstrate the ecological interconnectedness between nations within the Asia–Pacific region and throughout the world. In this way, transboundary environmental problems act as another dimension of international linkages and globalisation. Also, it is important to note that pollution spillovers and the existence of shared resources are an ecological fact that cannot be ignored.

The increased regional and global integration of Asia–Pacific economies has also led to increased pressure on Asia–Pacific nations from the major importing markets to conform to trade-related and investment-related environmental standards. This increased pressure to meet the environmental standards of international importing markets is not very surprising. Just as increased globalisation and integration of economies leads to some loss of autonomy over fiscal and monetary policy through the imposition of external conditions set by international markets (Cooper 1994; Greider 1997), some autonomy over domestic environmental policy will be lost as well (Anderson 1995; Rodrik 1997). However, exporting nations express concern that the imposition of rigorous environmental standards by importing nations may actually be motivated by protectionism. In truth, it can be difficult to determine which environmental standards are legitimate and which are designed to protect domestic industries (for a discussion on how to separate bona fide environmental policies from those that are protectionist, see Esty 1994a: 117-27).

Responding to transboundary environmental problems

Classifying environmental harms by geographic scope aids in determining the level of government that should respond to various environmental issues. The optimal level of governmental intervention necessary to remedy environmental harms depends on the geographic range of the harm (Esty 1996).[4] The scope of governmental response should match the scope of the environmental harm at issue. From an economics point of view, this 'matching principle' demands that the scope of the regulating government's authority or jurisdiction match the scope of the externality being addressed or public good being provided. Accordingly, local/national harms, such as pollution of local/national watersheds and airsheds, will perhaps best be handled by national governments or their subdivisions. In contrast, the appropriate scale of response to regional and global transboundary harms cannot be merely local or national but should also include regional or global efforts.

The international action necessary to address transboundary harms (global or regional) can involve complexities not at issue in local/national problems. In particular, the principle of national sovereignty makes it difficult to obligate nations to remedy spillover harms originating within their national borders. At the same time, the causes and impacts of some transboundary harms, such as global warming, are geographically spread across the world. Combining this geographic spread with the diversity of views

---

4 Numerous authors have discussed the issue of optimal allocation of jurisdictional responsibility for environmental harms at the national and regional level, particularly within the US federal system (see e.g. Fischel 1975; Gray 1983; Krier and Brownstein 1992; Oates and Schwab 1988; Revesz 1992; Stewart 1977).

on how (and whether) to solve environmental problems makes it difficult to co-ordinate solutions. As is often the case within international governing bodies, the ability of laggards (the least-committed nations) to determine the level of action taken to solve problems (Susskind 1994) further limits progress in dealing with transboundary environmental harms. For these and other reasons the current global environmental regime has been largely ineffective in dealing with transboundary environmental harms (Dua and Esty 1997).

Regional co-operation in the Asia–Pacific region has also not proven to be effective in coping with transboundary environmental problems. The haze problem that originated from forest fires in Indonesia and affected its neighbours provides an example of the ineffectiveness of regional co-operation to date. Despite a region's recognition of the problem and agreement to co-operate on issues and actions that need to be taken, the lack of implementation and political will makes the effectiveness of co-operation questionable. Given the lack of capacity to undertake national-level reactions, especially at present with Indonesia's economic and political crisis, regional assistance and action should play a role. The Association of South-East Asian Nations (ASEAN) Co-operation Plan on Transboundary Pollution and the ASEAN Regional Haze Action Plan set out the appropriate principles of prevention, mutual assistance and co-operation (see Box 3.1). However, these are not treaties with legally enforceable obligations and they contain no provisions for progress by different countries to be monitored and no provisions for regional institutions to assist implementation at the national level. The emphasis remains on the co-ordination of national plans rather than on providing regional responses or inter-country assistance (Tay 1999b).

The Indonesian forest fires suggest the need for action on several fronts. First, Indonesia's capacity to respond to the haze problem would be enhanced if a multinational task force were created to work closely with the Indonesian authorities to prevent and fight forest fires. Second, the co-operation agreement should be elevated to the status of a treaty, thereby creating a legal obligation to prevent forest fires and forcing signatories to take their environmental obligations more seriously. Third, implementation of a fire-control strategy and monitoring of compliance cannot be left to government officials and environment ministers. Broader societal engagement is needed, especially at the grass-roots level where the fires originate. Non-governmental organisations (NGOs) should also be galvanised and co-ordinated to provide early warning systems and to monitor the implementation of fire prevention measures (Esty 1998a). Finally, independent peer reviews and observations of the countries' progress should also be considered.

Points of leverage for clean growth

President Clinton in his 1998 visit to China urged the leaders of the world's largest nation to find a cleaner path to economic growth. 'China has a unique opportunity . . . to avoid

THE HAZE PROBLEM IN ASEAN HAS BECOME A REGIONAL PROBLEM OF CRITICAL magnitude because of the resulting regional tensions and costs in dealing with fires and smog. A preliminary study by the Economics and Environment Programme of South-East Asia and the World Wide Fund for Nature has estimated the damage for the 1997–98 period at US$1.4 billion of economic losses from disrupted production and transportation, health and ecological problems, declines in tourism and damaged crops and forests. The haze resulted from forest fires in Indonesia caused by slash-and-burn land-clearing by plantations, timber concessions and farmers, despite a ban on the use of fire for land-clearing. The fires were aggravated as a result of a prolonged drought, ground-area peat or coal beds and the slow response by the Indonesian government because of a lack of political will and capacity as well as a lack of co-ordination between different agencies and levels of government. The smoke and pollution has affected parts of Indonesia (Sumatra and Kalimantan), Malaysia (Sarawak), Singapore and Brunei.

The haze problem is not a new issue, as forest fires from Indonesia have affected the region before—in the 1980s, in 1991 and again in 1994. After the previous episode a regional agreement was reached: the ASEAN Co-operation Plan on Transboundary Pollution, which banned the clearing of land by fire in 1995. However, the problem became acute in 1997 until the beginning of 1998 and the Pollutant Standard Indexes (PSI) levels rose to between 400 and 600, well above the safe benchmark of 100. At the time of writing (April 1999), the problem appeared to be recurring, with the PSI approaching 100.

The ASEAN Co-operation Plan on Transboundary Pollution does address transboundary atmospheric pollution and there is recognition of the major issues, including: assessing the origin and causes, nature and extent of local and regional haze incidents; preventing and controlling sources of haze at national and regional levels by using appropriate technologies and capacity-building; and developing national and regional emergency response plans.

A regional action plan on haze was also developed in December 1997 with the following short-term and long-term strategies: timely detection, prevention of fire use for land-clearing activities, especially in dry periods, sharing of information and networks and allowing land that is susceptible to fire outbreaks—such as coal beds and peat fields—to be developed. Furthermore, recommendations of activities and steps were made to establish national focal points to enable the creation of an inventory, to create a mechanism for the dissemination and exchange of information, to develop a common air-quality index and harmonise air-quality sampling techniques and to craft a regional fire danger rating system.

However, there has been a lack of implementation because of the low level of commitment among the ASEAN governments. ASEAN environment ministers met in April 1999 and adopted a zero burning policy in member states to make plantation owners and loggers answerable for the fires on their land. The crux of the matter is enforcement. Indonesia is experiencing its worst crisis at the worst time—during elections, which necessarily signifies less implementation and enforcement. At the April meeting, environment ministers gave Indonesia until July 1999 to crack down on those starting the fires and to implement a zero burning policy.

Box 3.1 **Association of South-East Asian Nations (ASEAN) regional co-operation on the haze problem** some of the terrible mistakes we've made', he told the Chinese.[5] There are many ways to change the economic development trajectory of Asia—and many points of leverage that could be used to induce a clean industrial revolution.

Globalised markets, liberalised trade and investment, improved responses to international pollution resource issues, conditions on international assistance, enhanced private-sector environmental stewardship, improved national government policies, more environmental leadership from NGOs, international reform, and the potential influence of globalised communication, culture and the media all provide environmental opportunities.

The role of multilateral and regional trade agreements

The issues and potential for influencing environmental conditions

The globalisation of trade and capital flows and the increasingly strong economic interrelationships between nations that facilitate trade agreements create an important opportunity for environmental policy influence. Incorporation of environmental conditions into trade agreements represents a critical dimension of modern trade policy-making (Esty and Geradin 1997). Simplistic neoclassical economic theory argues that differences in national and local environmental standards are to be welcomed as variations in comparative advantage that can serve as the basis for gains from trade (see Esty 1996 in which Fischel 1975, Oates and Schwab 1988, Revesz 1992 and Tiebout 1956 are discussed). More nuanced recent analyses suggest that some differences in environmental policy choices will inevitably arise from variations in climate, weather, population density, background pollution levels, risk preferences or level of development (Esty 1996). These differences represent 'legitimate' bases for comparative advantage.

Other divergences in environmental standards, however, may arise from governmental regulatory incapacity, public choice failures (corruption, special-interest lobbying, campaign contributions or simply the short time-horizon of politicians) or market failures (uninternalised cross-border externalities or mismanaged shared resources). In these cases, the variation in standards is not 'legitimate' and competition on environmental variables may induce a welfare-reducing 'race to the bottom' (Esty 1996: 627-28).

How competition in environmental standards plays out in a globalised world is thus a critical pollution and resource management issue. Increasingly, government officials and business leaders recognise the need for some form of 'convergence' or harmonisation of standards between nations (Esty and Geradin 1997). This policy intervention need not—and should generally not—lead to uniform standards across all jurisdictions (Esty and Geradin 1997). Instead it may well entail the adoption of minimum standards. This set of standards can be multi-tiered so that the requirements are tailored to the needs of nations at different levels of development or so that requirements may be refined in other ways to make harmonisation more flexible.

---

5 President Clinton made this statement while visiting the scenic Guilin area along the Li River. *AP Online* 2 (7 July 1998) (1998WL6690183).

The European Union (EU)'s experience in this regard provides an object lesson. In support of its 'single market', the EU has issued more than 200 environmental 'directives' aimed at ensuring co-operation, and not harmful competition, among its member states in the environmental policy realm (Esty and Geradin 1997). Although not all of these directives may be necessary or advisable, some of them have helped to ensure the integrity of the EU single market by preventing competitive advantages from being established on the basis of market failures such as uninternalised transboundary externalities.

There remains doubt, however, in many quarters, particularly in Asia, about the wisdom of harmonised environmental standards, especially standards related to production processes or methods (PPMs). Many developing nations believe that harmonisation will lead to mandated high standards consistent with developed-country needs and thereby deprive them of opportunities for growth, block their products from export markets and perhaps even constitute an overt act of eco-imperialism. Industrialised nations, on the other hand, are concerned that a lack of harmonisation could lead to a regulatory 'race to the bottom' (see previous section) as developing nations compete to attract industries by promising lax controls on pollution. Developed nations fear that such a 'race to the bottom' could induce industrial activity that is dirtier than it needs to be, leaving all nations worse off. If properly done, however, harmonisation of PPM standards could permit legitimate regulatory differences to exist, thus allowing appropriate aspects of comparative advantage to be exploited without promoting an environmentally destructive 'race to the bottom' (Esty and Geradin 1997).

Harmonisation of standards could also produce a number of advantages beyond preventing welfare-reducing competition over environmental regulation. The convergence of process standards can encourage economies of scale in regulation (Esty 1994a). In the United States, for instance, national standards for vehicle emissions allow the city of Detroit, MI, to produce cars for the whole nation from a single production line, streamlining the production process and achieving substantial cost savings. At the same time, these regulations lifted the burden of identifying, setting and monitoring exhaust standards from the states.

Similar standards can also yield 'network' externalities (Dybrig and Spatt 1983; Katz and Shapiro 1985). If one jurisdiction assumes the pollution standards of another, it can avoid the costly and time-intensive process of developing and implementing its own standards. The creation of appropriate standards in the environmental realm is often particularly difficult because of the technical and scientific expertise required. In addition, governments may face diminished political opposition to regulation when producers do not face a plethora of unique standards.

Furthermore, the convergence of standards strengthens the incentives for companies to innovate. In small countries, companies can gain only relatively small benefits from researching and producing technologies that more effectively meet (or exceed) the local pollution requirements. When a single innovation has the potential to advance a company's position in multiple markets, a firm is far more likely to invest in research and development (R&D) efforts.

Harmonisation of environmental standards need not lead to inflexible, one-size-fits-all regulation. By utilising regulatory means that afford companies flexibility in meeting compliance standards, governing authorities can achieve many of the advantages of decentralised control (Esty 1996). Performance-based (or quantity-based) measures, such as tradable permits, allow jurisdictions to determine the level of pollution that is acceptable. Simultaneously, companies can decide on the compliance strategies that will yield the lowest possible total cost. Moreover, the use of ambient standards to account for geographic differences in pollution compliance can lead to more efficient, welfare-enhancing outcomes.

The consideration of environmental issues that occurred on a parallel track within the North American Free Trade Agreement (NAFTA) negotiations could serve as a model for the incorporation of environmental issues into trade liberalisation efforts elsewhere. This parallel track facilitated many meaningful accomplishments, including a Mexico–US Border Environmental Plan, an Environmental Review of the NAFTA and an Environmental Side Agreement. In addition, the NAFTA preamble makes explicit references to environmental goals by calling for the pursuit of free trade consistent with 'sustainable development' and in balance with 'environmental protection and conservation'. During the negotiations themselves, environmentalists were given prominent roles. For example, the Office of the United States Trade Representative included an influential environmentalist on the Advisory Committee on Trade Policy and Negotiation, and top US Environmental Protection Agency (EPA) officials were included on the US negotiating team (for further discussion of the integration of environmental considerations into the NAFTA, see Esty 1994c).

Increased focus on environmental issues in the trade context seems inevitable, especially in the wake of the failed 1999 WTO ministerial meeting in Seattle, WA. The pace at which actual progress occurs is, however, more uncertain. In March 1999 the WTO hosted a symposium on trade and environment issues in Geneva. The event drew more than 600 participants representing governments, businesses, environmental groups, research centres and universities from around the world. Although there was no consensus on a precise action plan, there was broad agreement on the need to better co-ordinate trade and environmental policy-making and on the risk that environmental disputes could disrupt efforts to open markets further (IISD 1999).

Potential for regional trade agreements and co-operation in the Asia–Pacific

Within the Asia–Pacific region, APEC has the potential to facilitate co-ordinated environmental policy-making in the context of trade and investment liberalisation, as well as part of its economic and technical co-operation programme. A glance at APEC's achievements in this area indicates that much has been said but little has been done. In 1993, APEC leaders gathered in Seattle, WA, and issued a call for environmental protection and sustainable growth. In 1994, the APEC ministers of the environment met in Canada and adopted an Environmental Vision Statement and a Framework of Principles for integrating economic and environmental issues. In 1997, in Vancouver, the APEC leaders reiterated that 'achieving sustainable development is at the heart of APEC's mandate', and they restated this commitment in 1998 in Kuala Lumpur, as they agreed 'to advance sustainable development across the entire spectrum of [APEC's] workplan'. In its various subcommittees and groups APEC has taken a few first steps toward making real its commitment to sustainable development. Initial efforts can be seen in the work in the standards area (Subcommittee on Standards and Conformance), in the programme 'Action for Sustainable Cities', in the strategy for cleaner production, in the Marine Resource Conservation Group, in the development of sustainable aquaculture in APEC and in the energy programme (see Box 3.2).

Much of APEC's effort has been focused on coming up with guidelines and principles pertaining to policy as well as technical issues, on conducting technical meetings and workshops and on facilitating exchange of information. Superficially, at least, there appears to be a recognition of the major issues, but, like the WTO, APEC has done little to address the environmental dimensions of globalisation (Dua and Esty 1997). Perhaps the intangible benefits of capacity-building through APEC meetings and the exchange of ideas that these sessions promote should not be underestimated. Concrete results, however, have not yet been felt in terms of actions or results.

ASEAN co-operation on the environment has taken place over a longer period compared with that of APEC, as ASEAN environment ministers have met regularly since the 1980s and have made various declarations and commitments on sustainable development and conservation. Again, the government officials enunciate the need to integrate the environment with development but in fact do little. Most of the agreements discuss co-operation on policies, information exchange, institutional development, technology co-operation, public awareness and specific programmes, such as the Haze Agreement. A consensus to adopt common stands and actively participate in international efforts to protect the global environment have also been advanced (see Box 3.3).

As noted earlier, the ASEAN programmes do not appear to be effective. For example, the agreement and action programmes in place failed to respond to the haze problem in 1997, and implementation and enforcement remains an issue today as the haze problem emerges again. The main problems with such 'soft' agreements are that there are no sanctions and that there has been a long-held, sacred principle of non-interference in the region. Given the seriousness of the transboundary environmental problem caused by the haze, some have advocated that two courses of action are needed. One is to make the agreement a treaty and do away with the ('Asian Way') principle of non-interference. The second is to rely on NGO groups and society pressure to push the government into action and to facilitate the provision of early warning signals and monitoring (Tay 1999b).

Foreign direct investment

The flow of capital across the world provides another point of policy intervention through which environmentally sound industrialisation might be promoted. Fundamentally, foreign direct investment (FDI) offers great environmental promise. Private-sector FDI flowing into APEC economies, prior to the economic crisis, exceeded US$70 billion annually (Dua and Esty 1997). FDI, especially in the form of joint ventures between established firms (multinational corporations [MNCs]) in the developed world and partners in the developing world, often results in the transfer of modern plant equipment that is almost always less polluting than what otherwise would be available

SUBCOMMITTEE ON STANDARDS AND CONFORMANCE
This subcommittee:
- Works on technical infrastructure development
- Creates guidelines for technical regulations
- Exchanges information on standards, regulations and labelling for products such as food
- Works towards mutual recognition arrangements (MRAs) for various products

ENVIRONMENT MINISTERS MEETINGS 1994
These created:
- An environmental vision statement and framework of principles integrating economic and environment issues, 1997
- A programme for Action for Sustainable Cities
- The APEC Action Plan for Sustainability of the Marine Environment (this integrates approaches to coastal management, prevention, reduction and control of marine pollution and promotes sustainable management of marine resources)

FISHERIES GROUP
This was set up to promote, among other things:
- The conservation and sustainable use of fisheries resources, the sustainable development of aquaculture and habitat preservation
- Resource management and enhanced food safety and quality
- Trade and investment liberalisation and investment in fisheries

Technical workshops have been held on many of these issues.

MARINE RESOURCE CONSERVATION GROUP
This was created to promote initiatives in APEC that will protect the marine environment and its resources and to maintain marine environmental quality. In 1997–98 workshops were held on various technical issues, such as:
- Preventing maritime pollution
- Promoting marine resource conservation
- Protection of the marine environment

ENERGY
This was set up to maximise the energy sector's contribution to the region's economic and social wellbeing while mitigating the environmental effects of energy supply and use. It has five expert groups, including groups on energy and the environment and energy efficiency and conservation. Their achievements are:
- 14 rational energy policy principles
- An energy ministers meeting in 1997 endorsed a set of non-binding principles for good environment proactive measures for independent power producers; they also endorsed a standards notification procedure
- A programme of seminars
- Training and exchange of information to promote new and renewable energy technologies
- The promotion of the more efficient production and use of energy
- Mitigation of environmental impact of energy production and use

Box 3.2 **Asia Pacific Economic Co-operation (APEC) co-operation and environmental issues**

THE CO-OPERATION HAS CONSISTED OF:

- Environment ministers meeting
- Jakarta Declaration on Environment and Development (18 September 1997)
- Bandar Seri Begawan Resolution on Environment and Development (1994)
- Singapore Resolution on Environment and Development (1992)
- Kuala Lumpur Accord on Environment and Development (1990)
- Jakarta Resolution on Sustainable Development (1987)
- The Agreement on the Conservation on Nature and Natural Resources (1985)
- The ASEAN Declaration on Heritage Parks and Reserves (1984)
- Manila Declaration on the ASEAN Environment (1981)
- The ASEAN Co-operation Plan on Transboundary Pollution
- The Regional Haze Action Plan
- The Working Group on Sub Regional Fire Fighting Arrangements
- The ASEAN Consultative Committee for Standards and Quality
- Technical Working Group on Standards and Information

The Technical working Groups on Standards and Information looks at the ISO 14000 series of standards on the environment to monitor developments in the environment area, including those in

- Certification practices
- Co-ordination of appropriate actions for ASEAN in this area
- The exchange of information, and implementation
- The exploration of possible joint co-operation between the United States and the USAEP to conduct joint seminars on the ISO 14000 series

Box 3.3 **Association of South-East Asian Nations (ASEAN) co-operation on environment**

(Esty and Gentry 1997; Gentry 1998). FDI-based joint ventures frequently further the transfer of environmental management systems and training programmes as well.

However, not all flows of private capital bring environmental gains. In China, for example, the drive for economic development, and particularly for new electricity-generating stations, has led to environmentally harmful competition among various Chinese provinces and the foreign suppliers seeking to build power plants (Esty and Mendelsohn 1995). The business and government leaders in many Chinese provinces and municipalities are quite eager for new sources of electricity to power further economic growth. In fact, these leaders are so eager that they insist that those supplying the plants strip out of their proposals pollution control devices in order to maximise the kilowatt-hours of generating capacity per dollar invested. Any company refusing to go along with this requirement is passed over in the bidding process, and, with a nearly endless supply of foreign companies seeking to build power plants in China, someone else is always willing to build whatever the Chinese authorities request.

These welfare-reducing competitiveness pressures in the context of private capital flows could be addressed by establishing baseline environmental standards for foreign investments. But the requests by environmentalists to build such standards into the Multilateral Agreement on Investment (MAI), under negotiation at the Organisation for Economic Co-operation and Development (OECD) from 1996 to 1998, were rebuffed by the trade officials involved. As a result, environmental groups around the world vehemently opposed the plan, and the push to complete the MAI faltered.

A future MAI could contain environmental standards for all projects funded through foreign investment. If all investors were required to meet a common set of standards, none would be competitively disadvantaged. Similarly, if all recipients of foreign capital had to ensure that their development projects contained basic pollution abatement elements, the competition for foreign investment would not be affected. At the very least, controls on transboundary pollution spillovers should be built into all FDI-funded projects.

It is not yet clear whether efforts will be made to revive a free-standing MAI. Some advocates of an investment treaty believe that such an accord could be developed through greater consultation with those parties who felt excluded from the prior effort (e.g. developing countries, environmental advocates, consumer interests). Others see the issue folded into a WTO millennium round, signifying that it will not become a central point of policy leverage for another four to six years.

Multilateral environmental agreements

Multilateral environmental agreements (MEAs) potentially provide another powerful tool for promoting environmental responsibility. Where nations agree on environmental standards—such as prohibitions on the production of ozone layer-destroying chlorofluorocarbons (CFCs) or the protection of endangered species[6]—these agreements should be seen as defining the acceptable bounds of behaviour in the international economic realm. Nations (or companies) that fail to sign on to widely accepted standards or that fail to comply with their obligations should be seen as 'free-riders' whose behaviour is disruptive to the smooth functioning of international economic relations. Any competitive advantage obtained by non-compliance should be deemed an 'unfair' trade advantage and therefore 'countervailable' by other nations or punishable through multilaterally defined trade measures.

Of course, there exists a need to avoid protectionism disguised as environmental regulation. For this reason, MEAs raise numerous contentious issues, especially from the perspective of developing countries (Reiterer 1996). First, MEAs may be seen as unfair when they penalise non-parties by forcing them to suffer trade penalties if they refrain from joining the agreement for legitimate reasons. Perhaps, for example, the non-signatory concluded that it had other priorities for its limited environmental budget. But this logic does not hold if the non-signatory was simply 'free-riding' on the efforts of others to address an international environmental harm.

---

6 The Montreal Protocol on Substances that Deplete the Ozone Layer prohibits the production of CFCs by developed nations; the Convention on International Trade in Endangered Species of Wild Fauna and Flora strictly regulates trade in endangered species by its signatories.

In judging the legitimacy of a decision not to sign on to an MEA, one variable is critical: the scope of the harm that is being addressed. Where harms are localised it is reasonable for individual nations to set their own course. Where, however, harms spill across national borders, the claim to a sovereign right to set one's own standards breaks down. Once an agreement addressing a transboundary issue is widely endorsed, hold-outs cannot claim to be acting 'legitimately' or defending their sovereignty.

Second, MEAs may be seen as a form of eco-imperialism used to advance the environmental agendas of rich nations outside of their borders. Who determines what level of pollution or resource exploitation is optimal? Who decides how fast a problem should be brought under control? Whose values are used to make judgements? Many environmental issues are marked by deep uncertainties and the need for political judgements in setting standards. Finding a fair and effective forum to resolve disputes arising from MEAs will often not be easy.

The Montreal Protocol provides a model for addressing many of these issues. Although it contains the threat of trade restrictions against those in non-compliance, the Montreal Protocol attracted the support of developing nations (Brack 1996) through subsidies for the purchase of CFC substitutes and a commitment to technology assistance. Moreover, with more than 160 signatories, the Montreal Protocol now represents an established international consensus.

Conditional adjustment programmes

Multilateral and donor agencies have been involved in providing aid (sometimes conditioned on 'structural' reforms) since the 1970s. The Asia–Pacific nations have been the recipients of much of this assistance (WRI 1998). Typically, structural adjustment programmes focus on appropriate macroeconomic stabilisation policies, including tight monetary policies to curb inflation, fiscal deficits management, flexible exchange rates, trade liberalisation, removal of subsidies and other microeconomic distortions, and institutional reforms to ensure effective implementation of policies. The first generation of adjustment programmes was much criticised by NGOs because they did not address the social and environmental impacts of these adjustment programmes (McAllister 1993; Reed 1993). The premise was that growth supported by appropriate economic policies would be sufficient to lead to sustainable environmental policies and, by reducing poverty, would significantly contribute to reducing environmental degradation, but, although adjustment programmes have generally had a positive impact on the environment by reducing poverty, removing price distortions and subsidies, making environmental services affordable and generating resources to manage the environment, the institutional capacity and political commitment necessary for success have often been lacking. In many cases, after some initial reduction in environmental degradation, further improvements in environmental conditions have not been achieved (World Bank 1998b).

Furthermore, given that reforms are undertaken in situations where market and regulatory failures still persist, the steps taken usually have serious negative short-term social and environmental impacts that are not fully considered. For instance, fuel price increases without concomitant adjustments to substitute products can lead to the increased use of lower-priced but more environmentally damaging fuels. Fiscal cuts can affect social and environmental services. Many adjustment programmes have not addressed the ability of the privileged élite to access natural resources on a preferential basis. Moreover, macroeconomic crises, which increase unemployment and poverty, can exacerbate environmental harms. In Thailand, for example, large areas of forest have recently been converted to agricultural lands to support impoverished people.

Multilateral institutions have recently placed a more direct emphasis on ensuring that the social and environmental dimensions are fully accounted for in the adjustment programmes (Graham 1994). There has also been greater emphasis placed on the institutional aspects of adjustment programmes. For example, the 1997 World Bank report on Indonesia, issued prior to the economic crisis, actually mentioned the need to address corruption and no longer made use of the euphemism 'high-cost economy' (World Bank 1997b). In recent years, most nations receiving International Monetary Fund (IMF) help have had specific environmental programmes built into their adjustment programmes as well as into the loans provided.

Asia's economic crisis offers opportunities to redirect environmental policies in several Asia–Pacific nations, as some of these countries are under comprehensive IMF reform packages. A close look at the various reforms and deadlines required by these programmes indicates that there has been an attempt to use the opportunity for structural reform to introduce desirable environmental policy changes. This is most evident in the IMF programme for Indonesia and less evident in the Thai and Korean programmes. In the Indonesian programme there are ten reforms aimed at creating improved forestry and agriculture policies which, if implemented properly, will mean a better environmental outcome in the long run.

The Indonesian case suggests that important steps have been taken that would previously have been difficult because of collusion between the private plywood industry and government. Disbanding the private plywood industry cartel, which previously had set prices and controlled export contracts, represents a major step forward both economically and environmentally. The other steps are typical of the reforms recommended to improve environmental sustainability: increased royalties, liberalised trade in logs, effective anti-trust policy and, most importantly, effective regulatory action to control illegal logging (Table 3.1).

The role of the market

The globalisation of markets for goods and services heightens the potential role that private companies can play in making Asia's industrialisation 'greener'. Increasingly, and especially as foreign aid declines, private capital flows are being recognised as the most promising engine for sustainable development across the developing world (Esty and Gentry 1997). China, for example, received about US$3 billion in foreign aid in 1998 both from multilateral sources (such as the World Bank) and from bilateral assistance (from Japan, the EU and the United States). More than US$40 billion in international private capital flowed into China in the same year, as overseas companies set up joint ventures and expanded businesses aimed at serving the Chinese market. How the more than US$40 billion of foreign investment is deployed—determining what factories, roads, dams, power plants, waste facilities, water systems are built, as well as what sort of

Policy Change	Effectiveness
Introduce new resource rent taxes on timber. Level: from 0% to 6% on timber sales (replaces the forest products fee).	5 May 1998
Reduce export taxes on logs and rattan to a maximum of 30% *ad valorem*. The aim is to reduce export taxes to 20% by the end of 1998, 15% by the end of 1999, and 10% by the end of 2000.	15 April 1998
Increase timber stumpage fees charge to forest concessions, implement auction system for new concessions, allow transferability of forestry concessions, and delink ownership from processing for new concessions.	30 June 1998
Reduce land conversion targets to environmentally sustainable levels and implement a system of performance bonds for forest concessions.	31 December 1998
Eliminate the Indonesian Plywood Association (APKINDO) monopoly over plywood exports.	30 March 1998
Transfer control over all government-owned commercial forestry companies from the Ministry of Forestry to the Ministry of Finance.	Early 1998
Incorporate the reforestation fund into the national budget.	Early 1998
Remove restrictions on foreign investment in palm oil plantations.	Early 1998
Remove the ban on palm oil product exports and replace it with an export tax of 40% to be reduced to 10% by 1999.	22 April 1998

Table 3.1 **Forestry-related policy changes under the IMF stabilisation and adjustment programme in Indonesia**

Source: Sunderlin 1998

infrastructure is constructed (i.e. whether pollution control investments are part of these projects)—will have a far bigger impact on China's future than how the US$3 billion in foreign assistance is spent.

In some cases, attention to environmental concerns improves competitiveness (Esty and Porter 1998). Where, for example, energy costs are an important factor in production, investments in reduced energy use will generate cost savings and environmental benefits. Just how far the corporate world can go with 'eco-efficiency' is a matter of considerable debate. Porter and van der Linde (1995b) and Flavin and Tunali (1996) see big opportunities. Others are more sceptical about how many win–win opportunities exist (Jaffe *et al.* 1995; Whalley and Whitehead 1994).

In other cases companies feel pressure to meet high environmental standards neither because they expect to save money nor because governments or international treaties require it but because market forces demand it. Both the EU Eco-Management and Audit Scheme (EMAS) and the International Organization for Standardization (ISO) have developed environmental management standards in response to customer requirements, reflecting a voluntary approach to improved corporate environmental performance. The ISO 14000 series, in particular, has become an environmental benchmark for many companies and an environmental quality threshold that all of their suppliers must meet (for more information on this series, see Pesapane 1998). A growing number of corporations today insist that their suppliers be ISO 14000 'certified'. Matsushita, for instance, plans to have all of its Indonesian plants conform to ISO 14000 standards by March 1999, exemplifying how the private-sector certification process and the pressures of the marketplace can help to align a corporation's self-interest with environmental aims (*New Straits Times* 1998).

Although the ISO 14000 series is aimed at corporate consumers, other policy tools have been developed that aim to harness the purchasing power of the public for environmental ends. Some consumers prefer environmentally friendly products and may be willing to pay a premium for them. In the United States one study has suggested that consumers would be willing to pay an average of 20% extra for products manufactured in a manner more sensitive to environmental concerns (Salzhauer 1991). Books informing individuals about the numerous small steps that they can take to improve the surrounding environment sell remarkably well. The unexpected success of *The Green Consumer Guide* (Elkington and Hailes) in 1998, which was a bestseller for nine months, is only one notable example (Salzman 1997).

Eco-labels represent another mechanism for steering the marketplace toward environmentally preferable products. These labels, which provide the consumer with information on the environmental attributes of a product, can come in many forms (Salzhauer 1991; Salzman 1997). Some focus on a single issue (e.g. dolphin-safe, recycled, recyclable), others address a number of criteria and distil the results into a single signal of environmental quality (e.g. Germany's 'Blue Angel', Scandinavia's 'White Swan', Canada's 'Environmental Choice'). Some labels are government-sponsored (such as the European and Canadian labels just mentioned), others are issued by private parties (such as the 'Green Seal' programme in the United States). Many focus on positive product qualities, but some highlight harmful environmental effects (e.g. California's 'Prop 65' labels on carcinogenic products). In addition, eco-labels can be mandatory or voluntary.

Eco-labels are attractive because they do not dictate consumer behaviour but rather educate buyers, allowing them to make more informed choices. The Marine Stewardship Council (MSC) has launched a labelling campaign to signal to fish buyers—in both supermarkets and restaurants—which supplies come from sustainably managed fisheries. Similarly, the Forest Stewardship Council (FSC) has developed a label for timber from sustainably managed forests.

Eco-labels can, however, be controversial. First, to eco-label a product accurately an analysis of a product's full 'life-cycle' must be undertaken. Unless the environmental impacts of a product's raw materials, manufacturing, transportation, consumption and disposal are all considered, it is possible that the overall environmental effect of the product will be inaccurately gauged (Salzman 1997). Currently, however, there are no agreed methods for life-cycle analysis.

Second, who pays for the analysis and who monitors compliance can be contested. Organisations that eco-label products may have problems raising funds, but to accept funds from manufacturers lends at least an appearance of bias.

Third, some businesses may not apply to be evaluated because of apprehension over a perceived relationship with environmental groups that are actively hostile to business interests (Salzhauer 1991).

Fourth, questions persist over how consumers will actually react to eco-labels. Although in surveys people express preferences for 'cleaner' products, it is not clear that they will act on those preferences or necessarily believe that an eco-label (or proxy) is a legitimate means by which to differentiate products. Moreover, there may be concerns that eco-labels might be used to promote eco-protectionism (Salzman 1997). Whether the policy underlying an eco-label is sound is often a matter of debate, and those who seek to eco-label their products may be subject to economic pressure. For example, a 1992 Austrian law requiring that timber products made from tropical timber be labelled as such was eventually rescinded after several ASEAN countries threatened to bring a GATT challenge against the law (Salzman 1997, citing Sucharipa-Behrmann 1994) and to prevent Austrian companies from entering their markets (Esty 1994b).

Additional consumer-driven environmental information initiatives include boycotts and other mechanisms to publicise environmental analyses and rankings. The Internet provides a particularly powerful new means by which to share product information quickly and cheaply. Likewise, as an information-intensive (not materials-intensive) service, the Internet is itself contributing to a world of reduced environmental harms.

The power of price signals should not be gainsaid. In fact, although not flashy and not fast, market-based incentives are having positive environmental effects. Incremental improvements driven both by corporate and by consumer pressures have sharpened the focus in much of the business world, leading to noticeable environmental results.

Government policies

Governments also shape how market forces play out in the environmental policy domain. In fact, ensuring that environmental externalities are internalised is a critical element in making Adam Smith's 'invisible hand' (Smith 1776) environmentally friendly. All environmental regulation should thus be geared toward promoting the polluter pays principle so that producers and consumers pay fully for the environmental harms that they cause.[7]

On a practical basis, governments can, and should, move to eliminate environmentally harmful subsidies to farmers, fishermen and certain energy suppliers. They can support eco-labelling schemes, mandate disclosure of corporate environmental perfor-

---

7 Not only does the polluter pays principle ensure that externalities are minimised, it also offers benefits in equity, as the costs of the harms are imposed on the individuals producing them.

mance measures (such as the US Toxic Release Inventory [TRI]), insist on accounting principles (Federal Accounting Standards Board [FASB]) that require disclosure of corporate environmental liabilities and demand Securities and Exchange Commission (SEC) filings (or their equivalent) on environmental results. All of these initiatives would inject more information on environmental performance into the marketplace, giving consumers, investors and communities a better basis on which to judge companies.

Role of non-governmental organisations

NGOs can help to ensure that environmental objectives are not forgotten in the context of economic development. NGOs have historically played an important role in keeping governments honest and are now making international organisations more efficient and effective (Charnovitz 1997). NGOs can provide services, mobilise public opinion, defend minority or inadequately represented perspectives, monitor enforcement agencies, advise governments, contribute to ongoing policy debates and bridge the divides between governments and grass-roots efforts (Bebbington and Farrington 1993; Esty 1998a; Princen and Finger 1994). Giving NGOs a structured role in the institutions managing globalisation and the processes of economic integration (e.g. WTO, NAFTA, EU, APEC) offers the promise of real environmental benefits. In fact, NGOs often play a critical role in the policy-making domain as 'competition' to governments (Esty 1998a).

Some commentators, however, remain critical of NGOs,[8] fearing they will act as manipulative special-interest groups. In general, the fear that NGOs will distort outcomes is overblown. Although NGOs may act as special interests to some extent, there are already many special interests at work within environmental policy-making bodies. Adding environmental groups to the mix may provide a valuable counterweight to industry views.

Government officials will feel a greater need to justify their decisions if they are subject to careful review. NGOs thus increase the accountability of regulators by providing the general public with a genuine idea of the factors influencing a given decision and the alternatives available. The value of having the prevailing wisdom constantly questioned is especially great in policy-making realms marked by high degrees of uncertainty, such as the environment.

The role of institutional reform

The forces unleashed by globalisation are often hard to control, either because they are very big and powerful (the demands of competitive markets), because they exceed the jurisdiction of any one nation (MNCs) or because they are hard to corral (financial flows given the fungibility of money). Global scale co-operation must be strengthened if these forces are to be prevented from leading to unnecessary environmental harms.

In fact, one of the central features of globalisation has been a broad-based international consensus in support of market economics and freer trade and investment flows.

---

8 The criticisms and responses outlined below draw on Esty's discussion of the role of NGOs in the WTO (see Esty 1998a, 1998b).

But the promise of a market economy will remain in doubt if there exists a threat of 'market failures', such as uninternalised pollution harms or poorly managed common resources. In the absence of a regulatory structure able to contain failures in the scope at which they arise (including the global scale), market systems tend not to deliver their full potential for improved social welfare, as market failures reduce the allocative efficiency of the international economic system, diminish the gains from freer trade and lead to environmental degradation.

The international bodies charged with protecting the environment at the global scale, especially the United Nations Environment Programme (UNEP), are largely ineffectual (Haas *et al.* 1993; Hurrell and Kingsbury 1992; Victor *et al.* 1997). The location of UNEP in Nairobi isolates it, and the lack of sufficient funding also limits its efforts. Although some successful initiatives have been launched, such as the Regional Seas Programme, UNEP has been largely incapable of leading more substantive responses to international environmental policy challenges. UNEP officials themselves recently admitted,

> global governance structures and global environmental solidarity remain too weak to make progress a worldwide reality . . . The gap between what has been done thus far and what is realistically needed is widening (UNEP 1997: 3).

A strong argument could be made for creating a global environmental organisation (Esty 1994a, 1994b). Such an entity would facilitate data and information exchange and develop and evaluate current policy efforts. The OECD, WTO, and World Health Organisation (WHO) provide relevant models for setting up such an organisation. In practical terms, however, such an effort is unlikely to occur in the near future because of sovereignty concerns.

Institutional reform and strengthening could, however, occur within some of the existing organisations. In moving toward a clean revolution in Asia, APEC could play an important role. To lead a meaningful effort, APEC itself needs to undergo some structural changes. The organisation would benefit from the creation of an environment committee, an external environmental advisory group, and an environmental dispute mediation service (Dua and Esty 1997). By developing this structural capacity and a set of environmental performance indicators and measures, APEC could provide further leverage for a movement to ensure clean growth in Asia.

Conclusions and policy implications

The past 30 years have changed life in Asia dramatically. Although substantial economic development has occurred throughout the region, lifting millions of Asians out of poverty, it has been accompanied by serious environmental problems. In the context of deepening economic integration pollution control and resource management concerns must be given attention as well. As we attempt to reach sustainable outcomes, economic and environmental issues can mutually reinforce and support one another, but policies must be explicitly designed to do so.

In trying to promote a clean revolution in Asia, points of leverage can be found in many policy realms. Institutional strengthening of the capacity of the WTO, UN, APEC and ASEAN to address environmental issues would be a positive step. Impetus for 'greener' products can also stem from business activities and the desires of consumers. Simply put, there is no need to sacrifice environmental progress at the altar of economic growth. Despite many obstacles and complications that currently exist, real prospects for an economically and environmentally healthy Asia remain alive.

The promise of a greener Asia depends first and foremost on efforts at the national level. Nations in the region must have the determination to improve the quality of the environment and ecological sustainability at the national, regional and global levels. Governments in the region must have the political will to undertake and sustain those efforts. Forces from within the society itself, either in the form of organised activities by environment-oriented NGOs or in the form of greater awareness of environmental matters on the part of the wider civil society, are critical to the success of national programmes in improving the quality of the environment. At the national level there is thus the need to strengthen environmental decision-making by engaging a broader range of perspectives in the policy-making process. Opening up the policy process is not easy, but transparency and inclusiveness would help to ensure that national efforts to promote environmentally sound policies are successful.

In many Asian countries the national agenda that would result from a more democratic policy process would focus on what local communities regard as areas of greatest urgency, such as cleaning up rivers in heavily populated regions. The agenda must be understandable and manageable and should promise to bring early results to broaden the constituency for further investments in environmental improvements. It can then be extended to other areas in a progressive fashion. The environmental standards that are adopted should also be realistic and should begin with those that can be enforced. Adopting at the outset standards that are too high can be self-defeating. Those baseline standards should gradually be strengthened and improved.

With these conditions in place, regional leverage points can play a constructive role in strengthening the implementation of national programmes and in further improving those programmes over time. Global agreements help set benchmarks and principles that could guide national governments in their efforts to achieve internationally accepted targets. Regional efforts can help reinforce those national efforts through more intensive exchanges, mutual assistance and mutual encouragement. In essence, this has been the centrepiece of the regional co-operation that has evolved in ASEAN and APEC. At times, and on certain matters, this modality has worked—often it has not. Yet both ASEAN and APEC remain exciting experiments. As such, they must engage in institutional innovations. Given the dynamic context, ongoing assessments of why certain programmes do not produce results are critical. Rather, as the haze problem illustrates, the disappointing record of ASEAN and APEC co-operation on environmental matters is perhaps a result of the fact that regional, collective actions (or programmes) are formulated and undertaken without having put into place effective national and community-level action plans.

National plans are critical to the success of efforts at the national, regional and global levels. The importance of 'ownership' of the plan should not be overlooked. Many regional co-operation efforts have failed because of the absence of national ownership of the programmes. ASEAN and APEC should experiment with this institutional innovation to promote greater and sustained efforts to strengthen environmentally oriented policies in their member economies and in dealing with local, national and transboundary environmental problems in a co-operative fashion, both regionally and globally. In doing so they can more effectively contribute to the attainment of a greener and cleaner Asia.

4

PUBLIC POLICIES TO PROMOTE CLEANER SHARED INDUSTRIAL GROWTH IN EAST ASIA

Michael T. Rock, Ooi Giok Ling and Victor Kimm

The purpose of this chapter is to identify the policies for overcoming the historically unique sustainability challenge identified in Chapter 1. The hope is that discussion of the policy responses to this challenge will initiate a dialogue that mobilises governments, donors, communities and private enterprises in the developing market economies (DMEs) in East Asia[1] to act now to ensure that future industrial growth is substantially cleaner. This result is labelled 'cleaner shared industrial growth'.

The argument proceeds in three steps. In the first section, we develop a simple theoretical framework for identifying cost-effective regulatory and non-regulatory policies for cleaner shared industrial growth. We go on to examine the role of regulatory polices in cleaner shared growth. In the penultimate section, we focus on the specific industrial, investment promotion, technology and economy-wide polices most likely to contribute to cleaner shared industrial growth. In the final section, we identify the implications of the argument for those interested in promoting cleaner shared growth in particular countries and regions in East Asia.

Policy choices for cleaner shared industrial growth

How might public policy be used to promote cleaner shared industrial growth? The underlying theory can best be demonstrated by a simple diagram (Fig. 4.1). Let the length

1   The DMEs of East Asia include several low-income countries (China, Cambodia, Laos People's Democratic Republic, Mongolia, Myanmar and Vietnam), a number of middle-income countries (Indonesia, Papua New Guinea, the Philippines, Thailand, Malaysia and Korea) and several high-income economies (Hong Kong, Taiwan and Singapore). The argument that follows can easily be extended to South Asia.

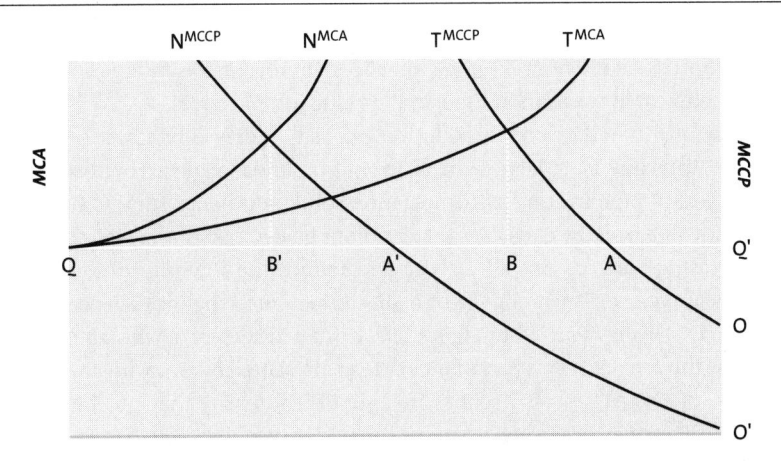

Figure 4.1 **The costs of clean production: marginal cost of abatement (MCA) and marginal cost of cleaner production (MCCP)**

of line QQ' equal a desired reduction in the pollution intensity of industrial production for a firm, industry (sector) or economy. The amount QQ' might reflect either an absolute reduction in pollution intensity (measured in weight of pollution per unit of value added) or a percentage reduction in pollution intensity needed to sustain a given level of ambient environmental quality.[2] The vertical axis on the left-hand side measures the marginal dollar cost of reducing pollution intensity through traditional post-pollution abatement (end-of-pipe expenditures), the marginal cost of abatement (MCA). Curve T^{MCA} reflects the traditional rising MCA associated with increasing reductions in pollution intensity through post-pollution treatment.

The vertical axis on the right-hand side measures the marginal dollar cost of reducing pollution intensity by reducing the intensities of energy, water and materials use of industrial production. This is often referred to as pollution prevention, cleaner production or what industrial ecologists call 'dematerialisation' (Warnick *et al.* 1996). The axis is labelled MCCP to refer to the marginal cost of cleaner production. The curve associated with this, T^{MCCP}, like that for MCA, is reflected in a rising (but from left to right) marginal cost of reducing pollution intensity by use of cleaner production.

There are several important differences between the MCA curve and the MCCP curve. First, to reiterate, MCA reduces pollution intensity by treating pollution after it has occurred, whereas MCCP prevents pollution by reducing the intensity of energy, water and materials use by substituting less polluting inputs for more polluting inputs, by improving the efficiency of energy, water and materials use and by recycling energy, water and

---

2 If the scale of industrial activity increases, the size of QQ' may have to be expanded to sustain a given level of ambient environmental quality.

materials. Normally, these cleaner production alternatives are brought about by some combination of better 'housekeeping' practices, minor process modifications or fundamental technical innovation in industrial production processes.

Because of this, reductions in pollution intensity achieved by lowering the intensity of energy, water and materials use are different from those achieved by abating pollution through end-of-pipe treatment (for instance, some pollutants, such as carbon dioxide gas, cannot be abated by end-of-pipe technology). First, end-of-pipe treatment is always cost-increasing whereas not all activities to reduce the intensity of energy, water or materials use are cost-increasing (but it should be noted that not all clean production pays either). This is depicted in Figure 4.1, with an MCCP curve with an origin, O, that lies below the zero axis. This part of the curve (represented by OA, with an area 'beneath' the curve of OQ'A) reflects declines in pollution intensity that can be attributed to declining intensities of energy, water and materials use that 'pay'. Second, end-of-pipe treatment is almost always a derivative of environmental regulatory policy. Although reductions in intensity of energy, water and materials use can flow from regulatory policy, they can also flow from changes in the relative prices of energy, water and other material inputs, from industrial and investment policies and from the pace, pattern and rate of diffusion of technological change saving on the use of energy, water and materials. This means that reduction of intensity of energy, water and materials use need not be driven solely by regulatory policy. As will be argued below, understanding this and appreciating how regulatory and other policies can reinforce these effects is critical to the design of cost-effective public policies aimed at reducing the intensity of energy, water and materials use. One example of this should suffice.

In the context of the DMEs of East Asia, dematerialisation and pollution prevention effects that 'pay' might well represent declines in intensity of energy, water and materials use associated with new (and cleaner) investment. Given the volume of expected new investment relative to the size of the existing industrial capital stock in the DMEs of East Asia (see Chapter 1), these effects could be substantial. This suggests that governments in East Asia might consider industrial, investment promotion and technology policies that encourage firms and plants to adopt and rapidly diffuse cleaner technologies.

For heuristic purposes, assume the 'starting point' for the MCCP curve is T^{MCCP} and that of the MCA curve is T^{MCA}. The most cost-effective strategy for reducing pollution intensity in a plant, firm, industry or economy by an amount QQ' requires reductions in pollution intensity through end-of-pipe control by an amount QB and reductions in pollution intensity through a decrease in the intensity of energy, water and materials use by an amount BQ'. Note that, as drawn, most of the reduction in pollution intensity (amount QB) comes from conventional end-of-pipe control.

From a policy perspective, four questions must be asked about this outcome. First, what environmental regulatory policies contribute to this outcome? Second, what role do non-regulatory policies have in promoting this outcome? Third, is the outcome described above the most cost-effective way to reduce the intensity of pollution (and energy, water and materials use)? If not, what might an alternative set of cost-effective policies look like (such as those depicted by the alternative MCA and MCCP curves, N^{MCA} and N^{MCCP}, respectively, in Figure 4.1, which would alter the two reductions from amounts QB and BQ' to amounts QB' and B'Q', respectively)? Each of these question is taken up in turn in the following discussion.

Environmental regulatory policies

Environmental regulatory policies for cleaner shared growth can best be understood by reference to the environmental policies currently in use by countries in the Organisation for Economic Co-operations and Development (OECD). Those policies either impose legal limits on emissions from major point sources of pollution, encourage facilities and firms to prevent pollution before it occurs or reward firms for 'superior environmental performance'. In terms of Figure 4.1, policies that impose legal limits on emissions (most often referred to as command-and-control policies) work on MCA; pollution prevention policies work on MCCP; and policies that reward firms for superior performance affect the magnitude of the amount QQ'.

Until recently, environmental protection agencies within the OECD relied heavily on command-and-control policies to meet mandated pollution intensity reduction goals such as that represented by amount QQ' in Figure 4.1. Even now, command-and-control policies are the base on which pollution prevention policies and superior performance policies rest. How command-and-control policies promote pollution reduction goals such as the reduction QQ' is fairly well understood. To begin with, they are almost always rooted in comprehensive environmental legislation that vests legal authority in environmental regulatory agencies to protect the environment. Landmark environmental legislation enables environmental protection agencies to set ambient standards and facility-specific emissions standards, to monitor and report to concerned publics on ambient conditions and on the compliance status of regulated facilities with emissions limits and to impose penalties on regulated facilities that fail to meet pollution discharge requirements. Without clear legal authority to do these things, it is virtually impossible for regulatory agencies to define ambient standards and facility-specific emissions standards, to clarify expectations for the regulated community or to promote equity in the burdens placed on similar point sources of pollution.

Because ambient air and water quality standards are critical for the protection of public health and ecosystems, environmental protection agencies typically get actively involved in setting ambient standards. Doing this right depends on reliance on 'mainstream' science, peer review and on an open, participatory and transparent standard-setting process that gives major stakeholders input into the setting of ambient standards. After this, numerical concentration limits can be set for air and drinking water quality and for surface water, based on intended uses.

Regulatory agencies can complement ambient standards with procedural requirements for handling solid and hazardous wastes. If it is not possible to meet ambient standards without imposing undue hardships on regulated facilities, regulatory agencies can set interim goals with attainable milestones. These interim goals are often reached following arduous consultation and negotiation with the regulated community and the public.

Environmental protection agencies can also take responsibility for monitoring and reporting ambient conditions and changes in those conditions. Reliable information on ambient conditions and changes in them can be an important way for these agencies to generate public and political support for pollution control. Because the public and regulated facilities are usually quite interested in the impact of command-and-control policies on ambient environmental quality it is critical that both have sufficient respect for the institutions charged with ambient monitoring. This is obtained by conducting ambient monitoring in accordance with widely accepted professional standards and protocols, and by using reliable monitoring equipment.

In addition to setting ambient standards, regulatory agencies also set facility-specific emissions or discharge limits on major point sources of pollution. Although it is recognised that facility-specific discharge limits should be set on the basis of expected impact on ambient environmental conditions, this is difficult to do in practice. Because of this, discharge limits are most often set on the basis of what best available technology can obtain without imposing undue hardships on regulated facilities. Most environmental protection agencies also differentiate between new point sources of pollution and existing sources. Emission limits for new sources are often more stringent than those for existing facilities.

To ensure facility-specific compliance with discharge limits, environmental protection agencies require major point sources of pollution to monitor emissions, record outcomes, report serious violations immediately and periodically to report compliance information to regulatory agencies and the public. This is often complemented by periodic monitoring by regulatory agencies and by unannounced inspections of regulated facilities. If it is economically difficult for a facility to meet emissions standards, regulatory agencies can offer compliance assistance and work out formal compliance schedules with regulated facilities that will bring them into compliance over time.

Since environmental protection agencies are legally entrusted to ensure that the regulated community is in compliance with established emissions limits, facilities found to be in substantial violation of discharge standards are subjected to a range of sanctions designed to enforce compliance. Thus regulatory agencies routinely issue administrative warnings, order improvements, suspend operations and occasionally shut down operations of facilities found to be in persistent violation of emissions standards. The agencies also rely on the courts to try to impose civil and/or criminal penalties.

Available evidence (World Bank 1992) suggests that these command-and-control policies are highly effective at de-linking growth from environmental degradation. They also contribute to notable improvements in ambient environmental quality. Despite this success, technology-based standards have not been sufficient to meet desired ambient standards. This is not the only criticism of technology-based command-and-control regulatory policies. Economists argue that these policies ignore efficiency considerations in the way facilities meet emissions limits. The regulated community has echoed this view and has complained that command-and-control policies impose onerous administrative burdens on regulated facilities and result in heavy-handed use of enforcement discretion by regulators. More recently, the regulated community has opined that increasingly stringent emissions limits impose high costs on regulated facilities (that is, they are forced to operate very high up on the steepest part of the MCA curve in Fig. 4.1) while yielding small or insignificant improvements in ambient environmental quality.

Others have criticised command-and-control policies for emphasising the cleaning-up of pollution after it occurs rather than preventing it in the first place (that is, for failing to recognise the MCCP curve in Fig. 4.1). Still others have criticised command-and-control policies for failing to reward firms for 'beyond-compliance' performance (that is, for failing to recognise that some leading firms may be willing to go beyond the statutory level of reduction of emissions [amount QQ' in Fig. 4.1]).

Because of these criticisms, regulatory agencies in the OECD began experimenting with 'market-based policy instruments', pollution prevention policies and 'superior performance' policies. These new policies were complements to, not substitutes for, the basic command-and-control policies that essentially required major point sources of pollution to abate pollution by investing in end-of-pipe pollution control equipment (represented by the MCA curves of Fig. 4.1). The major impact of this shift in regulatory policy was that regulated point sources were given greater flexibility in how they met required reductions in emissions (amount QQ' in Fig. 4.1). Market-based instruments were designed to take efficiency considerations into account in the meeting of emissions standards. In terms of Figure 4.1, market-based instruments were designed to lower and move the MCA curve 'to the right'. In the case of tradable permits, this was accomplished by allowing facilities with high marginal costs of abatement to purchase the right to increase emissions above those of facilities with lower marginal costs of abatement. The net effect of trade in permits to emit was that overall reductions in emissions (amount QQ') were met at a lower abatement cost (the overall MCA curve shifted 'down' and 'to the right'). Other market-based instruments include pollution charges, performance bond schemes and deposit–refund systems. Although attractive in terms of gains in economic efficiency, in practice the effective use of market-based instruments has been limited. In the United States, market-based instruments have been successful in reducing lead in gasoline and in reducing sulphur dioxide emissions from large power plants (for reviews or the strengths and weaknesses of such approaches, see Jaffe *et al.* 1995).

Unlike traditional regulatory programmes and market-based instruments that work on the MCA curves shown in Figure 4.1, pollution prevention policies encourage point source facilities to prevent pollution before it occurs. That is, they impact the MCCP curves in Figure 4.1. The ultimate goals of pollution prevention policies are to avoid or reduce the quantity and toxicity of waste-streams and to reduce or eliminate the need for end-of-pipe treatment. Although the initial focus of pollution prevention policies was on small batch-type production processes that resulted in especially toxic wastes, over time pollution prevention policies have been expanded to deal with all types of interventions designed to reduce pollution and conserve energy, water and raw materials.

In the United States this approach was incorporated in the Pollution Prevention Act of 1990. Among other things, the Act postulated a hierarchy for waste reduction activities including: (1) process changes to limit or reduce the toxicity of the waste-streams; (2) re-use of raw materials; (3) recycling of process streams; and (4) if all else fails, complete treatment prior to disposal. Pollution prevention policies also sparked collaborative efforts between industry-specific trade associations and regulatory agencies to find less costly ways to meet tighter emissions standards. This proved to be important as more stringent emissions requirements significantly raised the unit costs of abatement. Because there was some evidence that pollution prevention activities 'paid', as depicted by the area OAQ' in Figure 4.1, by actually reducing costs of production and by reducing or eliminating the costs of abatement they came to be seen as potentially attractive alternatives to cleaning up pollution after it occurred.

Advocates of policies for pollution prevention or cleaner production have argued that, because of information, policy, market and co-ordination failures in clean technology markets, the outcome to decrease MCA-type and MCCP-type reductions by amounts QB and BQ', respectively (Fig. 4.1) is not cost-effective. They argued that markets failed to convey to polluters both the real lower marginal cost of clean production (denoted by the lower-cost N^{MCCP} curve) and the real higher marginal cost of abatement (denoted by the higher N^{MCA} curve). If the real costs of a cleaner production environmental management strategy are given by the N^{MCCP} curve and the real MCA curve is N^{MCA}, several important differences result. First, the range of pollution prevention or activities to reduce the intensity of energy, water and materials use that 'pay' expands from the area represented by OAQ' to that represented by O'A'Q'. This provides more win–win opportunities for polluters. It may also convey Porter-like 'competitive' advantages to firms that shift in this direction (Porter and van der Linde 1995a). Second, cost-effective pollution reduction requires more clean production (an increase in the reduction of intensity of energy, water and materials use from B to B') and less end-of-pipe expenditure (a reduction from B to B'). Third, except in the case where the real marginal costs of cleaner production are less than the real marginal costs of abatement for all levels of pollution reduction, firm-level and plant-level cost-effective industrial–environmental management requires identifying the optimal combination of end-of-pipe and clean production.

But why might existing policies and market forces generate pollution intensity reduction outcomes such as QB (by MCA) and BQ' (by MCCP) rather than the more cost-effective outcome given by QB' (MCA) and B'Q' (MCCP)? There are two answers to this question. First, traditional 'command-and-control' technology-based industrial–environmental management systems favour end-of-pipe pollution intensity reduction strategies over clean production strategies. Because technology-based standards underlying existing 'command-and-control' industrial–environmental management systems identify the range of pollution intensity reduction possible with best available end-of-pipe technologies, they are easier and less risky both for regulators and for polluters. This biases pollution intensity reduction strategies in an end-of-pipe direction. If this bias is combined with increasingly stringent emissions standards, this provides incentives for the end-of-pipe pollution control industry to search for cost-reducing end-of-pipe technological change. In terms of Figure 4.1, this has the effect of pushing the curve N^{MCA} 'down' and 'to the right'. The use of market-based instruments reinforces this shift. Assuming no change in the curve N^{MCCP}, this biases cost-effectiveness toward more pollution intensity reduction by abating pollution after it has occurred.

If, in addition, markets for cleaner production are characterised by information and co-ordination failures and/or high risks and high transactions and learning costs, curve N^{MCCP} may be 'higher' and 'to the right' of that shown in Figure 4.1. This reinforces the end-of-pipe policy bias. But why should clean production markets be characterised by information and/or co-ordination failures or high risks and high transactions and learning costs?

There are several answers to this question. To begin with, implementing a firm-level or plant-level clean production industrial–environmental management strategy raises several new problems for manufacturing firms and plants. Several examples should suffice to demonstrate this. Substitution of a less toxic input for a more toxic input may be perceived to change, or may actually change, the quality of the final product (Laughlin and Corson 1995). Although it might pay to make this substitution, firms may be unwilling to take the risk of a negative customer reaction to this 'new' final product. The same might be said about basic process modifications that 'pay'. In addition, before firms make these switches, they may have to invest scarce managerial and engineering time and even scarcer capital to identify clean production alternatives (Kiesling 1994). Unless these expenditures have known or expected pay-offs that are better than the alternatives, firms may be reluctant to make them (Panayotou and Zinnes 1994). That is, it may simply be prudent to stick with well-known end-of-pipe abatement alternatives.

If current 'command-and-control' policies, including use of market-based instruments, bias industrial–environmental management strategies in an end-of-pipe direction and if risks, information failures, and transactions and learning costs undervalue the benefits of clean production alternatives, governments can intervene to correct these policy and market failures. This is precisely what regulatory policies that promote cleaner production do. Information, technical assistance and demonstration projects about pollution prevention opportunities are designed to overcome information failures. Tax breaks, such as accelerated depreciation for cleaner production investments and subsidised loans, are meant to 'level the playing field' between pollution abatement and cleaner production alternatives for reducing intensities of pollution, energy and materials use. They are also meant to compensate firms for the risks and learning costs associated with cleaner production alternatives. If these programmes are successful in overcoming policy, market and information failures, as well as high transactions and learning costs, the real marginal cost of abatement in Figure 4.1 will be given by the curve N^{MCA}, and the real marginal cost of cleaner production will be given by the curve N^{MCCP}. With this, more of the reduction in pollution intensity comes about by reducing the intensity of energy, water and materials use and less will come about by abating pollution after it has occurred.

What about policies that champion 'superior environmental performance' by leading firms that voluntarily commit to pollution reductions that exceed the sum of regulatory restrictions on facility emissions? (In terms of Fig. 4.1, this means that emission reductions are larger than the amount QQ'.) In most instances, the chief incentive offered by regulatory agencies is some form of public recognition for credible beyond-compliance performance. When environmental reputation matters, public recognition can spur senior management of large, leading and highly visible firms towards superior performance.

Non-regulatory policies for cleaner shared growth

Several researchers (Wheeler and Martin 1992; World Bank 1994a, 1997c) have suggested that newer industrial plant and equipment developed within the OECD tends to be cleaner than existing industrial plant and equipment in East Asia. Because manufacturers in East Asia are dependent on firms in the OECD for plant, equipment and technology, it may be technically and economically possible for them to import, adopt, adapt, modify and innovate on an industrial capital stock that will be cleaner simply because it is newer. Given the expected increase in the size of the industrial capital stock in East Asia over the next 20 years, this could be an important avenue for cleaner shared industrial growth. Some (World Bank 1997c) have suggested that, because of the openness of countries in this region to trade, foreign investment and foreign technology, this will happen almost automatically.[3]

What are the implications of this possibility for a cost-effective cleaner shared industrial growth outcome such as QB' (MCA) and B'Q (MCCP) depicted in Figure 4.1? There are two answers to this question. If openness is sufficient to promote a cleaner industrial capital stock, the effect of openness will be to push curve T^{MCCP} 'down' and 'to the left' so that it moves toward curve N^{MCCP}. This results in more pollution intensity reduction through reduction of energy, water and materials use and less through post-pollution abatement. This suggests large win–win effects for the environment and the economy.[4] But it is important to ask if this possibility is inevitable or whether it is dependent on other policies. If it is dependent on other policies, it is important to identify those policies.

There are several reasons to suspect that openness, by itself, may not be sufficient to generate win–win outcomes such as QB' (MCA) and B'Q (MCCP) in Figure 4.1. First, win–win outcomes such as this will be less likely the more the 'new' investment consists of older and dirtier industrial capital. Second, win–win outcomes will be less likely if policies elsewhere in the economy discourage efficient use of energy, water and materials. Last, as will be argued below, even if new investment is cleaner and resource pricing policies are efficient, unless firms have the capacity to manage plant and equipment efficiently they may not be able to achieve such cost-effective pollution intensity outcomes.

What do we know about each of these? To begin with, there is little doubt that some of the 'new' investment in the second-tier (Indonesia, Malaysia, Thailand and the Philippines) and third-tier (China, Cambodia, Laos People's Democratic Republic and Vietnam) newly industrialising economies (NIEs) of East Asia consists of older and dirtier capital in 'sunset' industries. Several of the first-tier NIEs (Korea, Taiwan and Singapore) have encouraged the export of low-technology labour-intensive industries such as textile dyeing, leather-making and simple electroplating to China, Indonesia, Malaysia and the

---

3   The World Bank (1997c) has suggested that access to cleaner plant and equipment in Indonesia and China should lead an industrial capital stock 20 years from now that is 25%–30% cleaner.

4   Win–win effects for the economy are manifest in the larger area of gain (O'A'Q') in pollution prevention that 'pays'.

Philippines. Some researchers (Rock 1996b) have suggested that this is the natural outcome of shifting comparative advantage. This suggests that openness alone might just as easily promote dirtier industrial outcomes. This tendency can be and has been exacerbated by inappropriate pricing policies for energy, water and other materials in some of the NIEs. Sometimes, as in China, energy price policy favours dirty fuels over cleaner fuels. Sometimes, as in Indonesia, energy prices are kept well below international prices. Similarly, water and other materials (such as wood and primary metals) are also often underpriced.

How do the import of older and dirtier capital equipment and the underpricing of energy, water and materials affect the pollution intensity reduction outcomes depicted in Figure 4.1? The import of older and dirtier capital equipment has at least two effects. On the one hand, it forces firms and plants to rely on end-of-pipe treatment (MCA). It may also provide opportunities for plants to engage in better housekeeping practices and minor process changes that reduce intensity of energy, water and materials use (MCCP). But how much of each of these plants engage in will depend on the degree to which regulatory policies encourage both end-of-pipe treatment and cleaner production. It will also depend on energy, water and materials price policies. If regulatory policies emphasise end-of-pipe treatment and if energy, water and materials price policies discourage efficient use of energy, water and materials, outcomes will look more like that depicted by curves T^{MCA} and T^{MCCP} than curves N^{MCA} and N^{MCCP}. But if regulatory policies encourage clean production alternatives and if end-of-pipe treatment and energy, water and materials prices reflect at least international prices, pollution intensity reduction could move more toward outcomes represented by curves N^{MCA} and N^{MCCP}.

That being said, before industrial plants and firms in the DMEs of East Asia can take advantage of either end-of-pipe or cleaner production opportunities, they must have the capability to efficiently manage plant, equipment, technology, technical change (especially technology acquisition) and technical know-how. If industrial firms lack the capability to do these things, there may be significant limits to their ability to respond to regulatory, economy-wide and industrial policy incentives designed to push them in a direction that lowers the intensity of pollution and of energy, water, and materials use. Lack of capabilities in these areas might also limit the ability of firms to take advantage of new imported technologies that are cleaner.

What do we know about the capabilities of firms in the DMEs of East Asia to manage production efficiently, to improve production capabilities and to carry out technical change? There are several answers to these questions. First, there is enormous variability in the existing capabilities of firms to do this well (Hill 1996; Kim 1997; Roberts and Tybout 1996; Rock 1999). This capability varies by country, firm size, sector and ownership. Firms in North-East Asia appear to be better at this than their counterparts in South-East Asia (Hill 1996; Kim 1997). Large firms appear to be better at this than small firms (Lall 1992). This is easier for firms to do in supplier-dominated capital goods sectors (textiles) than it is to do either in scale-intensive sectors (automobiles or aircraft) or in science-based sectors (such as chemicals or electronics where a strong capacity for reverse engineering is needed) (Bell and Pavitt 1992). Firms engaged in joint ventures with large foreign firms appear to be better at this than domestically owned firms (Harrison 1996).

Second, because much of the acquisition of these capabilities is tacit—that is, it can be gained only from direct experience—variability also depends on a firm's willingness to invest in learning-by-doing in each of these areas (Bell and Pavitt 1992). There appears to be enormous variability in the willingness of firms to make these learning-by-doing investments. Moreover, this willingness is strongly influenced by country policies. A stable high-growth environment appears to be particularly conducive to firms' willingness to invest in technological capability acquisition (Lall 1992). Export-oriented industrialisation policies that require firms to reduce costs, raise quality and introduce new products help (Lall 1992). When trade policy is tethered to lucrative export incentives, it can be a powerful stimulus to technical capability-building within firms (Kim 1997; Rhee et al. 1984). State policies that favour and reward local firm technical capacity acquisition over reliance on foreign capital (direct foreign investment) can reinforce (and have reinforced) these effects (Mardon 1990).

Third, because there are significant externalities in the accumulation of production, technology and technology capabilities, government policies are needed to accelerate the process by which firms acquire new technical capabilities and diffuse them throughout the economy. Experiences in North-East Asia suggest that two distinct sets of issues affect the speed with which firms acquire new technical capabilities. The first concerns the influence of government policy on firm size. The second concerns the need for government to invest in the provision of public goods that accelerate acquisition of technical capabilities in industrial firms.

With respect to the size of firms, two distinct patterns have emerged. In the Republic of Korea, one aim of government policy was to promote the development of very large firms (*chaebols*) that could internalise, and hence appropriate, many of the externalities associated with technological learning (Jones and Sakong 1980; Lall 1992). When this was combined with stable and high growth, an export orientation and an administrative structure that rewarded performance, the consequences for technical capabilities acquisition were enormous (Kim 1997). Government support for the development of equally large industrial conglomerates in Indonesia, Thailand and Malaysia suggests that something similar may be at work in those countries (McVey 1992; Rock 1995, 1999). There is one other benefit to government policies promoting the development of large, diversified, industrial conglomerates. Some of those firms are likely to become leading firms. As experience in OECD countries shows, leading firms appear to be particularly susceptible to incentives designed to reward superior performance (i.e. to get them to reduce pollution, energy and materials intensities by more than QQ' in Fig. 4.1). Alternatively, in Taiwan, industrial development policy promoted the development of a large number of small firms (Wade 1990). Because not one of these in any industry was capable of internalising the externalities associated with all facets of acquisition of technical capabilities, much of this was done either in government-funded industrial technology research institutes or in public–private-sector programmes co-ordinated by government (Lall 1992; Wade 1990). When this happens, it is not surprising that the public sector rather than the private sector takes the lead in clean production and superior environmental performance.

Beyond this, public-sector investments in national technical capability-building also matter. As the experiences of Korea and Taiwan demonstrate, large investments in literacy, in secondary education and in tertiary education, particularly engineering training, make it easier for firms to acquire technical capabilities (Tan and Batra 1995). A technology infrastructure that provides information (including information on cleaner technologies), that tests materials and inspects and certifies quality control standards (including the ISO 14000 series) and that calibrates measuring instruments (Tan and Batra 1995) facilitates acquisition of technical capabilities, particularly in small and medium-sized enterprises (SMEs).

What are the implications of all of this for the pace and scale of diffusion within and between firms in the DMEs of East Asia of production and technological capabilities in the reduction of intensity of pollution and of energy, water and materials use? There are three answers to this question. First, policies that promote firm-level technical learning and capabilities acquisition are likely to be good for the reduction of intensity of pollution and of energy, water and materials use. They should make it easier for firms to engage in better housekeeping practices and minor process innovations that prevent pollution. They should make it possible for firms to 'stretch' existing plant and equipment by substantially modifying it to reduce pollution as well as energy, water and materials use. They should also make it easier for firms to evaluate 'new' imported plant, equipment and technology in terms of their ability to reduce these quantities. Each of these lowers (shifts to the left) the marginal costs of cleaner production (MCCP) and contributes to greater reduction of pollution intensity by increasing the efficiency of energy, water and materials use.

Second, because the reduction in intensity of pollution and of energy, water and materials use is or will be a relatively new activity for industrial firms in the DMEs of East Asia, industrial firms there are likely to need industry-specific and technology-specific information (and specialised technical training) on how to do this. This is just the kind of information and specialised training that institutions that are part of the national technology infrastructure (such as industrial technology institutes or standards agencies) are good at providing. They should be encouraged to provide such information and training to overcome information failures and the high transaction costs associated with reducing the intensity of pollution and of energy, water and materials use. This is most likely to be true for SMEs.

Last, existing SME–multinational corporation (MNC) linkage programmes aimed at technological upgrading of SMEs might well be modified to include MNC 'greening' of the supply-chain programmes (Battat *et al.* 1996).

Summing up and next steps

Our arguments suggest that getting policies right in three discrete but overlapping policy arenas—in environmental policy; in trade and resources pricing policies; and in industrial, investment promotion and technology policies—are critical to the success of cost-effective reduction in intensity of pollution and of energy, water and materials use. How might individual economies and sub-regions, such as the Association of South-East Asian

Nations (ASEAN), in East Asia use these insights to design and implement cost-effective pollution intensity reduction policies? To begin with, virtually all of these economies can gain by pricing energy, water and materials closer to their real scarcity values. Each of these economies can also gain by maintaining and increasing openness to trade, foreign investment and foreign technology and by policies that encourage firms to engage in high-speed technological learning and capability-building. Public investments in national technological capability-building and incentives that reward individual firms for engaging in high-speed technological learning should also help firms move toward cost-effective pollution intensity reduction. Beyond this, policies need to be tailored to take advantage of differences in existing conditions in each of the economies of East Asia.

At least three patterns of differences are visible. One group of economies (Korea, Taiwan, Hong Kong and Singapore) has relatively strong command-and-control environmental agencies, economies that are nearing the end of their industrial revolutions and firms with strong technical capabilities. A second group of economies (China, Indonesia, Thailand, Malaysia and the Philippines) has much weaker environmental protection agencies, economies that are in the midst of their industrial revolutions and firms with weaker technical capabilities. A third group of economies (Cambodia, Laos People's Democratic Republic and Vietnam) has extremely weak environmental protection agencies, economies that are at the start of their industrial revolutions and firms with extremely limited technical capabilities.

Economies in the first group face four problems/opportunities. To begin with, economies in this group are nearing the end of their industrial revolutions. This means that the intensity of pollution and of energy, water and materials use is likely to grow slower than income. It also means that most of the industrial capital stock that will be in place 20 years from now is already in place. Because of this and because economies in this group have relatively successful command-and-control environmental agencies, clean-up is either just about complete (as in Singapore) or well on the way to being completed (Korea and Taiwan).

Moreover, because environmental agencies in this group of economies are oriented toward command-and-control policies, pollution intensity reduction has been biased in favour of end-of-pipe solutions (MCA in Fig. 4.1). As we know from experiences in the rest of the OECD, there are rapidly diminishing returns to this strategy. As ambient environmental standards and facility-specific emissions standards are tightened, firms in these economies will be forced to move further up the MCA curve. This will undoubtedly create pressures, as it did within the OECD countries, on regulators to 'ease up' on the regulated community. Because of the close relationship between business and government in these economies, this could contribute to regulatory reversals. To counter this, regulatory agencies in this group of economies need to develop market-based instruments, pollution prevention and superior performance complements to command-and-control policies. This means that regulatory agencies in these economies are likely to be particularly open to policy initiatives that work on MCCP strategies (i.e. prevent pollution) and expand beyond what regulations require, QQ' (i.e. reward superior performance). Regulatory agencies in these economies also need to develop stronger relationships with and more support for their actions with political leaders, the public and the regulated community.

This may be necessary to prevent regulatory backsliding. Because publics, communities and environmental non-governmental organisations (NGOs) in this group of economies tend to be distrustful of governments, this may not be easy to do.

Because firms in this group of economies have made a habit of engaging in high-speed technological borrowing and learning, it should be relatively easy for them to engage in high-speed technological borrowing and learning in environmental management. Tough, competent regulatory agencies have, no doubt, already contributed to this, at least with respect to end-of-pipe solutions to pollution. Now is the time to extend firm-level learning to cleaner production and superior performance solutions to pollution. How this might best be done is likely to vary by country. In Korea, where large vertically integrated and conglomerated firms dominate, much of the new learning is likely to take place within the firm. Thus policies designed to promote technical–environmental learning in cleaner production and superior environmental performance must take account of this. One way to do this is by linking corporate leaders and environmental management units in these large firms with their counterparts in 'leading' firms in the United States. In Taiwan, where small firms dominate, the public sector is likely to be the primary conduit for learning about cleaner production and superior performance. This requires working with industrial policy agencies (such as the Industrial Development Bureau of the Ministry of Economic Affairs), science and technology institutes (such as the Industrial Technology Research Institute) and standards agencies.

Last, governments in several of these economies (particularly Korea, Taiwan and Singapore) are actively engaged in selective industrial policies that promote the development of indigenous environmental goods and service industries. In each instance, nascent domestic environmental goods and service industries are expected to become export-oriented. In some economies (Korea and Taiwan) government agencies expect this industry to capture a significant share of the market for environmental goods and services in countries such as Malaysia, Thailand, Indonesia and the Philippines. It would be unfortunate if firms in this industry in these economies end up successfully promoting and exporting only end-of-pipe solutions to pollution. To avoid this bias toward end-of-pipe solutions, efforts should be made to ensure that capabilities-building in this nascent industry in these economies includes learning about cleaner production and superior performance policies

Economies in the second group face more difficult tasks. First, their environmental regulatory agencies are much weaker. In some economies (Thailand and the Philippines) these agencies operate without landmark environmental legislation that empowers them to set ambient and emissions standards, monitor performance and enforce compliance. In others (Indonesia and Thailand) regulatory agencies have no authority to monitor, inspect or enforce facility-specific emissions standards. In virtually all of these economies regulatory agencies lack both sufficient technical capacity and sufficient resources to manage national environmental protection programmes effectively. Weaknesses in environmental protection programmes are exacerbated by the looming sustainability challenge outlined in Chapter 1. Because the economies in this group are in the midst of their industrial revolutions, they are poised for substantial and massive increases in industrial output over the next 20 years. This combination of weak environmental protection agencies and large expected increases in industrial output is particularly noxious.

What can or should governments do under these circumstances? First and foremost, substantial efforts must be made to enhance the capabilities of environmental protection agencies to set, monitor and enforce facility-specific emissions standards. Experiences in Singapore, Korea, Taiwan and within the OECD countries suggest that this will take time and resources. In Singapore, Korea, Taiwan, and in the OECD countries more broadly, this required building the capability of environmental protection agencies to implement and manage traditional command-and-control policies. Only after this was done did regulatory agencies introduce pollution prevention policies and superior performance policies. This raises an interesting question. Should the nascent environmental protection agencies in this group of economies follow this path or should they try to simultaneously develop command-and-control, pollution prevention and superior performance policies or should they attempt even more innovative alternatives such as integrated pollution control? Since pollution prevention policies and superior performance policies are complements to and not substitutes for sound command-and-control policies, we suspect that environmental protection agencies in this group of countries would be best served by developing the capacity to manage rigorous command-and-control programmes.

What might these agencies do in the interim while command-and-control capacities are being built? There is a simple and straightforward answer to this question. Environmental protection agencies need to be both opportunistic and strategic. That is, they need to look for opportunities where they can intervene to make a difference and where they can learn by doing. This suggests taking a problem-specific approach to capability-building. This can mean taking action that either builds on or galvanises public opinion and/or community pressure. There are several examples in East Asia of how this has already been done. The Department of the Environment in Malaysia took advantage of growing community and public dissatisfaction over unabated pollution from crude palm oil mills to fashion a highly effective intervention strategy that successfully de-linked palm oil production and exports from water pollution. This included development of a highly productive relationship with a quasi-public, quasi-private science and technology research institute. A local environmental agency in Indonesia (Aden and Rock 1999) did much the same when it used a highly publicised pollution case to mount a small-scale monitoring and inspection programme that worked. Indonesia's national environmental impact agency, BAPEDAL, has gone one step further by developing a simple environmental business rating programme, PROPER (see Chapter 7), which relies on public disclosure and shame to get plants to clean up pollution.

The export orientation of firms in these economies opens an additional opportunity for strategic intervention. There is growing evidence that external environmental market pressure can influence the environmental behaviour of manufacturing plants that export. Sometimes this takes the form of greening the supply chain programmes, sometimes it takes the form of international voluntary environmental standards (such as the ISO 14000 series) and sometimes it takes the form of industry codes of conduct (such the chemical industry's Responsible Care programme).

Nascent environmental regulatory agencies in this group of countries can take advantage of the opportunity created by the export orientation of industry by working with industrial policy agencies (with ministries of industry, science and technology institutes and with standards agencies) that provide assistance to local firms so they can meet these requirements. This might take the form of co-operation between an environmental protection agency and a national standards agency on development of policies for ISO 14000 certification of local firms. It might take the form of adding an environmental supply-chain programme to linkage programmes between local SME suppliers and multinational buyers, or it might take the form of development of a green labelling programme between environmental protection agencies and respected domestic environmental NGOs.

There are three potential advantages to these kinds of partnership programme between environmental protection agencies and industrial policy agencies. Because they place some of the implementation burden on others, they limit demands on nascent environmental protection agencies. They also encourage productive relationships between environmental protection agencies and industrial policy agencies. This can work to the benefit of the latter, particularly as the former learns that they can help their clients meet some of the external environmental demands these clients face. Finally, they actively engage industrial policy agencies in environmental protection.

Countries in the last group (Cambodia, Laos People's Democratic Republic and Vietnam) face the most formidable challenges. These economies are largely agrarian, they have very small industrial bases and they have even smaller export-oriented industrial bases. Their current comparative advantage in industry is in low-skill, low-wage, labour-intensive dirty industries such as textile dyeing, leather-making and low-skill electroplating. These are relatively footloose industries and the very industries that others in East Asia, particularly Korea, Taiwan, Singapore and Hong Kong, are losing comparative advantage in. Because of this loss of comparative advantage, many of the 'plants' in this industry are relocating to this third group of countries. Plants in this industry are also moving to other low-wage countries such as India, Bangladesh and Sri Lanka.

Comparative advantage in these dirty industries, high levels of poverty, low levels of education and great weaknesses in institutional capacity in government, generally, and in environmental protection, in particular, provide few obvious opportunities for effective intervention. Countries in this group might have much to gain from a regional (ASEAN-based) investment code of environmental conduct that binds foreign investors to a commonly agreed set of environmental practices. This could be particularly helpful if foreign investors from elsewhere in East Asia and from elsewhere in the OECD abided by a set of environmental requirements similar to those of investors' home countries or economies. Export-oriented industrial plants in these countries might also gain from greening the supply-chain programmes and other external environmental market pressures (such as ISO 14000 certification and green labelling programmes), particularly if they are managed either by foreign investors or by donors.

5

INDUSTRIALISING CITIES AND THE ENVIRONMENT IN PACIFIC ASIA
Toward a policy framework and agenda for action

Michael Douglass and Ooi Giok Ling

The purpose of this discussion is to develop a policy framework for addressing Pacific Asia's[1] urban–industrial environmental crisis. The economic importance of the twin processes of urban and industrial growth is not in doubt. Taken together, the benefits of urbanisation and industrialisation are manifested in rising incomes and levels of material welfare with a reduction in shares of population below basic-needs poverty lines throughout the market-oriented economies of Pacific Asia.[2]

These benefits have been achieved at very high environmental costs, however, and one of the most challenging issues of the present and future of Pacific Asia is how to drastically reduce the already serious environmental consequences of urban–industrial growth and concentration. Urban–industrial development represents a quantum increase in the uses of energy and natural resources, with the appropriation of the resources to service them and the impacts of the waste they create, now taking place on a global as well as a local scale. In the Pacific Asia context of dense human settlements—more than

1   In this chapter the term 'Pacific Asia' may be read as 'East and South-East Asia'.
2   Just as industrialisation spurs the growth of cities, cities provide necessary spatial structures to support industrial production and distribution. With industrial growth bringing quantum increases in material goods, employment and income, urbanisation provides high agglomeration economies for industrial performance and sites for the concentration of supporting services for producers and consumers. Cities also form the built environment, raising levels of household consumption, including access to higher-order services and amenities. At a higher spatial scale, national and international systems of cities form the spatial matrix for communications, transportation and trade that facilitate and accelerate globalisation.

30,000 people per square kilometre in many cities—these increases have translated into high concentrations of pollution and environmental degradation that are affecting the lives of the inhabitants of cities, rural hinterlands and the world as a whole. Yet only a fraction of the industrial production expected to be occurring in Pacific Asia in just two decades is in place today.[3] With most of the mega urban regions that are the foci of industrialisation as well as the development of expanding service sectors continuing to double in population size every 15–20 years, the use of resources for city-building, business and daily living adds to industrial pollution to far outpace capacities to cope with the growing environmental stress and degradation.

In the next section, to address the issues of environmental management under accelerated urbanisation and industrialisation, that is, the rapid growth of burgeoning cities as well as the growth of industry and manufacturing, we present an overview of the Pacific Asia urban transition and a summary of the ways in which both industrial and urban growth jointly create impacts on a wide variety of environmental issues. After this we move into the policy arena by focusing on raising institutional capacities around new forms of urban governance that include local government, private-sector and civic organisations in collaborative initiatives for environmental management.

The Pacific Asia urban transition

The urbanisation process under way in Pacific Asia is the most dynamic in the world. Over the next quarter of a century about one-third of the world's total increase in urban population will accrue to this region, and by the end of this period approximately one-third of the world's urban population will also be found in Pacific Asia cities. The accelerated urban–industrial growth that is under way is bringing an equally rapid transition from rural to urban societies. From levels as low as or less than 20% urban in the 1960s, more than half of the population of this region will be residing in cities within the next two decades. In the already highly industrial economies, notably Japan, Korea and Taiwan, the urban transition is near completion, and rural populations are declining in absolute terms, with similar expectations for some of the second round of newly industrialising economies (NIEs) in South-East Asia in the coming years.

The greatest shares of urban population increases are gravitating to extended metropolitan 'mega urban regions' that have populations now moving beyond 15 million and up to 80 million (the Pearl River Delta) residing in complex urban fields of interaction (Ginsburg *et al.* 1991; McGee and Robinson 1995). These 'science-fiction-sized' agglomerations of human activity are the cities of tomorrow. Their sheer scale and complexity challenge all conventional concepts of cities and methods of planning and managing them.

---

3 In the case of Indonesia, for example, 80% of the industrial capacity projected to be in place in the next 15 years is not yet installed (*Economist* 1993).

The urban transition in not only manifested in new wealth and higher incomes for many. Along with the emergence of urban middle and working classes has come the rise of civil society as a political force that, over time, has gone beyond a singular focus on economic benefits to push for a variety of political and social reforms. Urbanisation is not merely a movement of people into more densely constructed forms of habitation but also represents social transformations that are championing democratic political systems, transparent relations between government, business and citizens, and more liveable habitats.

Urban–industrial impacts on the environment

Much of the environmental stress in expanding urban regions—as well as in resource-providing regions—is directly related to industry and industrial processes, including:

- Water, air and ground pollution, including hazardous and toxic waste, heavy metals
- A shift from organic to non-biodegradable wastes in production
- Use of chemicals and other resources affecting workers and conditions at work sites
- Depletion of energy and environmental resources (water, timber, minerals) directly used in industrial transformative and assembly processes
- Energy and environmental resource requirements for product use (e.g. leaded gasoline for automobiles)
- Noise, smells and other neighbourhood disturbances such as high volume of trucking for industrial production and distribution from production sites

Most of these forms of environmental degradation and pollution have point sources that are readily identifiable. They also generally involve private-sector operations that respond to incentives and regulations related to profit motivations and market forces and are thus amenable to performance drivers, as discussed by Angel *et al.* (1999).

There are, however, more varied and far-reaching impacts that are indirectly related to industrialisation that are embedded in urbanisation, and often neither has easily identifiable individual point sources nor responds to market incentives to improve the environment. These include:

- Poor environmental conditions of worker housing and the communities in which those workers live, including slum formation and the intensive environmental degradation associated with it
- Air pollution from increased traffic between residence and workplaces, schools and shops
- Appropriation and depletion of environmental resources for urban use, such as rural reservoirs for urban water supplies

- Water and land pollution from multiple proximal sources, including household sewerage and solid waste disposal
- Noise from traffic and construction
- Loss of prime agricultural land, coastal ecologies, regional forests and upland vegetation around (larger) cities targeted for resource extraction, industrial location, infrastructure (such as airports and multimedia super-corridors) and housing development

As these examples show, the urban–industrial matrix of interrelationships is manifold and so intertwined that the dividing line between urban and industrial sources of environmental pollution and degradation is ultimately an arbitrary one. Also, although environmental degradation is occurring in cities that are not rapidly industrialising, the reality is that urbanisation in the open economies of Pacific Asia is being substantially driven by industrialisation, and, where industrialisation has taken off, local and global environmental risks are exceptionally high from interdependent urban and industrial sources.

Policy framework, drivers and directions

The structural–spatial intertwining of industrialisation and urbanisation asks for a more comprehensive, strategic policy framework to parallel those of the more typical sectoral and single-issue-oriented approaches. Environmental impacts of single activities are compounded in geographical space as industries and urban communities alike contribute to air, water and land pollution. These activities also compete for land use and development, which not only calls for finding appropriate land-use mixes but also raises important questions about how to keep urban–industrial growth from expanding into fragile ecological zones and scarce stocks of highly productive agricultural land or into an environmentally destructive and inefficient urban sprawl.

In a situation in which environmental degradation is already severe, but also in which 90% of the expected industrial capacity is not yet in place, harnessing this transformation at this juncture in history is of crucial importance. In a potentially more positive vein, given that most of the urban and industrial space of the future has not yet been created, there are also significant opportunities for improving urban environmental management as cities expand outward into rural hinterlands.

The principal question in this regard is how to scale up environmental management capabilities to reach the level and pace of environmental pollution and degradation occurring in Pacific Asia. A feature shared by Asia–Pacific cities is that environmental deterioration and pollution are worsening at rates that are in excess of population and economic growth rates. Economic growth alone has not been found to ameliorate environmental problems. Although there is some association between sectoral shifts during the course of economic growth toward cleaner technologies in some industries, when seen from a larger urban–industrial perspective, levels of energy and resource consumption continue to increase exponentially. Inefficiencies in energy and resource use, often associated with the failure of the market to internalise the true costs of environmental resources, add to the quantum jumps in their consumption and production of waste. Such conditions define an environmental crisis that, in lieu of fundamental shifts in policies and actions, is already producing unliveable cities and is reaching to larger scales that threaten the environmental sustainability of this world economic region and beyond.

The urban–industrial transformation now under way in Pacific Asia is an encompassing social, political and economic process involving not only quantum increases in the use of energy and natural resources but also fundamental changes in governmental and societal institutions. Command-and-control styles of planning associated with the 'developmental state' of the 1960s and 1970s are yielding to more democratic and inclusive forms of governance. Policy tools have, however, lagged behind the rise of new urban and national politics and potential management institutions. Most are based on segmented, often adversarial, relations among government, private enterprise and civil society. Likewise, environmental management suffers from attempts to institute regulatory controls in settings where enforcement capacities are weak and where non-compliance is routine. The principal question in this regard is how to develop new institutional capacities in parallel with ongoing political reforms toward a broader sharing of decision-making in public policy and implementation.

It should be recognised that in the longer-term transformation under way, urban issues and urban environmental problems will outlast and will become greater than those directly associated with the manufacturing sector in the region. Higher-income countries in Asia, such as Korea and Japan, have already passed the point at which manufacturing is a growing percentage of the gross national product (GNP) and employment, with the service sector now dominating expansion of both. In the case of Korea this sectoral transition occurred in less than three decades. The Seoul metropolitan region continues to grow in scale and population even though manufacturing has already reached its peak and producer services are replacing lower-end manufacturing that is being relocated to lower-income countries. Although industrial pollution will continue to be a major dimension of pollution, there is a need to go beyond attending to industrial sites as the principal focus for environmental management and instead to look at expected sources of pollution and environmental degradation in the context of the urban transition that has no automatic turning point away from quantum increases in the consumption of natural resources and energy or the production of waste.

With these perspectives in mind, and understanding the environmental issues related to manufacturing and industry do not subsume those of urban development, the focus here is on the particular historical period now being witnessed in several Pacific Asia countries: namely, one marked by simultaneous accelerated processes of urbanisation and industrialisation. The spatial overlap of these two processes is self-evident. Most industry gravitates to larger urban regions, and many of the concerns of urban governments are related to locational patterns of industry, particularly resource-intensive and polluting industry, near other urban activities. Labour must be housed, and producer services that add to urban pollution also appear in great numbers in cities. Large shares of transportation, notably the increase in use of fossil fuel-powered vehicles, within and among cities is similarly devoted to supporting industrial production and distribution of industrial products. The concern here is thus not with industrial or manufacturing sites alone but more generally with the impacts of manufacturing and industrial activities on the urban matrix of living and working.

Though environmental conditions are worsening almost everywhere, cities show significant variations in terms of sectoral mix and severity of environmental problems. Political commitments to addressing these problems and the degree of engagement of businesses and citizens in mobilising social and political energies toward addressing environmental concerns vary substantially as well. Capacities to move intentions to action in government and in society at large are equally diverse. These observations, along with the realisation that neither world powers nor central governments can be expected to effectively manage environmental questions of local cities and regions across a national landscape, suggest three cornerstones for a policy framework for redirecting current trajectories away from their environmentally destructive course.

First, there is a manifest need for a localisation of capacity-building at the urban and sub-national regional scale. This involves enhancing capacities to prioritise, to develop policies and to embark on innovative programmes for more efficient and reduced resource use, for pollution reduction and for improved environmental monitoring.

Second, the principal focus of localisation is collaborative governance. An inclusive term involving civil society and private enterprise as well as the state, collaborative governance is a means to create societal synergies for vastly increasing capacities for innovation and problem-solving (Evans 1997).

Last, forming and strengthening inter-city networks is needed for technical exchanges of innovations and for political–economic efforts to stem the 'bidding down' process that includes lowering environmental standards as a means of attracting global industrial investment.

Localisation of capacity-building

Several reasons can be put forth to call for a localisation of capacity-building to deal with urban–industrial environmental issues. First, variations among localities in constellations of types of industry and industrial organisation are high. These variations emanate from international as well as local scales. At a global scale, some segments of trans-national production are undergoing radical transformations, whereas others continue to be characterised by older forms of organisation, technology and production. When seen in real geographical terms, the general tendency has been for more polluting industries to be located in lower-income economies, not only because of relatively low environmental management capacities in these countries but also because of the older, more polluting technologies being transferred with the investment.

In terms of indigenous enterprises, manufacturing in many cities, especially but not only in South-East Asia, is characteristically found in shop houses and older commercial areas associated with particular ethnic groups. Although each enterprise may be small in size, the collective levels of pollution from them can be substantial yet difficult to monitor and even more difficult to regulate. Thus, in Asia's cities the range of industrial and manufacturing enterprises run from local and global sweatshops, which expand work by adding labour rather than technology, to higher-technology Fordist assembly-line operations and post-Fordist systems of flexible specialisation. Increasingly, too, commodity systems reaching from agricultural fields to biotechnology-based industrial processes are making intersectoral integration more vertically complex at all spatial scales.

In environmental terms, the implications of these trends, which are realised in actual industrial location patterns in cities and regions, are that policies need to be tailored to locationally specific types and ranges of technological and organisational possibilities. A setting, such as that of Hong Kong, which continues to be composed of very small enterprises and has moved toward a service economy, is different from that of Korea, where the industrial structure is dominated by the huge *chaebol* conglomerates that engage in heavy industry as well as light manufacturing; therefore, the strategy for dealing with each of these must also differ.

Second, although the world may be concerned about, for example, Bangkok's environmental problems, action taking is fundamentally a localised process. Variations in environmental situations are great among cities even within the same country, and, historically, central governments have not revealed a capacity either to show flexibility in adjusting policies and standards to local variations or to distribute resources among cities equitably. Further, taking action is not simply a matter of governments realising the need for improved urban and industrial environments but is vitally a political process that emanates from local pressures and interests.

To the extent that these sources of validating environmental policies are not included in policy formulation and implementation, the gap between policy rhetoric and local possibilities for action remains wide. Localisation of decision-making is the main channel for bringing civil society and other local stakeholders into the political process. Generally, the failure to localise decision-making has also meant that local communities have been unable to engage businesses in their areas in public discussions on environmental concerns. In lieu of a local government with substantive political authority, national political solutions tend to bypass or ignore local community inputs, and neither communities nor businesses are able by themselves to organise efforts to mutually work out solutions to environmental problems at the local level.

Much of the discussion on localisation of capacity-building is being furthered under the banner of decentralisation, and, almost without exception, governments throughout the region have adopted policies to decentralise decision-making powers and management capacities to local urban and regional scales. In some cases, such as in Korea and Taiwan, progress in this direction has been impressive. In most other instances, however, decentralisation in practice remains a form of deconcentration of administrative tasks still tightly controlled by central bureaux rather than authentic devolution of effective governmental capacities.

A process of decentralisation to build local capacities would entail at least three aspects: (1) increased local decision-making and management authority; (2) greater financial resources and autonomy at the local level; and (3) vastly increased manpower and personnel skills to engage in partnerships, streamlined regulatory processes and monitoring of the environment. In each of these aspects there are tremendous needs in cities throughout Pacific Asia. Even with increased authority to make and carry out environmental policies, initiatives may quickly fail as a result of acute shortages of skilled workers or as a result of the financial management capacities of local governments. Although the speed of decentralisation and local capacity-building will depend on national and local political processes, international and other supportive organisations could greatly assist environmental management efforts through the promotion of local workshops and worker training in this area.

Collaborative governance

Democratisation and governance

If policy-making on the environment is to be channelled toward more inclusive, problem-solving engagement in the public domain, mechanisms must be expanded to enlarge political communities through participatory processes. Moving in this direction focuses attention on the question of governance. In contrast to the term 'government', which centres on the organisation of the state apparatus, governance casts a wider net to capture the processes by which public decisions are made and put into practice. This includes relationships among government (the state), civil society (including non-governmental organisations [NGOs]) and economic enterprises (private sector) (Friedmann 1998). Focusing on governance is not for the sake of resolving environmental questions alone but rather to link solutions to environmental questions to the promotion of more open, transparent and democratic political processes. Analysis of environmental movements in Asia shows that democratisation and social mobilisation for proactive engagement in resolving environmental issues go hand in hand. Advancement toward more inclusive public decision-making also carries the potential to incorporate questions of environmental justice associated with urban poverty more directly onto policy agenda.

In this light, the emphasis here is placed on collaborative governance: namely, the capacity to go beyond reactive, sometimes violent, confrontations by building a broad societal process for proactive agreements for joint action among government, civil society and the private sector. One of the major trends over the past decade in Pacific Asia in this regard has been the outright rejection of autocratic regimes in several countries and significant political reform toward more open politics in others. The return to elected government in the Philippines more than a decade ago was a signal event in this regard. More recently, Korea and Taiwan have instituted direct election of governments under multi-party systems. In Thailand, following major social protests in 1992, the military has moved to arm's-length distance from government. The ousting of the Suharto New Order Government in Indonesia in 1998 also serves notice of a new, although as yet unresolved, relationship between the state and civil society. Although in several countries one party political machine dominates, elected governments are no longer the exception in the

Asia–Pacific region, and even non-elected governments are having to widen the scope for public discourse over political affairs. All of these trends suggest a much greater possibility for collaborative governance than ever before. In most cases, however, a further strengthening of the capacity and capability of civil society is needed before the citizens can effectively engage the state and for-profit business interests in addressing environmental issues.

Following fundamental reforms that allow for direct elections and accountability of public officials, the rise of civil society in some countries has already become a potent force in promoting environmental improvements and more liveable cities (Evans 1999; Friedmann 1988). In settings where political reform and more flexible approaches toward urban management are combined, policy arenas are shifting out of bureaucracies and toward collaborative forms of management among government, private-sector and civil society organisations. Although democratisation remains elusive in others, a trend toward reducing the presence of government in command planning and regulation of urban activities, including land use and the environment, is readily observed throughout the region. In many cities, municipal governments are realising that bearing the costs of environmental regulation on their own is unsustainable. Command-and-control mechanisms require growing amounts of resources, both financial and human, challenging the budgetary limits of local governments (Ooi 1998).

This phenomenon entails the emergence of both a broad urban middle class and the organisation of a growing wage labour force—and a widening unemployed population in countries now experiencing economic reversals. These, in turn, are intricately linked to the globalisation of urban economies through export-oriented industrialisation as well as to greater access to information afforded to the general public about domestic and international political affairs and to education and living experiences abroad. Urban populations in Pacific Asia are now able to know much more about their own governments and alternative perspectives—even in situations in which governments continue to try to suppress such information.

As a consequence, popular sentiments have moved beyond the desire for higher material welfare to include aspirations for accountable governments, democratic practices and a translation of economic gains into more liveable urban habitats and socially just societies. Citizens everywhere are also becoming more effective in challenging the ways in which cities are being planned and managed. Environmental movements emerged in significant numbers during the 1990s, and many have gone beyond demonstrations and protests to form their own organisations for longer-term planning around environmental issues (*AJEM* 1994; Ho 1997; KNCFH 1996; Lee 1995; Lee 1998). Conflict over such issues as the location of environmental infrastructure and services, industry and mega-infrastructure projects such as airports, railway lines and highways is increasingly common. The countless projects involved in creating industrial spaces and constructing mega urban regions have become principal sources of political mobilisation and confrontation throughout Pacific Asia. It is fair to say that environmental issues and conflict among special-interest groups, classes and communities will drive much of the future of urban politics in Pacific Asia. How to transform this social energy into a positive source of collaborative planning is of fundamental importance to the future of Pacific Asia cities.

At the same time that state–civil society relations are being transformed, private-sector interests are becoming more effective in insisting that governments move away from overt, red-tape regulation of the city and toward state-sector–private-sector 'partnerships' and privatisation as the new mechanisms not only for urban concerns but also for sustaining economic growth. The current economic crisis in Asia has further pointed toward the need for basic reform in state-sector–private-sector relations, both in terms of eliminating collusion between state and privileged enterprises and in terms of less cumbersome forms of regulation. As with civil society, there is a manifest need for greater transparency in regulatory processes related to economic enterprises.

Governments are thus increasingly being drawn into creating more inclusive forms of governance. As part of a process of democratisation, this trend can be seen as being desirable in its own terms. It is also a necessity for practical reasons: without collaboration with citizens and business interests, programmes and projects meet resistance or are undermined by non-compliance.

Variations in government responses to the rise of civil society and to calls for more transparent relations with the private sector are great. In several countries NGOs are flowering and have taken on positive mediating roles between state and society. As noted, locally elected governments have also been instituted for cities in, for example, Korea, Taiwan, Thailand and the Philippines. In other countries democratic reforms remain limited, but even in these cases there is growing popular resolve for political reform and more participatory governance. Where political reforms have been most cavalierly ignored or countered, such as in Burma, political crises continue to overwhelm and forestall improvements in all other social and economic issues.

Building synergistic capacities

In an era of slowed economic growth and severely compromised government resources, a key dilemma in Asia is how to drastically raise the capacity to manage environmental problems without incurring public costs that are far beyond the revenue base of government. To achieve longer-term benefits at the lowest cost, governments now need to do more with regard to providing environmental infrastructure and services. Under current economic reversals throughout the region, expectations for increasing levels of expenditure are, however, unrealistic. Evidence over the past decade of structural adjustment and economic crisis suggests that, in fact, the opposite trend of declining capacities to spend on environmental management is common, especially in the countries of South-East Asia where industrial growth is incipient and where per capita incomes are still significantly below those of Japan and the first generation of newly industrialised economies.

Unless such dramatic enhancements are achieved with relatively low public costs, neither the economic nor the environmental future of the region is bright. The thesis here is that accomplishing this scaling up of capacity can be done, but not by any one set of actors alone. It must instead be found through creative synergies among them.

Left to their own motives and methods, government, civil society and private enterprise are often more likely to produce counterproductive or limited outcomes than they are to address environmental management problems successfully.

A major hindrance to the search for means other than simply spending more on the environment is the lack of collaborative institutional arrangements among principal actors—government, civil society and private enterprise. As summarised in Table 5.1, these actors have different motives or drivers that lead them to use and manage environmental resources. Government, for example, is driven by the twofold, but often contradictory, need to sustain its revenue base from the economy and to secure its political base by responding to, for example, popular protest movements.

Table 5.1 **Non-collaborative drivers, actions and tendencies among government, civil society and the private sector**

Source of action	Driver	Type of action	Tendency
Government	Need to secure state revenues Need for legitimisation	'Corporatist' command planning through regulations and/or subsidies influenced by social pressures for environmental improvement	Ineffective or selective enforcement; non-compliance high; ad hoc responses to citizen demands; democratisation and rise of civil society bring environment more clearly onto political agenda.
Civil society	Desire for more liveable habitats	Confrontation, vigilante action	Reactive solutions: 'Nimbyism'; socially uneven sharing of burdens
	Access to environmental resources, infrastructure and services	Self-reliance in environmental management	Management limited by scale of response
Business/ private enterprise	Prices and profit motive	Systematic undervaluation of environmental resources; technological advances in resource conservation dependent on price signals	Inefficient use of environmental resources, large-scale pollution and degradation
	'Good neighbour' ethic	Voluntary use of recycled material, for example	Estimated 10% of US firms

Its principal powers are to regulate political, economic and social behaviour, which in most Asian countries is carried out in a command style of policy-making and implementation with little consistent participation of business or citizen groups. Because tensions are rife among various groups of businesses and strata of citizens, the general tendency of government has been to selectively enforce regulations and to respond to only the most overt forms of social mobilisation and protest. In many other instances, the capacity of the state to enforce its regulations is in doubt not only because of the limited pool of trained personnel but also because of the support facilities to enable them to do their work. While some business sectors complain of favouritism and citizens become cynical about corporate–government relations, the condition of the environment deteriorates as a result of non-compliance and sporadic enforcement. Even when they are evenly applied, regulatory approaches can also show a lack of appreciation of the need for private enterprise to remain profitable in a globally competitive world where the location of industry in one country over another is sensitive to costs imposed by environmental regulation.

In contrast to government, the drivers for action among citizens emerge from desires for more liveable communities and cities and, especially among the poor, for access to such basic environmental resources as clean water that invariably entail supporting infrastructure and services. Without government support many of the actions take the form of protest movements and reactive drives to force government to, for example, compel industries to stop polluting, desist from locating undesirable facilities in local areas or achieve environmental justice in the form of clean environments for slum areas. As the general process of economic growth is also a socially stratifying one marked by the appearance of an increasingly politically vocal urban middle class, a strong tendency emerges to focus on reactive NIMBY ('not in my back yard') protests that, in effect, seek to locate such environmental projects as locating new landfill sites away from richer to poorer neighbourhoods or out to rural areas. In some cases, such as in Taipei, such efforts to beautify central urban areas resulted in the demolition of slums to make way for large city parks. In such cases, the results are not an increased capacity to manage the environment but rather a shifting of impacts from one area to another.

The principal driver of business interests is well known: namely, a profit motive that responds to price signals. As noted, because the market is generally incapable of internalising the true costs of appropriating finite environmental resources and of polluting air, water and land, the tendency to undervalue resources and pollution impacts is high, environmental resources are inefficiently used and thus urban–industrial growth becomes a major cause of environmental degradation. With some enterprises better situated than others to avoid government regulatory sanctions, state–business relations tend not to be transparent and appear to be subject to significant levels of corruption and rent-seeking behaviour. Adopting policies to use price incentives as a driver for getting industry to use less resources and improve end-of-pipe treatment has great promise in many areas related to industrial sites and production processes, and environmental services have themselves become a source of private-sector expansion.

The major problem, then, is how to identify drivers that do not rest solely on state (government) regulatory powers but can instead be more positive in engaging civil society and for-profit enterprises in environmental management. Thus, although the reduction of environmental resource use at industrial, manufacturing or assembly plant sites through market-oriented price incentives is laudable and has substantial promise, it must be seen as a partial approach to the quest for cleaner urban environments.

The state, civil society and the business community are not monoliths. Within each there is a range of discouraging and encouraging agencies and people. In the private sector, for example, some percentage—thought by some to be 10% in the United States—of enterprises have manifest 'good-neighbour' ethics that result in conscious efforts to recycle waste materials and promote environmental consciousness among employees (Henderson 1997). Nonetheless, as long as such positive forces remain exceptional rather than commonplace, current trends toward deepening environmental crisis cannot be expected to be attenuated or reversed.

Creating collaborative synergies

The absence of authentic collaborative mechanisms among major stakeholders in the urban and industrial environment leads to limited, insufficient responses to growing environmental problems. In contrast, where energies of stakeholders can be combined through collaborative efforts, they can begin to resolve these problems in surprisingly positive ways. Drivers in this sense become combinations of drivers of each actor that are transformed from reactive, single-focus and adversarial contests to proactive, multi-stranded and co-operative outcomes. The drivers of each do not change, nor should they be expected to change in the real world. They are instead harnessed and redirected for mutual benefits for improved environmental outcomes. The following four cases show the benefits of substantially increased capacity through collaborative efforts involving citizens, government and business.

Waste collection in slum settlements. A common problem for slum and squatter settlements is that government-run refuse and salvage collectors cannot enter these communities because lanes are too narrow for their vehicles. By organising households to bring refuse to collection points at the edge of communities where trucks have access, community and government can combine energies to resolve what is a major environmental problem in most Asian cities. Case studies show that chief bottlenecks in this regard are seemingly simple, but in practice often difficult, to overcome. They include government refusal to provide pick-up services to communities of questionable legal status, which is often a secondary effect of governments not wanting to appear to be validating the existence of slum and squatter settlements (Douglass 1992). In such cases, the larger issue returns to questions of citizen rights and inclusive governance. On the community side, bottlenecks include the inability of households to organise routine refuse collection along lanes or to obtain even a small parcel of land as a refuse collection point. In the case of either government or community action the principal point to be made is that both parties must act together. One side acting alone will lead to failure. Yet, where they have joined in partnerships, as in a recent case in Bandung, the successes have been remarkably positive and catalytic (Dharmapatni and Prabatmodjo 1994).

Land readjustment. Another common problem in Asian cities is that the spatial configuration of the privately owned parcels of land in a given area do not include or readily allow for the provision or alignment of land needed for roads with drainage, sewerage or environmental infrastructure and services. Government and landowners working together can, however, pool land and redistribute it among the owners to allow for this infrastructure. Although owners receive less land than before, windfall increases in property values from infrastructure improvements yield a winning outcome for private land interests and improved environmental management. Whereas government can seek to directly regulate land developers by, for example, setback rules or land-building ratios, such actions are not able to guide larger-scale community development. Similarly, whereas governments can exercise rights of eminent domain to appropriate land for roads and other public goods, in many Asian cities the land prices are so high that government budgets are severely stressed just to compensate owners for road rights-of-way. In Tokyo in the late 1980s, for example, 90% of the cost of road construction was absorbed by the cost of land purchase (Douglass 1993). Partnerships in the form of land readjustment thus make providing infrastructure more affordable and less of a constraint on improving the environment, particularly at the urban fringe where land is for the first time being developed for urban expansion.

Private–public–community collaboration for community improvement. Slum upgrading is often attempted by government agencies with communities acting as passive recipients that perhaps supply free labour power but not much more. The general results are that improvements are expensive on a per household basis, reach very few communities because of costs and bureaucratic procedures and do not match the priorities of residents. Among the most successful efforts is the Kampung Improvement Program (KIP) in Indonesia, which is credited with upgrading hundreds of low-income communities through site and service projects, including drainage, lane construction and communal toilets (Douglass and Lee 1996). But, even with this record of success, only long-established settlements were included in the programme, leaving the newer settlements appearing at the edge of cities unserved and, through its top-down style of planning, neglecting many of the poorest sections of communities. No private-sector involvement is included except as paid subcontractors for government projects.

An alternative form of collaborative planning involving private sector, public university professors and a slum community emerged in Bangkok in the late 1980s. Bangkok has more than 1,000 slums, with a common feature among them being severely deteriorated environments. In 1990 two professors from Mahidol University secured an agreement from Citibank to provide a pool of US$50,000 for interest-free loans for community development (Ard-am and Soonthorndhada 1994; Douglass et al. 1999). One community, Wat Chonglom, agreed to be the recipient of loans from this fund. Although the community was unable to raise such funds on its own, once they were received it initiated its first project: cleaning up the rubbish and waste that for years had been collecting under the community, which was built over a low-lying area subject to frequent flooding. The success of this campaign was a source of pride for the community and encouraged it to adopt a continuing string of projects, including completing piped water connections to all houses, building a two-storey community hall complete with a children's day-care centre and making improvements in individual houses. The cumulative positive effect of the success of this project, which led to immediate improvements in the environment and personal health, is illustrated in Figure 5.1. Every new activity added to the potential of the community both to improve the wellbeing of its households and to generate new income-earning opportunities. With lanes safe to walk and the ambience decidedly improved, small shops started springing up in the community. Beauty shops, video rental stores, food stalls and mini dry-goods stores appeared in the front of homes along the pathways. Many were owned and run by women. Reversing the conventional belief that income gains must come before environmental improvements, the Wat Chonglom story shows how environmental improvements create economic growth (Douglass et al. 2000).

The community was subsequently identified as 'Community of the Year' in Bangkok, with government adopting it as a model for other communities. What is of interest here is how three forces—private-sector resources, university-based community activists and the desire of a community to improve itself—created a synergy that transformed a slum into a clean, healthy and thriving community within the space of about two or three years. The community was able to pay every loan back on time without a single default; the estimated cost to Citibank for being a good neighbour was thus just the foregone interest on the loan of less than US$10,000 per year. With such a small amount of support, a community of about 1,000 people was able to reverse a process of steady environmental deterioration and create a virtuous process of steady environmental improvements and community vitality.

Solid waste recycling. Large segments of the urban population in Pacific Asia are directly involved in the recycling of solid waste. In most cities these activities are highly organised and include a hierarchy of waste pickers, petty buyers and brokers, small shops that transform waste into new products, and larger buyers that include industries such as paper manufacturing. Ignored as 'scavenging', these activities are crucial to environmental management in cities throughout the region. In Jakarta, for example, they are estimated to reduce urban refuse by at least a third (Douglass 1993). In large cities in Vietnam the estimates are even higher; waste recycling has, in fact, become a significant portion of the urban economy (Digregorio 1993). By recycling waste, resource depletion is reduced through re-use of resources.

Despite the benefits to the environment of the city, these activities remain limited as income sources for the poorer waste pickers. They also tend to decline rapidly as governments invariably expand city-wide closed-container dustcart services that prevent waste 'harvesting' between kerbside sources and landfill sites. Although more orderly from an urban household point of view, the closed-cart system thus diminishes recycling and adds to the daily amount of waste that is chronically in search of new landfill sites. Income from lowest levels of waste picking is also limited because of the lack of space and capital available to waste pickers to transform collected waste into saleable commodities.

To address these issues, in the 1970s the City of Bandung initiated a programme of assistance to waste pickers. Called Jati Dua, the programme found land and provided for

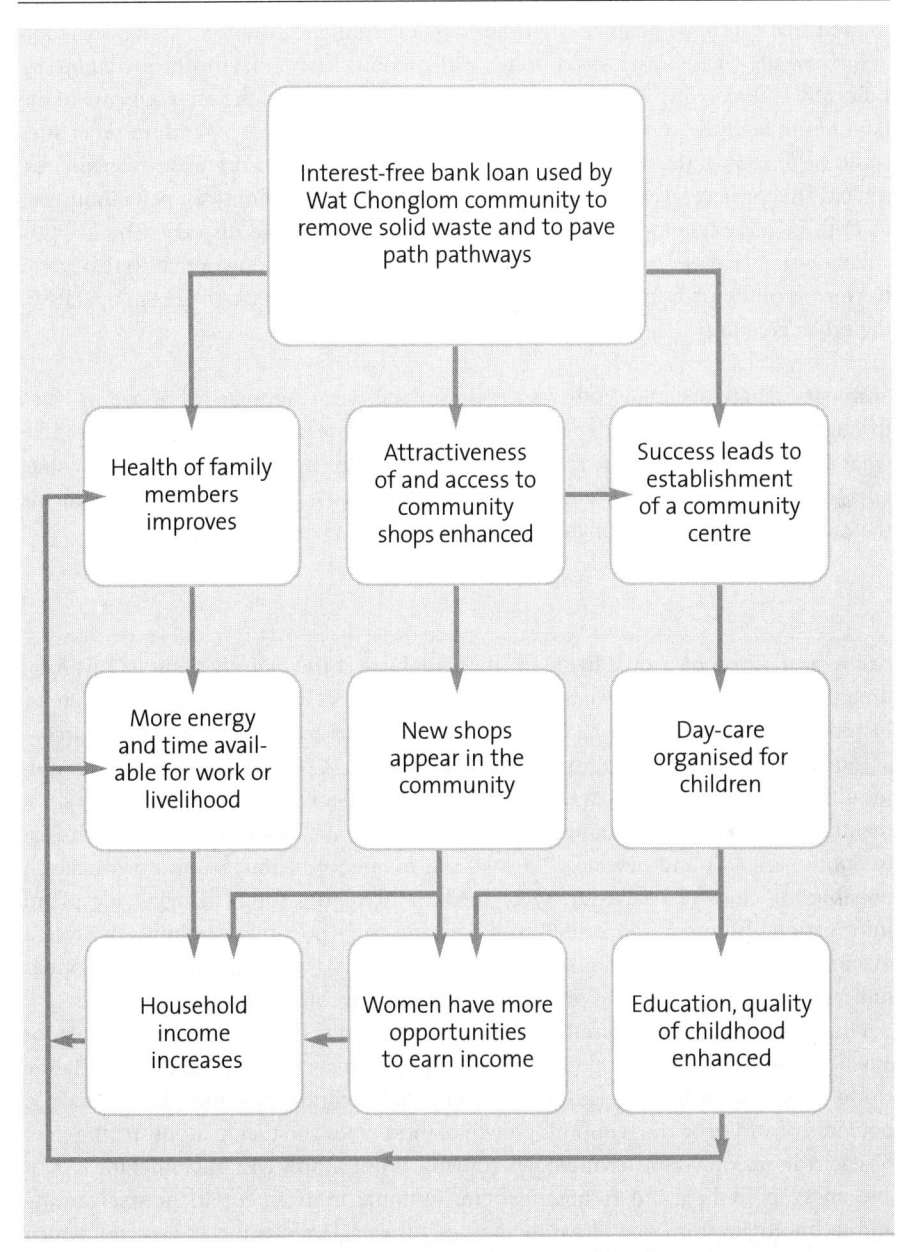

5. INDUSTRIALISING CITIES AND THE ENVIRONMENT *Douglass and Ling* 119

```
                    ┌─────────────────────────────┐
                    │ Interest-free bank loan used by │
                    │ Wat Chonglom community to    │
                    │ remove solid waste and to pave│
                    │ path pathways                │
                    └─────────────────────────────┘
         ↓                      ↓                      ↓
┌──────────────┐        ┌──────────────┐        ┌──────────────┐
│ Health of    │        │ Attractiveness│       │ Success leads │
│ family       │  →     │ of and access │  →    │ to establishment│
│ members      │        │ to community │        │ of a community│
│ improves     │        │ shops enhanced│       │ centre        │
└──────────────┘        └──────────────┘        └──────────────┘
         ↓                      ↓                      ↓
┌──────────────┐        ┌──────────────┐        ┌──────────────┐
│ More energy  │        │ New shops    │        │ Day-care     │
│ and time     │        │ appear in the│        │ organised for│
│ available for│        │ community    │        │ children     │
│ work or      │        │              │        │              │
│ livelihood   │        │              │        │              │
└──────────────┘        └──────────────┘        └──────────────┘
         ↓                      ↓                      ↓
┌──────────────┐        ┌──────────────┐        ┌──────────────┐
│ Household    │   ←    │ Women have   │        │ Education,   │
│ income       │        │ more         │        │ quality of   │
│ increases    │        │ opportunities│        │ childhood    │
│              │        │ to earn income│       │ enhanced     │
└──────────────┘        └──────────────┘        └──────────────┘
```

Figure 5.1 **Synergistic outcomes of collaborative community development**

Source: Douglass and Zoghlin 1994

modest credit to these people, who in turn organised their own production association at the new site where waste was collected. Early reports found that incomes went up significantly and recycling was further encouraged. Unfortunately, the private owner of the land terminated the project by not extending permission to use it, and no other sites could be found for the low-income recyclers to continue their enterprise. Nonetheless, the Jati Dua project provided a window on the types of opportunity for petty industries to both increase incomes and reduce the need for solid waste disposal. The key was collaboration between government assistance in providing resources that were beyond the means of the waste pickers but then leaving the industry to the pickers to operate in the urban economy.

Summary. There are many other successes of the types summarised above in cities throughout the world. Their common trait is that barriers to collaboration caused by what are often seen as opposing interests and competing agendas have been overcome, releasing a synergistic capacity for rapid and substantial environmental improvements that, in many cases, were thought to be improbable in the pre-existing climate.

Inter-city networks

As the proportion of people living in cities climbs over the halfway point in East Asia, urban areas also account for increasing shares of national economic growth. Even now, in almost all countries, one or a few mega urban regions account for very large shares—as much as half or more—of the gross domestic product (GDP). Organisational, communications and transportation networks among these cities have grown significantly since the mid-1980s and are beginning to span the entire Pacific Asia region from North-East to South-East Asia and beyond. The ease and frequency of interaction now makes it possible for cities to use a variety of media and venues for exchanging ideas and information about environmental problems and their possible solutions. The recent Mayors' Asia–Pacific Environmental Summit held in Honolulu was an important indication of government and private-sector interest in collaboration among cities.

What was also evident from the conference is the array of technologies available for ameliorating environmental pollution in key problem areas facing Asia's cities. As the world's fastest-growing industries, environmental technologies and services have a potential of adding to the economic growth of these cities and their nations. Yet the chief obstacles in making these technologies available is not simply cost but rather the lack of awareness about them and the absence of institutional arrangements to transfer learning and technologies from one locale to another. Also neglected is the NGO sector, which can bring important insights on urban environmental problems to bear on private-sector–government interaction. NGOs such as consumer groups can also have a great impact not only by engaging the market through businesses in improving environmental performance but also by influencing consumer behaviour.

On all of these matters, concerted attention on fostering inter-city networks can have high pay-offs in terms of raising awareness of common problems, availability of technologies and opportunities for collaborative innovations across urban and national boundaries. Building such networks is also consistent with 'performance-focused' environmental management. As summarised by Metzembaum (1998: ix), performance-oriented environmental management, which calls for a local-level setting of performance measurements and supporting information systems, can benefit greatly from exchanges through inter-city networks:

> The dynamic capacity of performance measurement is unleashed when it is used to compare the performance of one performer to another because it creates a mechanism that automatically updates performance expectations whenever new measurements are taken, thereby motivating continual performance improvement without necessitating a complicated and often lengthy decision-making apparatus to update targets. This is an approach that nearly a hundred localities, working with the International City/County Management Association, are testing.

In other words, cities can facilitate improvements in other cities through sharing of performance achievements, which can be used to stimulate improvements among all cities. Such achievements can be further spotlighted by such practices as giving national and international awards to 'best cities' in various environmental management categories.

A perhaps more problematic issue that can be addressed through inter-city collaboration is that of inter-city competition for economic investment in manufacturing and services. Given that their economic health is part of an intensive international competition for investment, cities face a number of difficulties in sustaining their economies. One difficulty arises from the changing nature of comparative advantage in an increasingly globalised system of production, distribution and finance within corporate networks rather than among nations. Tremendous reductions in transportation costs have radically changed the notion and reality of comparative advantage away from being anchored in natural-resource endowments toward competition to attract globally footloose industrial capital through 'created assets'. In the broadest sense, cities themselves are the spatial arena for these created assets, which range from specific industrial requirements for reliable electricity and water supplies, diverse labour supply and transportation infrastructure to higher-order functions such as hospitals and universities, parks and amenities and centres for hosting international conferences and spectacular world events.

As part of this process, inter-city competition is shifting world economic integration away from nation-states and toward inter-city networks through which the global economy is articulated. This includes a shift toward cities as the basic units of policy-making about global economic interaction. Thus municipal governments are now charged with devising their own strategies for globalisation and to engage in overt campaigns to attract world investment. As the stakes in the competition increase, so does the scale of projects to win the competition. In Pacific Asia this translates into mega infrastructure projects such as world hub airports, high-speed trains, 'Manhattan' skylines, large new towns with gated communities, and a plethora of amenities such as golf courses, museums and convention centres.

Funding such ventures has driven some cities, such as Osaka, close to bankruptcy under the weight of high project costs and diminished revenues from investors. A related concern is that cities will become trapped in a bidding-down process of competition for investment that will entail relaxing environmental measures and their enforcement beyond the point at which urban environments can be sustained. This short-term horizon on economic growth may prove disastrous in the longer term for both the environment and the economy. Paradoxically, as investment seeks out high-amenity locations but cannot be readily tapped to provide these amenities, a city's comparative advantage may deteriorate. As concluded by the World Bank (1993a), urban regions that cannot successfully sustain their environments may begin to be abandoned in favour of other 'new' regions where environmental deterioration is not as great. In some cities, such as Bangkok, Manila and Jakarta, there is already growing awareness and concern that deteriorating environments have begun to erode the capacity to continue to attract investment.

The problem is how to maintain a long-term capacity to host citizens and businesses alike while also meeting the short-term, highly competitive demands for deregulation, subsidies and other give-aways to would-be investors that seem to threaten capacities for improving environmental management and keeping industrial investment. This dilemma in the relationship between industrialisation and the urban environment is arguably the most important of all the issues confronting cities. One response is to develop capacities in inter-city networks to reach agreements about environmental management with regard to global investors. Again, the use of performance-oriented approaches could offer a means for cities to raise their environmental quality while flexibly allowing investors to develop their own means of reducing resource use and limiting pollution.

Toward an agenda for action: process and performance

The concept of 'policy' in the context of urban environmental management concerns actions taken in the public domain. This does not mean that governments are the sole or even principal actors but that decision-making and action are open to public discourse and scrutiny and the participation of diverse private, civic and governmental interests. In effect, the policies that emerge are implicitly, if not explicitly, negotiated agreements among actors having different interests and incentives (drivers) for environmental management. Based on this understanding, the proposal for action here joins in an ongoing paradigm shift away from centring public policy on government regulation toward more participatory forms of self-regulation, responsibility and positive incentives as well as regulatory controls. As detailed above, the focus for this paradigm shift is on localisation of governance to environmentally guide Asia through its urban transition.

Even without collaboration, each type of actor has a role to play in environmental management. In most cities, effective regulation by government of land use and environmental resource use is needed. Though sometimes violent, the mobilisation of citizens against the state and against flagrantly polluting industries has brought significant changes in government policy and practices. Price signals and the 'good-neighbour' ethic are also important drivers of voluntary business actions that result in improved environmental management practices of private-sector firms. Yet each has severe limitations in what it can accomplish. State regulation without broad willingness to comply, self-reliant communities trying to improve their environments without government or private-sector support, and dependence on price signals or voluntary action by business to reduce resource use has each failed on its own to create a process of environmental improvement sufficient to the monumental tasks at hand. In many instances, it can be said that they may have had contrary effects that have led to a worsening of environmental quality.

Having made the case that synergistic associations between state–society–business are the keys to making substantial improvements in the environment, the question arises about how such associations emerge. Although there is no single point of entry for initiating this process, what is instructive from the experiences of several cities, such as Seoul and Taipei, is that the rise of environmental movements and collaboration around environmental issues went hand in hand with democratic reforms. In the case of Korea, for example, environmental issues were the source of the 'moral high ground' for popular mobilisation and the recipient of increased policy attention once democratic reforms were undertaken.[4] In other words, there is ultimately little separation between the tasks of improving the urban environment and progress toward localisation and more democratic forms of governance. In this context, the challenge is to go beyond reactive conflict and toward proactive collaboration in coming to terms with the environmental deterioration of Asia's cities. The information revolution, the rise of civil society, the widening global marketplace for environmental technologies all suggest a greater potential for cities to learn from within and with each other.

In the first instance, the focus of the discussion here is on how to translate the needs for localisation, collaborative governance and inter-city exchanges into an agenda for action. The recommendations emerge at two levels. The first is concerned with supporting the creation of institutional structures and organisations. As discussed, this entails support for localisation of government capacities to manage the environment, broadening the scope of governance at the local level and promoting inter-city exchanges among all actors, including government, private-sector and civil society organisations. All of these actions are associated with what can be summarised as establishing the process for environmental management.

4 Environmental movements in Seoul were largely underground and led either by university radicals or by Christian religious organisations, until the late 1980s when fundamental democratic reforms created openings for a flowering of environment organisations in civil society. By 1993, at least 175 groups had been formed around environmental issues in Korea, and newspaper reports of incidents of citizen mobilisation on these issues increased from fewer than 500 in the 1982–86 period to almost 9,000 per year by the early 1990s (Lee 1995: 12). In 1995 the consolidation of various environmental groups led to the foundation of the Korea Federation for Environmental Movement, with 20,000 members and 21 local chapters. Along with local elections held for the first time in the same year, two environmental candidates won mayoral elections. Environmental movements were also able to compel the national government to reorganise environmental planning bureaus and adopt an environmental impact assessment system. The overall result has been the adoption of stricter standards by the government, and more information is now made available to citizens about environmental conditions (Kim 1994).

The second is concerned with performance dimensions of environmental management. It focuses on how a collaborative process can facilitate and be enhanced by performance-driven forms of environmental management that avoid the extremes of command-style planning and unregulated development. Both process and performance orientations are needed, and each can help to reinforce the other.

Policy and implementation process

As detailed above, for environmental management to begin to reverse trends of environmental deterioration related to urban and industrial growth, one of the most critical needs is to create a more open, inclusive and locally aware process of decision-making and implementation. Supportive actions include:

☐ Localisation

☐ Collaborative governance

☐ Inter-city networks

Localisation calls for support for (1) enhancing the revenue base of local governments for environmental management,[5] (2) training of government personnel in areas of environmental management, particularly with regard to setting up much needed systems of monitoring various kinds of pollution and environmental degradation and (3) supporting policies at the central government level for political decentralisation and increased capacity for decision-making at local levels.

Collaborative governance calls for devolution of governmental powers to municipal levels and institutionalising democratic practices. An area of much needed support is the training of government, community and business personnel in skills of mediation, negotiation and facilitation. Creating openness and transparency of government–business–community relations through legal rights and rights of access to information is another important area for support. Information systems such as geographical information systems (GISs) to map environmental problems over the city, open public forums on environmental issues, education programmes and celebrations such as civic awards for best practices are all considerations for supporting enhanced awareness and collaboration.

The creation of inter-city networks entails establishing new 'horizontal' linkages among cities that can more directly set in motion a process of exchange of ideas among those people from local governments, businesses and communities who know best about urban problems and who are also more likely to commit themselves to improving the urban environment. Sponsoring national and international urban conferences, perhaps with rotating city secretariats, that explicitly include local government, business and

---

5 Comparative studies show that an average of about 90% of public revenue is collected and spent by national governments in developing countries; for high-income countries, the level averages about 65% (Oates 1993). In many instances, municipal governments in developing countries have no capacity to engage in such routines as collecting property taxes or recouping expenditures through locally initiated cost-recovery programmes.

community organisations is one major way to promote these networks. Rotating responsibilities over different areas of expertise among cities could assist in capacity-building as well. Setting up bulletin boards and other forms of Internet-based information and discussion forums is another way to support and sustain such networks. Within the category of information exchange, there is a manifest need for support to create baseline data across an array of environmental issues, including the human health impacts and implicit costs to local economies of environmental degradation.

A process emanating from these three types of action enhances environmental management in a somewhat paradoxical way because the drivers of each are different and push for different dimensions of performance. To be effective, collaborative agreements will include elements of each. Thus, whereas government might be driven toward trying to induce companies to recycle waste, industry will want to try to ensure that profits are maintained and citizen groups might be concerned with the location of waste-recycling plants. Together they represent a potential for integrating environmental management with economic growth and more liveable cities. Alone, they represent single-driver, single-dimension objectives that, as discussed above, have been shown to have equally partial outcomes.

Performance orientation

How do governance and other aspects of the process summarised above relate to the idea of 'performance-focused' environmental protection? The most direct answer is that to the extent that stakeholders have a genuine interest in outcomes and that power is distributed among them the older form of command-and-control approaches toward environmental management will give way to a mutual concern for collaboration to continuously improve performance. Business, citizens and governments working together from city to city will challenge each other to adopt innovative ways of reducing resource use and pollution. There is, however, a need to consider a reorientation of information-gathering concerning performance indicators in ways that will enable cities and citizens to be informed of how programmes are performing.

Since such collaboration works best when decision-making systems are transparent, decentralised and shared among many actors, it also directs attention toward accountability through, in part, calling for accessible performance measures. That is, instead of insisting on a single 'standard' way of improving environmental management, a performance orientation more flexibly allows for a multitude of ways to achieve improvements (Metzembaum 1998). A focus on performance calls for a substantial increase in information, including its creation, dissemination and accessibility. This is also consistent with the three overarching needs for building social capacities for environmental management, namely, localisation, inclusive governments and networks among cities within and among nations. As stated by Metzembaum (1998: viii):

> An effective performance focused system can improve the way we address public problems in several complementary ways—by boosting outcomes, strengthening accountability, and enhancing the transparency of processes and decisions that affect the public's wellbeing.

The aim of improving performance focuses attention on innovation and on enhancing the capacity for innovation, which is the core of what has been called 'learning societies' which, in their basic form, thrive from the meshing of local or experiential knowledge with more specialised knowledge of environmental 'experts'. Such mutual learning in a continuous process of problem-solving is also an outcome of building high levels of social capital through co-operative action (Woolcock 1997).

Adopting a performance orientation can include three types of measurement of environmental management successes and failures (Metzembaum 1998). The most traditional is setting targets for achievement. They can also be used to establish benchmarks for similar activities in the same city or other cities within a country or internationally. Third, they can be used to reflect on the value of programmes and to help pinpoint bottlenecks and areas in need of improvement. Each has a role in collaborative governance and environmental management.

Although some enthusiasts of performance-focused management state that performance can be given precedence over process (Metzembaum 1998), process and performance orientations both need attention and both are ultimately interdependent. Process here is taken to mean building the institutional basis for undertaking four key tasks:

1. Assessing conditions and prioritising problems to be tackled
2. Analysing and understanding the sources of these problems
3. Developing policies and identifying policy tools to link solutions to priority problems
4. Devising organisational means for taking action and proceeding to implementation

Combining these aspects of planning and management creates an iterative process involving feedback loops that do not necessarily follow in a predetermined order. Each nonetheless constitutes a realm of knowledge and action that must be included in any environmental management process. All too often there is a leap from the description of environmental conditions to the formulation of policies, without a clear understanding of underlying forces creating the problems. Blaming the poor for creating environmentally degraded communities is, for example, a common policy stance that lacks clear foundations in actual cause-and-effect relations. The poor are often compelled by circumstances to locate on already degraded land or near industries that are the major sources of land and water pollution. They also expend hours of household labour time on obtaining basic environmental resources such as clean water.

Similarly, policies need to be linked to realistic expectations for successful implementation. This is where the idea of drivers becomes central. Without a clear understanding of motives and incentives of businesses, households or other actors who are expected to comply with policies, policies become little more than statements of intentions or hopes.

Finally, process is important in that it reveals who is and is not included in decision-making and taking action. It would be deceptive to argue that performance standards can be treated separately from who is included in setting the standards. Yet this question of inclusion is not readily addressed by a performance orientation alone. The results can be that closed decision-making—which is usually also associated with partial information—about performance criteria leads to poor understanding of problems and an élitist agenda for action. Process must be attended to as an integral part of environmental management.

Focusing on performance rather than on input standards opens environmental management to a wider degree of collaborative innovation. It also potentially allows for a shift toward a more flexible approach within what has been called the 'environmental management option hierarchy', which, in using the example of industrial production, rests on giving source reduction and elimination highest priority, followed by recycling and re-use, and with treatment of effluents and emissions considered as the least desirable waste-management strategy.[6] It also allows for more integrated waste management, which, in contrast to hierarchy approaches, allows a community to mix and match options according to its needs and resources.[7]

The promise of performance-based management is that it will allow the process to be in the hands of stakeholders who are committed to solving problems, who will respond to the drivers in a constructive way and who are willing to act. In this sense, it is not a substitute for but rather adds another layer to the process. It can be used in prioritisation by helping to assess gaps between existing conditions in one city and improvements achieved in comparable cities elsewhere. Its most important role, however, is in fostering a more dynamic process of implementation by sharing insights from high-performance areas to spur improvements elsewhere and, in this sense, become part of the drivers by continuously raising expected outcomes to higher levels and by providing actors with exemplary accomplishments.

---

6 Waste management can be seen to have a parallel hierarchy of options, with reduction of waste ranked at the top, followed by re-use and recycling, incineration with energy recovery, incineration of unsorted wastes and landfilling.

7 Whereas the integrated approach defines solid waste as 'rubbish that needs to be disposed of', the hierarchy approach considers solid waste as a potential resource tied into the production cycle. Consequently, the former tends to emphasise the development of new technologies and pollution-control systems, searching for capital-intensive high-technology programmes, whereas the latter is inclined toward labour-intensive, low-technology options.

6
CIVIL SOCIETY AND THE FUTURE OF ENVIRONMENTAL GOVERNANCE IN ASIA

Lyuba Zarsky and Simon S.C. Tay

The financial and economic crises that have swept East Asia since mid-1997 have brought the region to something of a crossroads in its policies in terms of both development and environmental protection. In many countries, the crisis has also changed the balance of power between government, business and non-governmental organisations (NGOs): that is, between the state and civil society.

To some analysts of Asian affairs, the crisis signalled the need and provided an opening for a change of course in development strategy for Asian newly industrialised economies and nearly industrialised economies. Central to this change is a new emphasis on the importance of 'good governance', not least to attract future foreign investment. A host of voices, spanning the International Monetary Fund (IMF), multinational corporations (MNCs), domestic business, NGOs and individual citizens, are calling for more transparent, accountable, efficient and capable government. The demand for better governance is taking place in the context of market-oriented policy reforms. If successfully implemented, these reforms will raise the economic profile of competitive businesses, both foreign and domestic, and generally increase the role of market forces in daily life.

This chapter examines a variety of conceptual frameworks and institutional models for the engagement of civil society in environmental governance in Asia. In the next section we examine the political and environmental landscape in the wake of the crisis. We then explore the character of civil society in Asia and argue that civil society is itself a contested concept. In the penultimate section we sketch six functional roles for civil society groups and develop three alternative models of environmental governance. Finally, we present some conclusions.

The fork in the road?

The crisis has given new strength and impetus to civil society in many countries in South-East Asia. The influence of civil society actors, such as NGOs, has grown dramatically, even in the most important political and policy decisions. With the crisis came the delegitimation of politically repressive governments whose popular legitimacy derived largely from high-speed economic growth. The clearest example is perhaps seen in Indonesia. After 32 years of rule President Suharto was driven from office, leaving behind a more plural and open political and social landscape. A more open politics is a positive trend in Indonesia, notwithstanding emergent concerns with law and order, including violent conflict in areas such as East Timor and Aceh.

Similarly, if less dramatically, the civil society of South Korea, Thailand and other countries has grown in the wake of the financial crisis. The increasing information demands of open markets, as well as the new opportunities and competitive pressures in a globalised economy linked by information technology, will further strengthen civil society. In the future, both business and NGOs will seek a greater role in societal governance (Tay 1999a). From this perspective, one fork at the crossroads will lead countries of the region towards more open markets and a more influential civil society. The interaction of open markets and civil society will, in turn, put strong demands on Asian governments to improve internal governance. Reforms in governance will aim both to facilitate market transactions and to promote social goals. The environment will be a major beneficiary.

The other fork of the road for East Asia is quite different. For some, the crisis has meant the end of the so-called 'Asian miracle'. Reliant on open markets and export-led industrialisation, the 'Asian model' touted by the World Bank was exemplified by the first-tier (Hong Kong, Singapore, South Korea, Taiwan) and second-tier (Malaysia, Thailand, Indonesia, the Philippines) newly industrialised economies (World Bank 1993). In the wake of the crisis, the 'Asian model' may no longer be seen as the best model to emulate, either for Asia or other developing countries.

Instead, the crisis has sparked increasing scepticism and concern about openness to trade, investment and capital and to globalisation in general. This doubt exists not just within the newly and nearly industrialised economies but also in other countries, such as Vietnam, that have opened up more slowly and now have seemingly good reason to postpone or curtail such opening. Throughout East Asia, there is a discernible rise in a narrow nationalism and a sympathetic sentiment towards autarchy, as first seen in the Malaysian experiment with capital controls (Mallet 1999; Tay 1999a). This is more than just a matter of economic policy.

On this path, there is also resistance to a greater role for civil society. Instead, the call for better governance is resisted or transmuted into a call for 'strong' government. Wary of the fate of the former Indonesian president, other Asian leaders may cling even closer to power and continue to deny demands for democracy and human rights championed by civil society (Tay 1999a).

The reluctance to open or, in other cases, the moves to partially close markets in some Asian countries is most often allied with relatively closed political systems. However, trends towards political closure also co-exist with economic openness. Indeed, the 'Asian way' has precisely espoused the value of relatively open markets and economies, combined with relatively closed political structures. Looking at Asia's 10–20-year future in the wake of the crisis, therefore, suggests a matrix of four possible scenarios in which the 'critical uncertainties' of open/closed economies and democratic/authoritarian political systems unfold in different ways. In such a matrix, economic and political regimes are in dynamic interplay to portend different paths: (1) open/authoritarian; (2) closed/authoritarian; (3) closed/democratic; and (4) open/democratic. Impacts on society and the environment differ in each scenario.

In a scenario characterised by an open economy and a relatively authoritarian political system, governments will seek rapid economic recovery and a return to old ways in government's relations to society and to business. The development path will probably continue with policies that put growth before environmental protection, rather than genuinely seeking a balance of sustainable development.

In the second scenario, characterised by a relatively closed economy and a controlled polity, environmental impacts are much the same or even worse. Without open markets, the likelihood of inefficient production and black-market transactions is even greater. The benefits of growth, moreover, will accrue to a few winners in a system that combines access to environmental resources with political power (Tay 1998a).

Whether the economy is open or closed, a politically controlling state is thus likely to engender high levels of environmental degradation. Even without political access, however, civil society is likely to make its influence felt. In the face of deteriorating environmental conditions, depletion of resources, and escalating health and other effects, government neglect is likely to strengthen discontent at the local level. Escalating environmental degradation and deteriorating 'environmental security' may thus be a source of civil and political conflict (Barber 1997; Homer Dixon 1999). Governments may also find themselves subject to external pressures stemming from consumer-led demands in developed and more environmentally conscious societies. In short, if the state in either of these scenarios chooses to ignore environmental problems, civil society, nationally and internationally, is likely to be one of the key drivers to make the state pay.

Future development paths based on relatively democratic governance—whether or not economies are open—are likely to have very different impacts on the environment. Via political access, civil society is likely to press for cleaner, healthier and more equitable economic policies. With a relatively closed economy, however, states and their natural environments may forego the benefits of foreign trade and investment, including newer, cleaner and more efficient technology and the benefits of economies of scale in production. With an open economy, on the other hand, consumer demands and per capita income are likely to grow more rapidly, triggering or exacerbating environmental stress. The dramatic rise in personal car ownership and the attendant problems of urban air pollution and traffic congestion are emblematic of this shift.

In the fourth scenario, which sketches an open economy and a democratic political system, the deepening of democracy may drive not only regional economic recovery but also a transformation in East Asia's development policies. A new paradigm would aim not simply to restart the old, pre-crisis patterns of growth but to transform them towards the path of more sustainable and equitable development. On the one hand, economic openness will require and enable efficient production in order to compete globally for export markets and foreign investment. On the other hand, the greater role of civil society could restrain government from self-serving policy excesses and encourage policies that promote the goals—economic, environmental and social—of other groups in society. With greater efficiency, equity and participation in policy-making, a better environment may well be one of the dividends (Tay 1998a).

It is not certain which of these scenarios will emerge in Asia over the long term. Perhaps all of them will emerge in different places, at least in the short term. With the intervention and surveillance of the international community and the IMF, the path towards open markets and some form of greater democracy seems most likely at present. However, in almost all societies there continue to be countervailing forces. Accordingly, more economic reforms have been promised than carried out in most of the crisis-hit countries. With political change, too, there is no guarantee that democratic forces will prevail in the medium to long term.

It is also likely that responses in various Asian countries will increasingly differ. While Asian countries clearly differed in their economic and political characteristics prior to the crisis, their responses and policies in the face of the crisis have sharpened their differences. In the Association of South-East Asian Nations (ASEAN), for example, the differences between the older and more developed member states (Brunei, Indonesia, Malaysia, the Philippines, Singapore and Thailand) and the newer members (Vietnam, Laos People's Democratic Republic, Myanmar and Cambodia) have grown sharply both in economics and in politics.

The ability of civil society to drive constructive changes in environmental governance would be enhanced, obviously, in a more democratic political regime. Even under conditions of political repression, however, the deterioration of basic ecosystems and the depletion of natural resources on which many depend for their livelihood is likely to stir up civil society protest and action. Civil society, in short, is likely to be a key variable in any scenario. The central questions revolve around whether it will be allowed to participate in charting Asia's future and what particular role or function it will play.

Environmental impacts of the crisis

Widespread, serious and highly visible before the crisis, the rate of environmental deterioration in East Asia is likely to increase as a result of the crisis. There are three reasons. First, the crisis is likely to reduce funds available for environmental protection and infrastructure and to weaken regulatory agencies and efforts to ensure compliance. Second, the crisis is likely to stimulate a resurgence of environmental problems associated with poverty. Last, there are likely to be negative environmental impacts from the economic policy responses to the crisis, especially attempts to restart 'business-as-usual' economic growth.

Decreased funds and weakened regulation

In the wake of the crisis, government revenues for almost any application have decreased, including the protection or improvement of the environment. Moreover, attention to environmental protection has declined, along with the wherewithal to meet environmental standards. This applies not only to governments but also to corporations and other producers, as well as to the members of the public that may otherwise be concerned with environmental issues.

Prior to the crisis, in response to growing public concern and external pressure, the East Asian newly industrialised economies moved to strengthen their environmental laws and regulations. During the 1990s, substantial investments were made to improve pollution control. Infrastructure investment in water supply and sanitation and, to a lesser extent, in mass transit systems brought environmental benefits. There were also signs of more effective regulation over industrial activity, with a strong record in some countries, such as Singapore, and improvements in some others, including Taiwan and Malaysia. The Asian Development Bank predicted that air and water quality would improve in the higher-income countries of the region.

In the short to medium term, the region will probably see a slowdown in the provision of such infrastructure (although some economic prescriptions now suggest the need to stimulate internal demand by, among other things, proceeding with infrastructure projects). Investment in improved and less polluting technology is also likely to be cut back. This is especially the case because most such technology has remained proprietary, despite the preferential terms of access envisaged for developing countries in international environmental agreements, including the Montreal Protocol on the Reduction of Ozone Depleting Substances and the Biodiversity and Climate Change Conventions.

There are similar concerns about whether environmental regulation and regulatory agencies will be weakened. The common wisdom is that the crisis is to be blamed, at least in part, on Asia's weakness in the governance of the public sector, financial institutions and industry. As a result, the regulatory reach of government is being weakened, with probable effects on environmental compliance.

Environmental regulation and, especially, enforcement was weak in Asia even before the economic crisis. The reasons generally stem from a lack of funding and weak, overstretched and often corrupt administrative and judicial systems (Barber 1998). The traditional command-and-control regulatory approaches to the environment are notoriously costly and are often difficult to implement effectively. The crisis may therefore worsen an already poor situation. In this context, the image of half-built, abandoned light-rail tracks in some Asian capitals is not only a portrait of stalled ambition but a surrender to the continuation of heavy vehicular traffic jams, with the attendant pollution costs.

The environmental dimensions of increasing poverty

The crisis has vastly reduced economic welfare for many individuals and has thrust a considerable percentage of people into poverty. The crisis has thus had and is likely to continue to have profound social, psychological and health as well as economic impacts on the people of Asia. Moreover, the decline in economic circumstances will have direct and indirect environmental impacts.

Prior to the crisis, the 'Asian miracle' displayed a number of negative social and human impacts, including an increasing disparity in the distribution of income and wealth between élites and the masses. Large companies, both domestic and foreign, drove rapid growth in export-oriented manufacturing, often bringing new technology and better-paid jobs. In many Asian countries, however, a domestic sector of medium-sized to smaller enterprises also grew rapidly. These small and medium-sized enterprises (SMEs) tend to be technology-poor and not competitive internationally. They tend to pay low wages, to be highly polluting and to be intensive in their use of natural resources.

The lower-level jobs and informal sector in Asia are most often associated with poor environmental and health standards. Largely, firms and individuals in these sectors have been unable to conform to higher environmental standards, especially as demanded by Western or international communities (Vossenaar and Jha 1996). If it leads to the loss of higher-level and middle-level jobs and an increase in either lower-level jobs or the informal sector, the crisis is likely to have a negative environmental (and human) impact.

Another discernible pre-crisis trend was the migration of people from rural areas to the cities, where they were often confined to low-level employment or the informal sector and lived in shantytowns (Rigg 1997). In the aftermath of the crisis, the reverse trend has become evident—urban poor are returning to the countryside. Although urban poverty has attendant problems, the reversal of this phenomenon is no solution. The push factors that initially triggered migration to cities, especially landlessness, are still extant. When the urban poor go back to the countryside, they do not return to their own land because they do not have any. Moreover, after a generation or more in the cities, the number of these landless urban dwellers has swelled. Consequently, those who return to the countryside have to find new land to cultivate.

Many will seek to survive by clearing forest or taking land previously set aside for agro-industry or luxury élite uses, such as golf courses. Others may turn to illegal hunting of endangered species of flora and fauna or to illegal logging. Such activities may be understandable where starvation is the alternative. The environmental impacts, however, may be considerable, especially where the landless use fire to clear land or drive out animals—a cheap and quick method that is polluting and often illegal. The situation is compounded when such land-takings are unauthorised and give no secure title to occupants. In such cases, the landless face strong short-term imperatives to clear and use land, but none to maintain and sustain it over the longer term.

Negative environmental impacts from policy responses

Current and future policy responses to the crisis may lead to negative environmental impacts. This may be intentional if the environment is seen as a luxury concern that can no longer be afforded in the crisis or be unintentional, given that policy responses are driven by economic agencies with little or no knowledge of environmental concerns and costs.

The environmental impacts from policy responses are hard to predict. One major reason is that economic policy prescriptions differ among different economists and agencies. Although there has also been a notable change in the tenor of the IMF and World Bank since the earlier days of the crisis, three general edicts remain operative. The first is the prescription for countries to export themselves out of the crisis. The second is the mandate to attract foreign direct investment (FDI). The third, and most general, is the aim to restart the economy and return to high growth rates. Such prescriptions trigger the concern of environmentalists that Asia's growth or, in this case, recovery will be at a high environmental cost.

This is especially so in the case of trade. Many of the exports in the region (e.g. oil, timber, minerals) are extractive or otherwise high in natural resource input. Many are also highly polluting in their production processes, especially given low regulatory standards and weak enforcement. The international trade system does not assist in this regard. The World Trade Organisation (WTO) has pointedly and repeatedly set aside environmental conditions imposed by importing countries in favour of unfettered free trade. Regional organisations such as Asia–Pacific Economic Co-operation (APEC) and ASEAN/AFTA (ASEAN Free Trade Agreement) have yet to conjoin their agendas for trade and investment with their environmental undertakings and may not have the political will to do so (Zarsky 1998).

As for investment, no international rules currently govern international investment. Such rules could potentially provide a framework for all governments in which to define and enforce the environmental responsibilities of foreign investors (Zarsky 1999b). Currently, it is up to national governments to screen foreign investment for environmental impacts. Most Asian countries have laws or policies to do so, especially requirements for environmental impact assessments before deciding on foreign investment. However, their track record has been mixed and exceptions or absences were notable even before the crisis. The political will to raise environmental performance may be even weaker now, given increased hunger for FDI and investor concerns about the competitiveness and stability of the region.

Last, blind to environmental (and social) costs, most economists see the task at hand as simply the need to restart stalled economic engines under business-as-usual social and environmental policies. If the progress of the 'Asian miracle' provided any consolation to environmentalists it was that greater affluence would eventually propel higher environmental standards. According to the hypothesised 'inverted-U pattern of pollution', environmental degradation first rises and then falls with growth in per capita income, with the turnaround point coming around US$5000 (Grossman and Krueger 1994). If the hypothesis is true, then continued rapid growth would eventually drive Asians to demand—and be able to afford—more environmental protection. With a drop in the per capita income of most countries it would seem that the region is being driven through another round of the lower end of the inverted U, when growth coincides with high levels of pollution. With the crisis, the search for environmentally sustainable development seems like a game of snakes and ladders: the Asian economies have been brought back to the bottom of the inverted U-shaped snake.

The potential for the environment to be sacrificed in the plans for recovery and restarting growth cannot be underestimated. This is despite some early reports from countries in the region which suggested that the environmental damage of the crisis has not been as bad as originally feared.

On the other hand, the crisis has triggered hope for better environmental protection in the future. In large part this hope is premised on the emerging importance of civil society, both a more competitive and efficient business sector and a greater voice for a myriad of community, religious, advocacy, professional and other non-governmental organisations. A significant increase in the role of NGOs *vis-à-vis* the state may bode especially well for the environment. Within more open systems of governance, NGOs may be able to better identify and articulate environmental and social concerns as well as to build the political will needed to enact and implement new policies.

The character of civil society in Asia

If it is presumed that Asian countries will continue to move down the path towards democratisation, we can expect civil society groups will in the future play a larger constructive role in political life. They may be especially central to the project of developing new and stronger forms of environmental governance. Indeed, the central impact of civil society may be to build the political will to drive significant policy reform.

To understand and anticipate the potential role of civil society in environmental governance, two key questions must be examined, both of which we grapple with in this section. First, what is the character of environment-oriented civil society groups in Asia? What does the concept of 'civil society' mean and what values, philosophies and issues animate it? Moreover, civil society in Asia is emerging in the context of globalisation, both of economies and of social norms. How will globalisation and 'international civil society' shape—and be shaped by—Asian NGOs?

The second key question concerns whether and how Asian governments will respond to the transformative potential of civil society, especially in relation to environmental (and social) governance. Given crisis-engendered fiscal constraints, as well as the turn towards greater openness to market forces, the logic of involving civil society in governance has become compelling. Government response will be shaped by a variety of cultural, political economic forces and pressures, both internal and external.

A contested concept

After a century of neglect,[1] the contemporary discussion of civil society was revived in the struggles against authoritarian socialist states in Eastern Europe and, to a lesser extent, military dictatorships in Latin America (Pelczynski 1988; Tismaneanu 1992). Although it was an elastic term that differed in nuance from one cultural context to another, 'civil society' emerged as a 'shining emblem' of resistance (Hann and Dunn 1996). In Eastern Europe, civil society was primarily a code word to demand rights that the Western liberal

---

1 Civil society was much discussed at the end of the 18th century, before disappearing into obscurity in the second half of the 19th century. Most scholars trace the term back to the work of Hobbes and Locke in England. Others suggest a starting point in the work of the Scottish Enlightenment and the work of Adam Ferguson (see Seligman 1992). For an overview of the history of the term, see Seligman 1992 or Keane 1988.

democracies already have as well as to delegitimate non-liberal regimes.[2] In this conception, civil society opposes the state or, at least, seeks to impose limits to an authoritarian and seemingly omnipotent state apparatus.

But this is not the only interpretation. Indeed, the term 'civil society' is a contested concept. Its meaning has become entangled with different political debates, such as the differences between Asia and the West, the defence of the welfare state against neo-conservative anti-statism, rights-oriented liberalism against communitarianism and élite against participatory democracy (Cohen and Arato 1997). These debates have added to and sometimes deviated from the East European idea of a civil society opposed to the state. Broadly speaking, two additional, emerging, conceptions of civil society can be discerned.

The first views civil society as a means to assist and trim back the state. The argument is that as civil society grows it should unburden the state of social and cultural duties such as the protection and promotion of religion, arts, families and education. It helps buttress neoconservative notions that reject the social welfare model of democracy and argue that the state should do less. 'Civil society' becomes an argument for a minimal state, and civil society itself is largely equated with the emergence and values of a middle-class, bourgeois community (Etzioni 1995; Walzer 1995).

There is another, quite different strand of the concept of civil society which emphasises the role of civil society in furthering democracy and keeping democratic culture vibrant. Proponents of this view—of which de Tocqueville was perhaps the first major theorist—consider civil society to be the lifeblood of political culture, essential to the socialisation of the citizen (de Tocqueville n.d.; Whitehead 1997). A lively civil society is the best safeguard for a stable democracy and the prevention of domination by any one group over the others. It is also what best brings the individual rights-holders together for common cause. This 'democratic' civil society does not seek primarily to oppose or assist the state. Rather, the role of civil society is to make the state more democratically accountable to the citizenry and to better enable the widest possible participation in governance.

Present proponents of this democratic concept of civil society reject the notion that civil society should be equated with or dominated by the bourgeois middle class. They call for civil society to be a pluralist entity, with fully representative participation that cuts across race, ethnicity, religion, age, gender and economic status.

The three concepts outlined—a civil society opposed to the state, a civil society assisting a minimal state and a pluralist civil society demanding an accountable state—are not merely academic conceptions. The discourse of civil society has been contested and shaped by different political agendas. The way we define civil society affects our expectations of what it means to be in favour of or opposed to nurturing that society (Tay 1998b).

In Asia, civil society is very much a received concept. There are, of course, native, indigenous ideas that correspond to the broad idea of community representation, self-organisation and action. After all, society long preceded the state in pre-colonial Asia.

---

2 A more 'radical' tendency that questioned the sufficiency of Western-style democracy can be detected in the writings of activists such as Vaclav Havel (see Havel 1988).

Each society and sub-group has its *banjar, kampung, kapitan* or other unit of community (Reid 1988). Yet the place of civil society on the agenda stems less from indigenous antecedents than from intellectual fashion.

Accordingly, there has been a growing literature on civil society in Asia. Riker (1995) characterises state–civil society relations as emerging in three differing waves. First, by the mid-1980s, civil society groups were seen by élites as having a complementary role in promoting development, especially in enabling the process of privatisation. The main tasks given to civil society in this minimal 'assist-the-state' concept were to promote production and market activities, deliver services to communities and groups beyond the reach of the state and to foster participation in the (state-driven) development process.

Towards the start of the 1990s, a second wave of civil society development emerged, one characterised as an autonomous and countervailing power to the state. Drawing from Western liberal thought, NGOs and civil society more generally began to play a greater role in calling Asian governments to account for a wide range of policies, including those on the environment, poverty alleviation, women's rights and human rights. Some also began to challenge development policies.

In some countries, the second wave of civil society has fostered new arrangements and forms of political organisation within existing political parties. New, cross-cutting alliances have emerged among different sectors of civil society, such as students, the media, the middle class and even business interests. A stark example was seen in the push for democratic reform in Thailand in the aftermath of the May 1992 protests. Even in the region's most politically stable countries, such as Japan, the growth of civil society is likely to change significantly the functions of government, business and civil society (Yamamoto 1999).

The third wave of civil society evolution in Asia is a reaction by governments to moderate growing pressures from civil society groups and to incorporate them as an instrument of state. Observers have noted the trend for governments in Asia to place legal controls on NGO activities and to try to co-opt and demobilise the more political, policy advocacy groups.

The three waves identified by Riker correspond somewhat to the different conceptions of civil society outlined earlier. The first wave follows from the minimal 'assist-the-state' concept of civil society. The third wave, in which states try to co-opt or control civil society, grows out of the 'oppose-the-state' concept. The second wave is a little murkier. The character of civil society in the second wave has elements of both the 'oppose-the-state' concept and the notion of civil society as a pluralist platform for democracy and accountability.

There is, of course, a difference between opposing the state and demanding that governments be publicly accountable. However, the line between them can become fuzzy. Certainly, in the minds of many governments unaccustomed to being questioned—and perhaps to civil society groups unaccustomed to questioning—the difference is often not felt. To those in power, a demand for public accountability is often viewed as a confrontation that can lead only to an erosion of their power. Indeed, sometimes it does.

Present and unfolding developments since the advent of the economic crisis in 1997 have revitalised interest in civil society. In South-East Asia, where the changes have been dramatic, especially in Thailand, Indonesia and Malaysia, some observers see vindication of their thesis that a democratic, pluralist and politically effective civil society will take root in the region. Others are concerned with the failure of governance and a rising tide of anarchy, especially in Indonesia. They are less sanguine about the strength and capacity of emerging democrats to govern effectively and therefore to deal with possible backlash.

Civil society roles in environmental governance

Since the 1980s an increasing environmental consciousness has emerged among the people of East Asia (Lee and So 1999). This consciousness stems in part from the ever-more-apparent environmental degradation and resource depletion that has accompanied rapid economic growth; and in part from the increasing political space occupied by NGOs, local community groups and other civil society groups, especially advocates for those hardest hit by ecological decline such as small farmers, fishers, the rural and urban poor, and indigenous groups. Business leaders have also voiced greater awareness, sometimes via their interaction with consumer or diplomatic demands from rich Western countries.

The central question for governments in Asia is—or should be—how to harness the transformative potential of this growing popular consciousness and burgeoning civil society towards greatly improving environmental performance. In essence, this means creatively bringing civil society groups squarely into the task of environmental governance.

Shifting to a development path that greatly reduces the energy, materials, pollution and waste intensity of urban–industrial growth will require significant investment in reshaping the structure of environmental governance. Policies will need to be developed that give the right market signals to innovators, entrepreneurs, managers, consumers and families. Rules and regulations will need to be established and enforced. Information about environmental performance and ecological health will need to be created and strategically disseminated. Trade-offs between economic, environmental and social goals will need to be debated and fairly resolved. Investment for large-scale infrastructure projects for transport, energy, water sanitation and waste management will need to be mobilised and projects will need to be implemented.

These are roles for government. Capable, credible, fair and efficient government is the bedrock of effective environmental governance. Carol Bellamy, Executive Director of the United Nations Children's Fund (UNICEF), argues:

> For all the importance of NGOs and other representatives of civil society, for all the potential value of development partnerships with business and the private sector, we must never lose sight of the fact that it is governments that remain the primary actors in . . . development (Bellamy 1999).

Governments face constraints, however, in Asia as elsewhere. In Asia, governments face four obstacles to significantly 'raising the game' on environmental performance: (1) a lack of political will; (2) fiscal constraints; (3) a lack of technical and regulatory capacity; and (4) the competitive pressures of globalisation, especially competition for FDI. As briefly outlined earlier, these constraints have been exacerbated by the crisis.

Business and civil society could play central roles in overcoming these obstacles. In the context of market-oriented, democratic (or democratising) societies, public opinion and popular demand play a large role in shaping public policy. An effectively mobilised and articulate citizenry is central to the task of building and sustaining strong, capable, publicly accountable governments with the political will to meet the sustainability challenge. Environmental quality and public health are issues in which most people feel themselves to be stakeholders. Concern for children in the strong family-centred societies of Asia is an especially potent mobilising force. In China, for example, the story is that Premier Jiang Zemin himself ordered improvements in Beijing's air quality after his grandchild was sent home from school gasping. In Japan, civil society as a whole emerged largely to demand protection against local industrial pollution (Yamamoto 1999).

Popular demand for better environmental performance is likely to be increasingly framed in the language both of efficiency and of human rights. The first—the pursuit of efficiency—is largely accepted as legitimate in Asia, though it is often cast aside if it entails the disruption of state or other élite monopolies. The second—human rights—remains controversial.

Although many Asian countries use the language of human rights there remain gaps in practice. Moreover, in the years prior to the crisis, many Asian states argued against Western impositions of human rights in favour of 'Asian values'. They argued that the definition of human rights in Asia had to differ from universal norms because of Asia's unique history, culture and stage of development. Many Asian analysts suspected such arguments to be self-serving fig leaves by authoritarian governments (Bell and Bauer 1999; Tay 1997b). The Asian-values argument is now largely discredited and the language of human rights is increasingly the idiom by which civil society groups demand better environmental protection (Boyle and Anderson 1996; Sachs 1995).

Human rights need not be viewed as being encompassed wholly by civil and political rights. Rather, they include all human rights in full recognition of their indivisible and interdependent nature. After all, Asian leaders have pressed for years the notion that the most important human rights are economic rights—the right to food and housing and to development itself. The right to health and to clean air and water can be seen as an extension of economic rights. Moreover, there is an increasing clamour in the international community for the articulation and embracing of a charter of environmental rights (Earth Council 1999).

In addition to building popular support and political will, the mobilisation of business, community and other civil society organisations can help to overcome fiscal constraints. Government revenues have dropped in the post-crisis era and, in many cases, have been reallocated away from environmental projects. Business and NGOs can fulfil a variety of functions that may otherwise fall to government or that governments simply cannot accomplish (see next section). Involving community groups in 'collaborative governance' can be both cost-effective and highly welfare-improving (see Chapter 5).

Business and NGOs are also potentially key players in designing, monitoring and enforcing the regulation of industry. Even if their capacity and will to regulate were greatly enhanced, governments could not adequately monitor thousands of companies and other agents to ensure regulatory compliance. Moreover, local community groups who live 'at the mineface' are often much better informed than government that a problem exists. Effectively mobilising community monitoring capacities and using them to spur better industry performance is central to effective environmental governance.

The fourth constraint on government capacity to govern stems from globalisation. Competition for foreign investment and trade can act as a force of gravity dragging down environmental commitments. Even if governments do not lower standards in order to attract FDI, they often do not enforce standards and are wary of raising them. Only international collective action can overcome this 'stuck-in-the-mud' problem (Zarsky 1997). International NGOs and coalitions between international NGOs and local groups are important players in the push to change the character of globalisation by incorporating environmental and social standards into the governance of the global economy.

Globalisation and social norms

The environmental and social impacts of globalisation are at the heart of the concerns of many Asian and international NGOs. In the wake of the crisis, critiques of globalisation have grown stronger. Even in the heyday of the Asian miracle doubts were voiced about the social and cultural aspects of globalisation. Many Asian leaders and intellectuals advocated globalisation for the economy but regional or national particularities for culture and society. They posited a world of convergence in economics but of essential (and essentialised) differences in social norms such as human rights and environmental protection as well as culture and politics.

Beyond Asia, environmental NGOs, labour groups and other advocacy groups throughout the world have raised concerns about the adverse costs of globalisation. In the West, many focus on the outflow of jobs from the more developed and more expensive economies to cheaper centres of production in Asia and elsewhere. Others voice concern about the accompanying social and cultural costs in environmental pollution and degradation, or the lack of protection of human and labour rights and the exploitation of vulnerable sectors of the populace, such as undocumented migrant workers, women and children. In this analysis, globalisation threatens a 'race to the bottom' in terms of social and environmental standards.

In the debate over globalisation, one focal point is the relationship between international trade and the environment. The United States has tried, unsuccessfully, to close its markets to goods that harmed species it wanted to protect, such as to tuna caught in ways that resulted in the accidental killing of dolphins and to imports of shrimp captured in ways that kill sea turtles. In human rights, the Asian-values debate has generated controversy, especially in relation to Western sanctions against China and Myanmar. Some predict the triumph of Western liberal democracy in Asia; others see a 'clash of civilisations'. Some view the choices starkly between the so-called 'McWorld' of convergence and the jihad of radical difference (Barber 1992). These debates about convergence and difference have not been resolved by the crisis; they have instead become more complicated.

In the West, a new triumphalism has emerged, especially with regard to the Asian-values debate in human rights and democracy. The United States has emerged from the

Cold War and the crisis not only as the sole superpower in military and strategic matters but also as a 'hyperpower' in terms of economic, political and cultural influence. US influence is associated with the urging of free and open markets and the promotion of democracy, human rights and (to a lesser degree) environmental protection. Critics have pointed to a new arrogance in US foreign policy, at times exacerbated by inexperience and incompetence in strategic leadership in a unipolar world (Zarsky 2000). To some it may seem like a repeat of the Western euphoria after the Cold War.

Yet differences between Asia and the West can be better understood and be made less controversial; areas of convergence can be better recognised and built on. Otherwise, the thin fabric of community and consensus in Asia–Pacific can be torn. Indeed, consensus about the benefits of globalisation itself can be shaken. Much depends on how processes of globalisation are shaped and governed, which, in turn, may depend largely on the emergence of what some have called 'international civil society' and what others see as 'Western interference'.

The international community is based on the idea that all states are sovereign and equal. No state has the right to compel another, and each state is free to order its internal, domestic affairs. However, newer trends have eroded this concept of sovereignty. With globalisation, increased interdependence has increased the need for international co-operation and supervision. The nation-state has not disappeared, but the nature of its sovereignty has changed. The ability of nation-states to govern unilaterally has diminished. Global law has arisen in tandem with globalisation, which intrudes into arenas traditionally belonging to a state's domestic jurisdiction.

In the economic sphere, global law, for example, limits the range of actions that a state can take against private investors and increases the rights such non-state actors hold in relation to the state. In international trade, the WTO regime binds states to observe the key principles of national treatment and non-discrimination, thus limiting what states can do unilaterally to promote domestic economic and commercial interests.

Many developing countries consider such global market rules to be inherently unfair to the WTO's poorest members. International rule-making on human rights and the environment is often even more controversial since it explicitly challenges the anarchic concept of national sovereignty and targets the internal norms and policies of states. Global concerns to conserve biodiversity and control climate change, for example, have propelled the concept that rainforests within national territorial boundaries are a global heritage not only of all peoples but also of all generations. Given their colonial histories, as well as the continuing imbalance of global power, many developing-country governments have resisted these aspects of global law as intrusions on their sovereignty.

Human rights and international environmental law are both strongly associated with NGOs. Environmental NGOs such as Greenpeace and the World Wide Fund for Nature (WWF) have become household names in Asia and elsewhere, as have Amnesty International and Human Rights Watch. A host of NGOs have organised around specific areas of human rights or environmental issues such as humanitarian emergencies, the women's movement, the anti-nuclear movement and the campaign against land mines. Such NGO movements have become increasingly strong and acknowledged sources of influence in the international sphere, with established roles in various United Nations forums.

But, although NGOs may be the core of an 'international civil society', a newer phenomenon seems to be emerging with globalisation. This is the rise of 'global public policy networks' which include not only NGOs but also networks of scholars, religious and other voluntary organisations, research institutes, media, international bureaucrats and government officials (Reinecke 2000). Often built around a specific policy goal, such as preventing the construction of big dams or reversing desertification, such networks intensively utilise information technology to create their own 'world wide web'. They counterpose the concept of 'globalisation from above' spurred by the movement of capital with 'globalisation from below', by which networks of citizens effectively organise themselves for common causes across borders.

These tensions in institutions, policies and conceptions of international order ignited at the 1999 WTO ministerial meeting, held in Seattle, WA. The mass and partially violent demonstrations against the WTO meeting displayed a frustration with an approach to the governance of globalisation that the demonstrators perceived as undemocratic, environmentally disastrous and bad for their individual economic prospects. Through their intensive use of the Internet, protesters from many sectors and many countries developed a sophisticated common critique and language.

The future role of civil society in Asia will take shape in this context of a growing global movement that rejects 'actually existing globalisation' (Dorman 2000). A host of groups and sectors from around the world is seeking new, ethically based approaches to global (and national) governance. This has two major implications for the potential role of Asian civil society in environmental governance.

First, it suggests that the evolution of environmental and social norms in Asia, both at the popular and at the NGO level, will be determined by external as well as internal forces. Coalitions of local and international NGOs will increasingly work together, targeting not only environmental degradation but also what the United Nations Development Programme calls the 'grotesque gap' between winners and losers (UNDP 1998). In the future, civil society in Asia is likely to call for more government accountability on issues of environment and human rights as well as for a new approach to development that promotes equity. There will be an ongoing debate between Asian civil society groups seeking to withdraw from and those seeking to expand and reform the processes of globalisation.

It is also likely that trans-Pacific environmental partnerships and NGO coalitions will blossom in the first decade of the 21st century based not only on ethics but also on ecological self-interest. In March 1999 a new report found that airborne chemicals from Asia—carbon monoxide, radon, aerosols, hydrocarbons and other chemicals—were reaching the west coast of the United States. 'The air that people breathe in Seattle today may contain chemicals that spewed from a factory in China last week', reported the *New York Times* (1999). Rising concern and activism in the United States is likely to give rise to new governmental and civil society action.

The second implication of the increasing globalisation of economies and social norms is that governments in Asia will feel increasingly pressed by contradictory external forces. On the one hand, they will be pressured not to raise industry environment standards for fear of losing foreign investment and trade competitiveness. On the other hand, they will be forced to accept higher standards set by North American or European states as conditions of market entry or demanded by Western consumers via free markets. Contentious environmental issues will continue to be on the agenda for global and regional trade diplomacy and NGOs will press hard to have their voices heard.

In many states in Asia, the primary reaction to international civil society and global public policy networks has often been denial. However, this will become increasingly difficult as such networks gain influence. Moreover, denial is counterproductive. Rather than interfering with state objectives, international civil society can be a resource for peace and development, helping the state deal with the negative consequences of globalisation and the present economic crisis. Indeed, NGOs, both domestic and international, can help states deal with the contradictory pressures of globalisation and find innovative ways to raise environmental and social standards while promoting economic development.

What is needed is more dialogue between states and international civil society. To make such dialogue constructive, it is not only states but also civil society groups who must change. In Asia, indigenous civil society and NGO networks will need to be strengthened and assisted both to enhance their work at home and to enable their participation in global networks. If global public policy networks are to claim legitimacy they must not reflect the interests of only one nation or a narrow group of nations and cultures, typically rich and Western. Rather, they must truly incorporate citizens and groups from all countries, finding common cause beyond borders. The participation of East Asians in such networks is likely to blossom in the next decade.

◢ Environmental governance and civil society: functions and models

Civil society groups, from both business and NGO sectors, can play a wide variety of functional roles in governance, including environmental governance. To some degree, the functional roles of civil society differ depending on the overarching governance model. In a 'command-and-control' approach, for example, government is the primary instigator, designer and enforcer of environmental regulation. In other models, citizen groups and market incentives are key agents in raising, designing, monitoring and enforcing environmental standards. This section first describes six broad functions that civil society groups can play in environmental governance and then outlines three, alternative governance models.

Six broad functions

Civil society groups can fulfil six broad functional roles in environmental governance.

1. **Intellectual and visionary.** Public policy think-tanks, as well as academic and journalistic writers, seek to define development paradigms and objectives and

to design and promote policy agendas. This independent source of creative intellectual input and visionary thinking provides an important channel for the development of strategic rather than reactive approaches to development challenges (Edwards and Hulme 1992).

2. **Advocacy.** Many groups are constituted around specific issues of social concern such as gender equality, labour rights, indigenous people, environment, public health, consumer rights, resource-dependent communities, etc. These groups, which mushroomed dramatically in Asia in the 1990s, help to bring issues to the public spotlight and to change social norms. This function of civil society groups has been the subject of considerable controversy and attention (Broad and Cavanagh 1993).

3. **Problem-solving.** A variety of professional associations, as well as community and advocacy groups, provide technical support and work with governments and businesses to develop solutions to specific environmental and social problems. In Thailand, for example, overseas Thai engineers helped the government write and implement its first environmental laws.

4. **Service provision.** Many NGOs, including religious and social service groups, provide direct services to the poor and other needy groups. Such services go beyond distribution of food and other basic needs to encompass capacity-building 'empowerment' activities, including the creation of community-based municipal public goods such as clean water, sewer services and refuse collection. In this capacity, NGOs often implement policies and programmes designed and promoted by government. This is an important function, especially where countries attempt decentralisation in response to environmental issues. In such cases, there is often a lack of local authority and resources to deal with environmental problems (Webster 1995).

5. **Critics and watchdogs.** NGOs, journalists and others can serve to monitor the activities of both government and industry. There is substantial evidence that community-group pressure is an important determinant of firm-level environmental performance in Asia (NIPR 1999).

6. **Financial support.** Although it is still relatively young, philanthropy in Asia is growing (Yamamoto 1995). Philanthropic foundations and individuals provide resources for independent think-tanks and other NGO activities, often stemming from their own visionary leanings and interest in solving problems. Philanthropic foundations also sometimes provide funds for government and business activity.

Many NGOs undertake multiple functions whereas others have a strong identity as serving one particular function. Some are national or regional and may have links with or be chapters of Western-based, international environmental NGOs (e.g. WWF-Indonesia). Others remain mainly local movements (Kalland and Persoon 1998). Moreover, a range of underlying values and philosophies guides environmental NGOs. In some cases, Asian perceptions of nature have a considerable influence on environmentalism (Bruun and Kalland 1995).

Additionally, some environmental NGOs trace their roots to political movements, such as anti-colonialism, Marxism or feminism. Some developed from a concern about the poor and the need for environmental justice. Environmental movements in India, the Philippines and Thailand in particular tend to focus on what have been called the environmental problems of poverty: the lack of access by the poor to environmental resources or their suffering from the direct impact of pollution (Lee and So 1999).

Other environmental NGOs and movements in Asia, in contrast, have their origins and sources of support from the emerging middle class. Like their Western environmental counterparts, such environmental NGOs and movements espouse postmaterialist concerns about over-consumption and the quality of life, as well as health and environmental impacts of industrialisation. Typically, such concerns are the focus of environmental movements in Japan, Taiwan, South Korea and other more developed economies in Asia (Eder 1996; Lee and So 1999).

Their different origins and sources of support tend to influence the functions and strategies that various environmental NGOs undertake. In Thailand and the Philippines, which have clearer self-identities as democracies, a thick web of NGOs is developing, with independent think-tanks and NGOs taking on one or (most often) more of the functions outlined above.

In the Philippines, civil society groups are described either as people's or non-governmental organisations to underscore their different relationships to local communities. This self-description suggests that NGOs wish to go beyond a membership based in urban, middle-class élites and reach out more at the grass-roots level. This is especially necessary given the expanse and stretch of the Philippines as an archipelagic state and parallels the devolution of many governmental functions to the local or *baranguay* level since the fall of the martial law regime of President Ferdinand Marcos (Silliman and Noble 1998; Solidarity 1989).

Yet another important feature of civil society organisations in the Philippines is their extraordinary ability to form broad coalitions, networks and umbrella groups. This affords them greater legitimacy, opportunities to share resources and a platform to exchange opinions among each other. These factors in turn allow NGOs within these coalitions to gain better access to government decision-making processes. Those who wish to deliver services to supplement the action (or inaction) of government also have access to a broad array of partners and resources. In so doing, NGOs begin to appreciate the benefits from co-operating with government when the occasion arises. Moreover, they are able to develop more sophisticated negotiation techniques that aim to seek settlements agreeable to all major stakeholders rather than to gain 'total victory' over the state or business (Noble 1998).

In Thailand, too, environmental NGOs have been on the rise. In many cases, NGOs have arisen out of rural development and environmental issues such as local access to and ownership of forests (TEI 1996; Poffenberger 1990). However, urban and largely middle-class groups concerned with the environment have also mushroomed. In the main, these civil society groups have not been welcomed by the government (Lohmann

1995). Recently, the Thai government of Prime Minister Chuan has made an attempt to accommodate NGO input and effort. This is necessary given the rising democratic ethos in the society and the government's democratic self-image.

Although civil society groups have great potential to assist Asia's economic recovery and to enhance the future of environmental protection, it is not certain that this potential will be allowed to develop. The response of Asian governments to civil society has been uneven and at times uncertain. In many cases, governments have not welcomed civil society as a partner but have been suspicious of it and have sought to control or co-opt civil society actors. Independent national think-tanks that wish to advocate public policy for the environment or contend for intellectual leadership on these issues are still rare in some countries in East Asia, such as China, Indonesia, Malaysia and Singapore. In part, this is because they are kept under control by wary governments (Ooi and Koh 1999).

However, times may be changing. In Singapore, there is a recognised role for NGOs to work in environmental education and to help provide services to supplement government action (Mekani and Stengel 1995). Indonesia, too, has witnessed a growth of environmental civil society organisations and NGOs. Before the crisis, NGOs tended to grow up around particular issues, such as opposition to dams, or for conservation (Hirsch and Warren 1998). However, as a result of the political and economic crisis in Indonesia, local environmental struggles have often been linked to broader politics.

Not all observers are optimistic about the possibility of an independent civil society developing under authoritarian regimes in Asia (Hewison 1996; Murphy 1996). Nevertheless, a growing web of national and also regional civil society organisations and NGOs can clearly be discerned. One example is the network of ASEAN Institutes of Strategic and International Studies (ASEAN-ISIS) which is a largely independent set of think-tanks in ASEAN. For more than 15 years this group has played a 'track-two' role in discussing international policy, especially in security and economic relations, exploring new policies and possibilities as well as providing input into official policy circles. They have had a good measure of success; for example, ASEAN-ISIS helped develop the ASEAN Regional Forum for security in the Asia–Pacific (Yamamoto 1995).

It is possible that, in future, similar regional networks of NGOs and civil society organisations may evolve for environmental issues, especially in response to transboundary environmental problems. The Indonesia fires, for example, which caused widespread smoke and haze pollution in South-East Asia, propelled Indonesian and regional NGOs to come together to advocate policy action by ASEAN (Tay 1999b).

Three models of relationships between state and civil society

The euphoria and the contest over civil society in Asia have often led to confusion in the terms of the debate. It has also bred conceptual rigidity about actual and potential relationships between the state and civil society.

For example, if the oppositional idea of civil society is rigidly held, then the idea of joint projects and policy development between civil society and the state must be read to mean that participating civil society actors have been co-opted or otherwise compromised by the state. Blindly followed, the oppositional idea of civil society demands political contestation and refuses co-operation. On the other hand, if one ascribes rigidly to the 'assist-the-state' idea of civil society, true partnership and co-operation are also impossible, although for the opposite reasons. In this formulation, civil society can and must be only a 'handmaiden' to state-directed policy, taking on only the tasks that the state assigns to it and that the state no longer wishes to fulfil. In this mind-set there is no role for civil society in policy formulation or advocacy.

To move beyond single and narrow formulations, it is useful to think in terms of a variety of models of relations between state and civil society. One model is not inherently better in all circumstances. Rather, the optimal model will shift depending on the particular context or objective, even within the same society and with the same civil society actors. The critical difference would be that the choice of model be dictated by context and policy goal rather than by ideology and habit.

The central idea implicit in the demands of civil society for a greater political voice is the concept that people who are impacted by decisions—either by government or business—are 'stakeholders' and have both the right and the need to shape those decisions. In the burgeoning corporate accountability movement, for example, the concept of being responsible to a broad range of company stakeholders, including workers and the local community, is supplementing a more narrow traditional focus on the financial bottom line. The idea is that stakeholder involvement strengthens the long-run performance of a company, including its profitability (Svendson 1998).

The stakeholder idea has been especially relevant to broad civil society demands for greater participation in environmental decisions. Along with social and economic welfare, environmental quality and public health are issues of concern to all members of a society. They are universal issues in which all are stakeholders, even future generations. The poor often have the largest stake in environmental improvement, since they suffer the brunt of pollution and resource degradation. Mobilising popular concern in order to chart an economically viable, socially just and ecologically sustainable development path—for themselves, for their families, for the people for whom they are advocates, for their societies—is a fundamental goal of many Asian NGOs. In their myriad forms, these NGOs constitute the 'voice of the community' as a stakeholder in government and business decisions.

Another key stakeholder sector is business. SMEs, MNCs and domestic big business, industry associations, chambers of commerce and others have particular concerns about environmental governance. As a whole, business tends to prefer stability and predictability in policy-making, including on the environment. Transnational companies find it useful for standards to be similar in different countries in which they operate and often claim to adopt uniform internal company-wide standards. Some business groups have embraced the principle of 'social responsibility' which entails a 'triple bottom line': namely, financial, environmental and social concerns. Others fear that raising the bar for environmental performance will disadvantage them in the marketplace. In either case, business is a major stakeholder in government decisions, as well as NGO activities.

The key issue is what model of environmental governance will best carry forward a new 'sustainable development' paradigm that promotes cleaner industrial production and integrates environmental and social concerns into economic policy-making. The traditional model of environmental governance puts government, usually national government, in the role of regulator and enforcer (as well as financier and often operator of public goods). Whether in the United States or China, the government has traditionally been understood to engage in dyadic relationships—regulator and regulated, provider and consumer—with both industry and the wider community. In this command-and-control model, government's role is to directly wield sticks and provide carrots and to be the 'good parent' in providing public goods.

Although it has achieved some success in raising environmental performance in the United States and elsewhere, the traditional model is being re-examined because it is expensive and rigid (Ruckelshaus 1998). On the regulatory front, the command-and-control model requires that substantial resources be devoted to enforcement. Moreover, there is no incentive for business to exceed standards. On the public-goods front, government services are subject to problems of corruption, capture by sectoral interests and political determination of prices.

In East Asia the command-and-control model has not been very effective. Although a spate of environmental legislation bloomed in the early 1990s, enforcement has languished, in part as a result of lack of funds (as well as political will). As providers of public goods, national governments have been constrained by ineffective tax systems, priorities for other kinds of spending (especially military) and corruption—and, more recently, by financial crisis. Moreover, many East Asian countries lack strong traditions of law.

With stakeholders seeking to play a greater role, what governments should, and in many cases are, asking is how best to harness stakeholders to the task of raising industrial and urban environmental performance. There are three, potentially overlapping, models.

Community partnership

A community-partnership model of governance rests on partnerships with NGOs and business in the undertaking of specific projects and implementation of policies and programmes. NGOs, for example, might work collaboratively with government or take charge of a host of urban improvement/management projects, including water and waste management, slum redesign and improvement, urban reforestation and creation of parks, etc. (see Chapter 5). NGOs can also spearhead environmental education programmes and develop school and workplace environmental training.

In this model, the role of government *vis-à-vis* stakeholders is to mobilise and co-ordinate business and community efforts. Government may also help to finance projects or help to leverage private funds. The emphasis is on the service provision and problem-solving roles of NGOs.

Public–private partnerships between government and business are a specific form of the community-service model. Governments act as organisers and co-ordinators, bringing together various private interests to undertake socially beneficial investment projects such as the development of clean energy sources or power plants (e.g. natural gas, clean coal, renewables). The government could help to leverage private-sector financing through innovative methods such as technology risk guarantees for technologies that are not commercially proven (for a proposal that the Global Environment Facility creates such a mechanism, see NISSD 1999).

The community-partnership model has three benefits. First, by using volunteer and low-paid community labour, governments can greatly stretch scarce revenues. Second, by providing opportunities for people to engage with and improve their own communities, they can encourage a greater sense of civic engagement. There is substantial evidence that strong civic association is an important component of good governance, which in turn positively impacts economic growth. Third, the government's mobilisation of business promotes projects that otherwise would have languished or which would have drained the public purse.

In this model, government retains its central role as regulator and enforcer and continues to structure its relationships with business and community groups bilaterally (and typically top-down). This may be attractive to those governments who are wary of the potential for social disruption or political challenge that might stem from a wider role for NGOs. In this sense, it could work as a transition model. Over the longer term it is likely that groups and individuals who work in close partnership with government will seek not only to implement but also to design policy.

Corporate self-regulation

The corporate self-regulation model is based on two ideas: first, that companies will respond to consumer demands for better environmental performance regardless of government regulation and, second, that companies know better than governments how to improve their environmental performance. There is also a related notion that large, transnational corporations, especially from countries in the Organisation for Economic Co-operation and Development (OECD), have superior technology and management. Encouraging them to invest in Asia is thus a strategy for environmental improvement (UNCTAD 1999).

In this model, governments provide not an ever-denser thicket of command-and-control style regulation but broad frameworks and guidelines, as well as open markets. It is a 'hands-off' approach to regulation wherein private ordering is given much more play. Not only is business given a greater role in governance but also the relationship of the community to business is influential. In the absence of formal government regulation, communities take on the task of 'informal regulation' (NIPR 1999).

The most widely used form of corporate self-regulation in Asia is ISO 14001, which sets environmental management standards for firms. ISO 14001 sets process but not performance standards. Companies commit themselves to auditing and monitoring their environmental impacts as well as complying with domestic laws. Proof that they are doing both enables companies to be certified which in turn eases or even gains entry for company products to a number of OECD markets. The hope is that the audits will uncover opportunities to save money by reducing waste and improving energy and materials efficiency. The gains in eco-efficiency will spur companies to make production improvements.

There is as yet little evidence as to the efficacy of the ISO approach. Environmentalists have typically been wary or downright cynical, suggesting that ISO means that the fox is guarding the hen house. A survey of US manufacturers, consultants and regulatory agencies found widespread concern on two fronts: first, that ISO 14001 certification provides no guarantee of an actual and continuous improvement in reducing environmental impact and, second, that certification could become more of a paper chase than an effective tool to promote managerial innovation (Marcus and Willing 1997).

For Asian-based manufacturers, the concerns are quite the opposite. There is suspicion that ISO standards are set by industrialised countries to provide a non-tariff barrier to trade, using a green excuse for protectionist intention. In this view, ISO 14001 could become primarily a hurdle to be overcome in order to gain access to OECD markets. The sentiment is more strongly felt because of the differential impact that the same standards impose on the bulk of firms in Asia, which tend to be smaller enterprises with limited access to technology, finance and know-how to change their methods of production to meet the new standards (Tay 1997a; UNCTAD 1999; Vossenaar and Jha 1996).

A second method of corporate self-regulation is the 'code-of-conduct' approach. Largely in response to strong community and/or international criticism, many US and some European transnational corporations (TNCs) have adopted codes of conduct that spell out their social responsibility (UNCTAD 1999). Public criticism damages the reputation, standing and 'moral capital' of TNCs, weakening their manoeuvrability and threatening market share. Some of the most heavily targeted companies include big oil multinationals such as Shell, criticised for its role in the Brent Spar incident and Nigeria, and Unocal, criticised for supporting the military regime in Myanmar/Burma (Schwartz and Gibb 1999; Wallace 1997). High visibility consumer-oriented companies such as Reebok, Levi-Strauss, Intel and Hewlett-Packard have also joined the code-of-conduct bandwagon. Intel joined in response to intense community criticism of the environmental impacts of its production expansion plans in the US south-west (SWAPO 1995).

Other groups that have generated codes of conduct include international organisations such as the OECD, business support groups such as the World Business Council for Sustainable Development and other social interest groups, such as consumer, environmental and church organisations.

According to one comprehensive survey, codes of conduct tend to focus more on social than on environmental issues, in part because they are more easy to specify than environmental issues (SCOPE 1999). Some have argued that such corporate initiatives might be an increasingly important way to enforce human rights (Cassel 1996; Gibney and Emerick 1996). Nonetheless, environmental issues are becoming more common.

Do voluntary codes of conduct work? There is little information either whether codes target the most significant environmental (or social) issues or whether TNCs comply with their own codes. One study under way is examining codes of conduct and the need for policy innovation to enhance social accountability in the context of the US oil and high-technology industries (CGCAP 1999).

According to a European study, the likelihood of compliance depends on the specificity of the issues in the codes and the inclusion of compliance mechanisms. The study found that most codes state very general commitments and do not have compliance mechanisms, suggesting that compliance is low. 'Firms might design codes for other purposes than for the sake of their own ethical behaviour and corporate responsibility', the authors explain. 'It is highly conceivable that codes adopted by firms are in essence meant to influence other societal actors: regulators, customers, communities, suppliers and contractors, competitors or shareholders' (SCOPE 1999: 5). Indeed, the authors find that codes of conduct 'are drawn up to anticipate or prevent mandatory regulation' (SCOPE 1999: 6).

Regardless of whether they comply with particular codes, the fact that TNCs feel themselves to be under public scrutiny may well help to improve their environmental and social performance. Some NGOs have taken their concerns directly to shareholders and have won resolutions that require changes in company practice. Many companies will mobilise a process of change immediately on the filing of a resolution, even if it does not ultimately succeed. One long-time 'green business' analyst describes the coming of a 'triple bottom line', wherein companies will be examined on not only their financial but also their environmental and social performance (Elkington 1998). It is doubtful, however, that companies will do so unless they are so required by governments. Without policies such as mandatory information disclosure, claims of social responsibility will not be able to be monitored and thus will not be credible. As a result, codes of conduct may have a short shelf-life.

If the jury is out on whether codes of conduct significantly improve company performance, there is substantial evidence that direct community pressure can improve firm-level environmental performance. A number of studies conducted by the World Bank demonstrate that communities can strongly influence firms to clean up (NIPR 1999). This process of 'informal regulation' might take a variety of forms, including agreements between citizen groups and companies, discussions between community leaders and managers, or public protest. The World Bank studies also found strong correlations between income and education levels and firm-level environmental performance (Dasgupta *et al.* 1997).

The final issue concerns the relationship between FDI and the environment. Whereas environmentalists have warned of 'pollution havens' and a 'race to the bottom', market enthusiasts have trumpeted 'pollution havens' and a 'race to the top'. The idea is that, on average, TNCs from OECD countries utilise and transfer clean technology and have better management systems than local firms in developing and transition economies. Increasing FDI will thus help environmental standards and performance to converge towards OECD levels.

The effect varies considerably according to the industry involved. FDI stock and flows in mining and other resource-intensive industries are high and negative environmental impacts have been well documented in many cases. On the other hand, there is some evidence that, in the manufacturing sector, TNCs have adopted higher environmental standards than have comparable domestic producers (UNCTAD 1999). For example, TNC manufacturing plants in Côte d'Ivoire, Mexico, Morocco and Venezuela were found to be significantly more energy-efficient than their domestic counterparts (UNCTAD 1999, citing Eskeland and Harrison 1997).

A recent review of the statistical and case-study evidence, however, paints a murky picture (Zarsky 1999a). Although there are cases where FDI has helped to transfer cleaner technology (e.g. the Chinese energy sector), there are other cases where it has been the vehicle for widespread pollution and ecological destruction (e.g. mining in the Philippines). The review concluded that if FDI creates pollution halos, they are pretty small. Moreover, it argued that

> There is no average, performance is context-dependent and other things are more important than [foreign versus domestic] ownership. If the goal is improvement in industry environmental performance, at both micro and macro levels, then what is needed is effective regulation utilising both governments and communities to monitor, reward and sanction firms (Zarsky 1999a: 2).

Some Asian governments think that they can escape imperatives to develop effective environmental governance and infrastructure by encouraging 'clean' manufacturing such as high-technology manufacturing. Although it may not belch out black smoke, the production of silicon chips and semiconductors is highly toxic, as well as intensive in its use of water and energy. A host of ever-changing chemicals used in production, essential to maintaining a competitive edge in a highly dynamic industry, bedevils regulatory oversight even in the United States (Mazurek 1999).

Strategic stakeholder engagement

The third model of environmental governance, that of strategic stakeholder engagement, incorporates elements of the first two models but is based on a fundamentally different concept of the role of government. Rather than dyadic, top-down relationships, or a 'hands-off' approach in favour of market forces, this model envisions government as one of three key agents in the governance process. Business and community are the other two.

In this 'multiple-agent' model, the government works in partnership with business and community on specific projects and programmes, but it also seeks, provides avenues for and listens to community input on policy design and broader development strategies. Moreover, the government plays two kinds of specific roles as a strategic 'enabler'. First, it seeks to enable communities to enhance their role in monitoring and improving the environmental performance of industry. The primary mechanism is the ability of communities to have access to reliable, user-friendly information about industry environmental (and social) performance. This means that companies need to collect the information in a standardised fashion—standards set by the government—and, most importantly, they need to disclose it. The heart of a community-based monitoring system is (mandatory) information disclosure.

A disclosure-based approach to governance can work in tandem with corporate self-regulation, including ISO 14001. Firms can retain a significant amount of managerial flexibility in how they move towards better performance, but they will have a greater incentive to do so. It can also strengthen government regulatory capacities; however, communities will be able to press firms towards compliance. A disclosure-based approach can also strengthen market-based approaches to governance, including product labelling. With credible information, consumers will be more likely to trust a 'green label'.

Second, in this model the government is a 'convenor'. It creates institutional interfaces that enable community and business groups to have ongoing conversations both with government and with each other to resolve differences and set performance goals. Collaborative governance, in short, is not just about partnerships on projects or service delivery but about an ongoing, round-table process.

The role of government as organiser, co-ordinator, regulator and arbiter is not obviated in this model. However, the government seeks to engage directly with each of the two other key agents bilaterally, as well as to strategically engage the community sector to monitor and regulate industry (and government as well). The central policy emphasis is on transparency, accountability and the creation of institutions that allow broad debate and consensus about the fundamental goals of development.

One of the requirements in this model is for government to enhance the technical and intellectual capacities of NGOs and business (and, indeed, for government itself) for collaborative governance. It will require an investment in education and training, including potential in mediation. It will also require a significant investment in information infrastructure—gathering, storing, disseminating, etc.

The stakeholder engagement model is gaining credence in many quarters, including the OECD. 'All stakeholders in society . . . must participate both in the design and the implementation of cleaner production processes', the OECD advised in a recent study (1996a: 13). Its list, however, which included 'businesses, industry associations, chambers of commerce, academia and the research community', fell short. Labour, community and other NGO groups must be part and parcel of the process.

One of the advantages of the stakeholder engagement strategy is that it allows for a much wider and deeper range of information and intellectual input to be part of the policy-making and governance process. Intellectual competition, in turn, can help to develop more flexible, responsive and dynamic governance mechanisms. The ability to learn and change carries a great premium in the dynamic age of globalisation.

The strategic involvement of NGOs can be crucial not only in enhancing domestic environmental performance but also in 'raising the bar' internationally. Likewise, their exclusion can render international environmental diplomacy barren. In the APEC context, for example, NGO advocacy and scientific groups were largely excluded from regional discussion of an environmental agenda. During the early 1990s, prospects of APEC environmental co-operation gathered steam, with three environment ministries, including one sponsored by the Philippines. The effort lost momentum, however, in large part because it was driven by the West and remained in the realm of bureaucrats (Zarsky 1998). Without the political passions and visions that NGOs bring to inflame government action, environmental diplomacy tends to be choked by commercial and strategic interests. The strategic stakeholder engagement model will need to extend towards the inclusion of community voices in regional and global forums.

Conclusions

The prospects for a paradigm shift towards ecological sustainability in Asia—as, indeed, in other parts of the world—rest largely on the emergence of effective environmental governance. Effective governance requires not only strong technical and regulatory capacities within government but also the ability of government to engage strategically the two other key social agents: namely, business and the community.

In the old 'grow now, pay later' paradigm, environmental and social objectives sat at the margins of economic policy-making—and advocates sat outside the corridors of power. Any new paradigm based on clean and equitable urban–industrial growth can take root only if it garners a deep and broad political and social commitment within Asian societies. Such a commitment can best be nurtured by a broad range of stakeholders. Both business and civil society—spanning from labour, environmental and community groups to professionals to policy-oriented think-tanks, advocates and critics—could provide a crucial driver for paradigm and policy change.

This chapter has provided a broad sketch of a variety of conceptual frameworks and institutional models for the engagement of civil society in environmental governance in Asia. The question of how best to do so is far from settled, even setting aside deep differences in ideology and interests. Asia itself is not culturally homogenous and different concepts and models are likely to resonate in different countries. Moreover, indigenous concepts and models of civil society engagement are likely to emerge given the political space to do so.

Nonetheless, there is reason to expect that NGOs and business in Asia will continue to press for a greater role in policy-making and that NGOs in particular will demand some form of a strategic stakeholder engagement model. In this model, the government seeks to include a wide set of community voices in the policy-making process. It also works in partnership with community and business to implement projects and programmes. It also serves as a convenor, creating ongoing institutional interfaces between government, NGOs and business and providing arenas in which community and business groups can interact directly.

Far more research is required into the feasibility and design of such a model in particular locales in Asia, at the level of both national and urban metropolitan governance. Indeed, the question of devolution and the role of NGOs within urban governance is a research topic in its own right (see Chapter 5). Moreover, much more research is required to determine what kinds of industry information are both useful and feasible to obtain and what kind of standardisation is most practical. It is likely, for example, that tracking and reporting will be more useful on a sectoral rather than an across-the-board basis.

More research is also needed about the character and politics of civil society formations in Asia. Whether and how civil society, especially NGOs, will be a driver toward environmental improvement in Asia will depend at least in part on their own evolution. The choices that civil society groups make in terms of strategic direction, coalition partners, targets for policy intervention, use of information technology and so on could significantly affect their ability to influence environment policy—and to shape the future of Asia.

PART 2
Case Studies in Innovation

David P. Angel and Michael T. Rock

THE SECOND SECTION OF THIS BOOK COMPRISES FOUR CHAPTERS THAT REPORT on innovative policy approaches to reducing the environmental impact of industrialisation in East Asia. The achievements reported in the case studies are important both as evidence of the progress that has already been made in improving the environmental regulation of industry and as a basis on which to implement the policy approaches laid out in previous chapters of this book. The progress made is also testimony to the capacity for policy innovation in Asia and to the opportunity presented within the region for achieving a different trajectory of development that is substantially less energy-, materials-, waste- and pollution-intensive.

Chapter 7, written by Shakeb Afsah and Jeffrey R. Vincent, reports on the PROPER programme in Indonesia. PROPER is a relatively simple and low-cost environmental reporting system for industry in Indonesia. Implemented in 1995 by BAPEDAL, the Indonesian Environmental Impact Management Agency, PROPER employs a five-colour rating scheme for rating the environmental performance of factories and for public disclosure of this performance information. The case study reviews the implementation strategy and initial results of the PROPER programme.

In Chapter 8 Jeffrey R. Vincent, Rozali Mohamed Ali and Kahlid Abdul Rahim explain how Malaysia managed to dramatically reduce pollution from palm oil production without blocking the palm oil industry's growth. In this case, the key policy instrument was the implementation of a system of pollution charges along with other incentives to explore alternative, cleaner, production technologies. The authors discuss the mix of regulatory strategies used and the impact of these strategies on pollution in Malaysia's palm oil industry.

Chapter 9, written by Michael T. Rock, describes the strengthening of environmental regulation of industry in Taiwan from the mid-1970s onwards. Taiwan has put in place a variety of policies to improve the environmental performance of industry, ranging from quite traditional command-and-control environmental regulation to the innovative integration of environmental concerns into industrial policy. These efforts are apparently a response to democracy, growing public concern over the environment as well as increasing awareness that superior environmental performance is emerging as one dimension of international economic competitiveness.

In Chapter 10, Daryl Ditz and Janet Ranganathan provide a range of examples of the use of environmental performance indicators as a tool for improving the environmental performance of industry. Of all the policy tools under discussion, performance metrics are attracting the greatest attention among corporations, investors and the regulatory community. Partly this reflects the possibility of a new generation of environmental policy, based on the development and disclosure of standardised performance information, that is less costly and more flexible than currently available policy tools, but it also partly reflects a sense that, in an increasingly interconnected world, information is emerging as an ever more powerful force of change. The impacts on economic development depend, however, on the accuracy, availability and transparency of that information and on the ways in which environmental performance information is used in private and public decision-making.

PUTTING PRESSURE ON POLLUTERS
Indonesia's PROPER programme

A CASE STUDY FOR THE HARVARD INSTITUTE
FOR INTERNATIONAL DEVELOPMENT 1997
ASIA ENVIRONMENTAL ECONOMICS POLICY SEMINAR*

Shakeb Afsah and Jeffrey R. Vincent

Prior to its current crisis Indonesia was one of Asia's 'miracle economies'. A poor, primarily agricultural, country two decades ago, it has emerged as a regional industrial powerhouse. Gross investment in equipment and structures grew at a median rate of 8% per year during the period 1985–94.[1] Manufacturing value-added grew even more rapidly, at 10% per year. Between 1980 and 1991, the nation's output of iron and steel increased more than fourfold, processed wood products more than threefold, and a host of other industrial products, such as textiles, footwear, paper products, glass, metal products and transport equipment, increased twofold or better. Industrialisation has created much-needed jobs in a country that is the fourth most populous in the world. It is a prime reason why the average Indonesian's real income has doubled since 1980 and the number of Indonesians living in poverty has fallen by more than half since 1970. But by the late

---

* This case is a teaching tool, not a report on original research. The authors drew primarily on public-domain materials available at a World Bank website: www.nipr.org. Afsah, who was seconded by the World Bank to BAPEDAL as a resident advisor during the implementation of PROPER, was author or co-author of most of those materials. The authors thank David Wheeler of the World Bank for helping arrange the case study; Joseph Stern, Tim Buehrer, Steve Radelet, Sian Wayt and Xiang Yu of the Harvard Institute for International Development (HIID) for help with obtaining data and materials; and US Agency for International Development (USAID) for financial support. Opinions expressed in the case study reflect the authors' views and do not necessarily reflect official views of the HIID, the World Bank, USAID or the Indonesian government.

1 Unless indicated otherwise, all figures in the text are in real (inflation-adjusted) terms.

1980s a less positive consequence of industrialisation was becoming increasingly apparent to Indonesians living in and around industrial centres: rapidly deteriorating air and water quality. Three-quarters of industrial facilities were located on the island of Java, which is one of the most densely populated portions of the Earth's surface. When Indonesia established its first air-quality monitoring station in Jakarta in 1978, airborne concentrations of suspended particulate matter already exceeded the World Health Organisation's (WHO)'s recommended standard by 40%. By 1988, they were double the recommended standard. Sulphur dioxide concentrations rose even more rapidly, doubling between 1981 and 1988. Rivers were increasingly fouled by industrial effluent. Biochemical oxygen demand (BOD) measured at water-quality monitoring stations rose from 3–6 mg/l in the early 1980s to more than 10 mg/l by the early 1990s. A study by the World Bank estimated that exposure of urban residents to airborne particulate concentrations above the WHO standard caused an additional 1,263,352 deaths, 26,609,033 emergency-room visits, 184,453,618 asthma attacks and 5.3 million–11.8 million lost work days in Jakarta in 1989 (Ostro 1992).

The government scarcely monitored the environmental performance of industrial facilities in the 1980s, and its enforcement efforts were virtually non-existent. The Ministry of Population and Environment had limited resources to regulate industrial pollution, and governors of provinces, whose principal concern was to increase employment and income by attracting investment, felt little incentive to do so. Increases in the Ministry's annual budget allocation were tiny compared with the rate of industrial growth, exceeding the inflation rate by just over 1%. The Ministry's pollution control activities were not simply failing to keep pace with industrialisation, they were falling further and further behind.

The PROKASIH programme

This situation forced the Ministry to experiment with approaches to environmental regulation other than Western-style 'command and control'. In 1989, the Ministry decided to focus its limited resources on a semi-voluntary programme for controlling the discharge of industrial pollution in waterways. It formally announced its Clean River programme, better known as PROKASIH, on 19 June 1989.[2] This programme established inter-agency teams within individual provinces. These 'PROKASIH teams' included representatives from a range of agencies, including the regional development planning board, the public works department, the health department, environmental study centres and environmental laboratories. They were responsible for several activities, including:

- Selecting specific rivers or portions of rivers where concerns over water quality were the greatest

- Identifying the industrial facilities that were the most significant polluters

2 Our primary source of information on PROKASIH is Afsah *et al.* 1995.

- Drawing up pollution reduction agreements to be signed by provincial vice-governors and polluting facilities

- Collecting data on pollution concentrations in facilities' effluent and in receiving waters and reporting those data to the Ministry

From 1990 onwards, the new environmental impact management agency, BAPEDAL, administered the programme. BAPEDAL reported directly to the president's office. It co-signed the pollution reduction agreements, jointly financed the programme with provincial governments and reviewed the data collected by the PROKASIH teams. Eight provinces agreed to participate in the programme initially. Participation by polluters was not voluntary: facilities selected by the PROKASIH teams were obliged to negotiate and sign the pollution reduction agreements. These agreements were not legally binding, however, and their details were not made public. In this sense, the programme was voluntary: facilities could determine the degree of compliance with the terms of the agreements without suffering any regulatory consequences if they simply ignored them.

At the inception of PROKASIH, BAPEDAL could not point to any evidence that similar programmes had worked in neighbouring countries, as none had tried such programmes. Yet, BAPEDAL's lack of funds and workforce made the agency willing to gamble that PROKASIH would at least make polluters aware that they were polluting—the first step toward getting them to change their behaviour. In some respects, PROKASIH was similar to the US Environmental Protection Agency's 33/50 programme, a voluntary programme aimed at reducing the release of toxic chemicals. But the US agency had introduced this programme only a couple of years before PROKASIH, and its success did not become clear until the early 1990s.

To BAPEDAL's pleasant surprise, PROKASIH induced several polluters to leap, not step, toward improved environmental performance. A World Bank study of 34 river basins included in the programme found that the aggregate BOD load discharged by participating facilities fell in 24 river basins by 1994 (Afsah et al. 1995). The median reduction in BOD load in the 24 river basins was 59%. BOD pollution fell in relative terms as well: the BOD discharge per unit of output fell by about 55%. Rising pollution control effort, not falling output, was thus the driving force behind the reduction in aggregate BOD load. Encouraged by the programme's success, BAPEDAL expanded it to 13 provinces by 1994, with the number of industrial facilities included in it rising by a factor of more than three.

Perhaps the most important thing BAPEDAL learned from PROKASIH was that pollution discharge varied tremendously across facilities. Many officers on BAPEDAL's staff were environmental engineers, used to thinking of pollution control in terms of end-of-pipe technologies. If facilities had the 'right' technologies, and actually used them, then pollution would be negligible; if they did not, then pollution would be proportional to output. Given the history of weak enforcement in the country, there was no reason to expect that facilities had invested in pollution control. Even if they had, there was no reason to expect that they were actually running the equipment. In either case, there was no reason to expect pollution discharge to vary much across facilities: it would be extreme in all cases.

But this is not what BAPEDAL found when it examined the baseline data on pollution discharge collected by the PROKASIH teams at the start of the programme. Instead, it found that a small number of extreme polluters were discharging most of the BOD load and that most facilities were discharging relatively small amounts. A later, more rigorous, analysis by the World Bank confirmed BAPEDAL's findings: in 1990, when PROKASIH was being launched, just 10% of the facilities were discharging 50% of the BOD load, and just 20% were discharging 75% (Afsah et al. 1995). In contrast, the 50% 'cleanest' facilities were discharging less than 5% of the BOD load.

This skewed distribution confirmed the wisdom of targeting regulatory efforts, but it raised a question without an immediately obvious answer: why did pollution discharge vary so much, when all facilities were facing the same (weak) regulatory environment? BAPEDAL's hunch, confirmed subsequently by a World Bank study (Pargal and Wheeler 1995), was that the conventional regulatory approach, which assumed that the only relevant parties in the regulatory process were industrial polluters and government regulators, ignored two other key parties: the communities in which industrial facilities are located and the markets in which the polluters purchase their inputs and sell their products. This observation was the starting point for the development of BAPEDAL's programme for Pollution Control, Evaluation, and Rating, better known as PROPER, which is based on public disclosure of facilities' environmental performance.[3]

Development and design of PROPER

Although the Indonesian parliament approved the country's framework environmental law in 1982, promulgation of regulations under it proceeded slowly until the 1990s. Few regulations were in place when PROKASIH was launched. In fact, Government Regulation 20 of the Year 1990, 'Concerning the Control of Water Pollution', was issued by presidential decree almost simultaneously with the presidential decree that created BAPEDAL.

A 1991 ministerial decree (KEP/MEN/03/1991) specified effluent discharge standards for 14 industries and delimited more general standards, linked to water quality objectives in the receiving rivers, for remaining industries. This decree also authorised BAPEDAL to enforce the standards under a programme dubbed JAGANUSA. Regulations for other environmental media and for comprehensive environmental assessment followed. Government Regulation 51 of the Year 1993, 'Regarding Environmental Impact Assessment (AMDAL)', came two years later, and Government Regulation 19 of the Year 1994, 'Regarding Hazardous and Toxic Waste Management', came three years later. Both were

---

3 Our principal source of information on PROPER is the report by the PROPER PROKASIH Team (BAPEDAL) and PRDEI (Policy Research Department, Environment and Infrastructure Division) (World Bank), 'What is PROPER? Reputational Incentives for Pollution Control in Indonesia' (Washington, DC: World Bank, www.nipr.org/work_paper/propwhat/index.htm). For a more concise account, see Wheeler and Afsah 1996.

presidential decrees. A 1995 ministerial decree specified air emissions standards for stationary sources.

Environmental agencies around the world have traditionally treated regulatory compliance as an 'either/or' proposition: to be in compliance, facilities must satisfy all provisions of the pertinent regulations. There is no middle ground. Facilities cannot be, say, 50% in compliance. They either comply, in which case environmental regulators take no punitive action, or they do not comply, in which case regulators levy fines or other penalties (e.g. suspension of operating licences). This approach struck Nabiel Makarim, Deputy for Pollution Control at BAPEDAL, as unnecessarily limited in scope and degree. It was limited in scope in that regulators always punished and never rewarded. This made the relationship between regulators and industry purely negative: regulators existed to find industry's mistakes and punish it for those mistakes. Regulators did not reward superior performance. Polluters, which all facilities inevitably are to a greater or lesser degree, had no incentive to identify themselves to regulators, even if they were reasonably good performers, as the attention of regulators could bring only costs and no benefits.

The traditional approach was limited in degree in that it ignored the range in actual environmental performance that BAPEDAL was finding characterised facilities in the PROKASIH programme. The traditional approach grouped facilities that were violating only a few (perhaps only one) of the provisions in environmental regulations, and thus might be among the 50% of facilities responsible for only 5% of the pollution, into the same non-compliance category as facilities that were flagrantly violating the regulations and discharging the great bulk of pollution. Having just two categories, in-compliance and out-of-compliance, gave a distorted picture of the industrial pollution problem and limited the progress of any environmental agency to address it.

As an alternative, in December 1993 Makarim proposed a colour-coded rating system for 'grading' the performance of facilities. He proposed subdividing the in-compliance category into blue, green and gold ratings, and the out-of-compliance category into red and black ratings. A blue rating indicated that a facility just satisfied all the provisions in applicable environmental regulations. A green rating indicated that its performance was substantially better than the regulations required. A gold rating, which Makarim thought should be awarded rarely, indicated that its performance was exceptionally good. Similarly, a red rating indicated that a facility was applying some environmental management effort but not enough to satisfy all the provisions, and a black rating was reserved for the worst performers, who were making no effort to control their pollution discharge. Makarim chose these colours because they had cultural connotations in Indonesia analogous to the environmental performance levels they signified.

The idea of a colour rating system probably would not have attracted much attention among senior policy-makers had Makarim not gone on to propose that BAPEDAL make the ratings public. In his view, the value of the system came less from improving the relationship and flow of information between BAPEDAL and industry, although these were important, than from the provision of information to communities and markets that interacted with the facilities. That is, he thought the ratings could influence the facilities' reputations and thereby honour and shame could be used to create reputational incentives for better environmental performance. From experience with PROKASIH, he and others at BAPEDAL suspected that variations in environmental performance reflected a combination of community pressure, at least when the media, local governments, non-governmental organisations (NGOs) and other community organisations (e.g. religious groups) could attribute pollution problems to specific facilities, and market pressure, when companies thought they could obtain a market advantage from good environmental performance.

Public disclosure of ratings would help expose polluters to these reputational pressures. It would also help direct those pressures to the right problems: the most harmful pollutants are not always the most obvious ones (heavy metals, toxic chemicals), and conversely the most obvious ones are sometimes relatively innocuous (certain organic wastes). Moreover, identifying the worst pollution sources is often not easy when, for example, several facilities are located in the same industrial estate. By helping communities identify the main sources of the most damaging pollutants, the rating system would help them apply pressure where it really counted.

Public disclosure of ratings was consistent with a prominent but rarely used principle in the 1982 environmental law: community participation in environmental management. What had not been anticipated in 1982, however, and therefore was not mentioned in the law, was that markets might also offer incentives for companies to improve their environmental performance. Although the domestic 'green consumer' movement was small and largely limited to the well-educated, suburban élite in Jakarta and other cities, Indonesian companies in some sectors, notably wood products, were facing increasing pressure from consumers, environmental groups and in some cases legislators in export markets. The Earth Summit in Rio de Janeiro raised the profile of green consumerism and the domestic and international NGOs that championed it, as did discussions over domestic environmental practices during the Uruguay Round of the General Agreement on Tariffs and Trade (GATT) negotiations. At least one Indonesian coal mining company, P.T. Adaro, had started aggressively marketing its 'Envirocoal', which was unusually low in sulphur and ash content, to Western utilities facing strict environmental regulations. Makarim reasoned that companies in other sectors might also respond to the market opportunities generated by green consumerism and clean up their acts in the process if a credible source verified their superlative environmental performance in a clear, easy-to-understand fashion.

Makarim succeeded in convincing Sarwono Kusumaatmadja, State Minister of Environment, to proceed with a pilot colour-rating scheme, with a proposed launch date of June 1994. But BAPEDAL faced several challenges in turning Makarim's colour-rating system into an implementable programme, and these challenges forced it to postpone the launch until June 1995.

From regulations to colours

The first major challenge was to translate the country's complex set of environmental regulations into the five colour codes. By the early 1990s Indonesia had in place numerous newly adopted environmental regulations. Discharge standards necessarily varied by pollutant, and, as in the case of standards in the water regulations, they sometimes also varied across industries and locations. One option was to give a colour rating for every individual standard and provision in the regulations, but that would have destroyed the ability of the system to communicate a facility's overall 'grade'. Given BAPEDAL's relative depth of experience with water pollution, first through PROKASIH and then through JAGANUSA, it decided to simplify the task by focusing initially on water pollution. Even then there was a complication, as Indonesia had provincial water pollution regulations in addition to national regulations. In some cases the provincial regulations differed significantly from their national counterparts. To build public understanding and promote the rating system as a national programme, BAPEDAL decided to use only national water pollution regulations (i.e. KEP/MEN/03/1991) in defining the initial ratings.

In early 1994, BAPEDAL mobilised a technical team of environmental experts from Australia, Canada and the World Bank, as well as from its own staff, to translate the regulations into colour ratings and to design a programme for putting the ratings into practice. Initial efforts, which included an extensive survey to collect factory-level data, produced an elaborate system that seemed inappropriate given BAPEDAL's limited resources and the limited amount and quality of regularly available data. The proposed system was dropped, and expatriate members of the technical team returned to their respective countries.

Undaunted, a group of BAPEDAL staff members continued to collect data and refine the methodology. From February 1995 onwards BAPEDAL began a concerted effort to launch the programme, with a core team now consisting of seven of its own staff and an advisor from the World Bank. This group succeeded in designing a short series of 'yes/no' questions that covered key provisions of the regulations and made it easy to determine which colour rating a specific facility deserved. Figure 7.1 shows these questions and the mapping from regulations to the five colours. Most of the questions relate either to the size of the pollution load relative to the effluent discharge standards specified in KEP/MEN/03/1991 or to self-monitoring provisions related to the installation of an effluent flow meter, daily measurement of the flow rate and monthly sampling and analysis of the effluent.

The mapping from regulations to colours was a conservative one: a polluter had to comply with all provisions in the regulations to receive a blue or higher rating. Good or excellent performance according to several provisions was not allowed to compensate for inadequate performance according to even one. BAPEDAL's worries that NGOs and their constituencies might be sceptical of environmental performance ratings from a government with an obvious commitment to rapid industrial development prompted this conservative approach.

Minimising the risk of ratings errors

BAPEDAL also worried about mistakes being exposed after the ratings had been made public. This would not have been a problem if the ratings were not publicly disclosed, as BAPEDAL could in that case simply notify the industrial facility of the error. But

164 ASIA'S CLEAN REVOLUTION

Does the plant meet discharge standards? — **No** →

↓ **Yes**

Is there a flow meter? — **No** →

↓ **Yes**

Is the monthly reporting at least 50% complete? — **No** →

↓ **Yes**

Is there any hidden bypass for waste? — **Yes** →

↓ **No**

Does the plant treat 60% of the waste load? — **No** →

↓ **Yes**

Blue   **Red**   **Black**

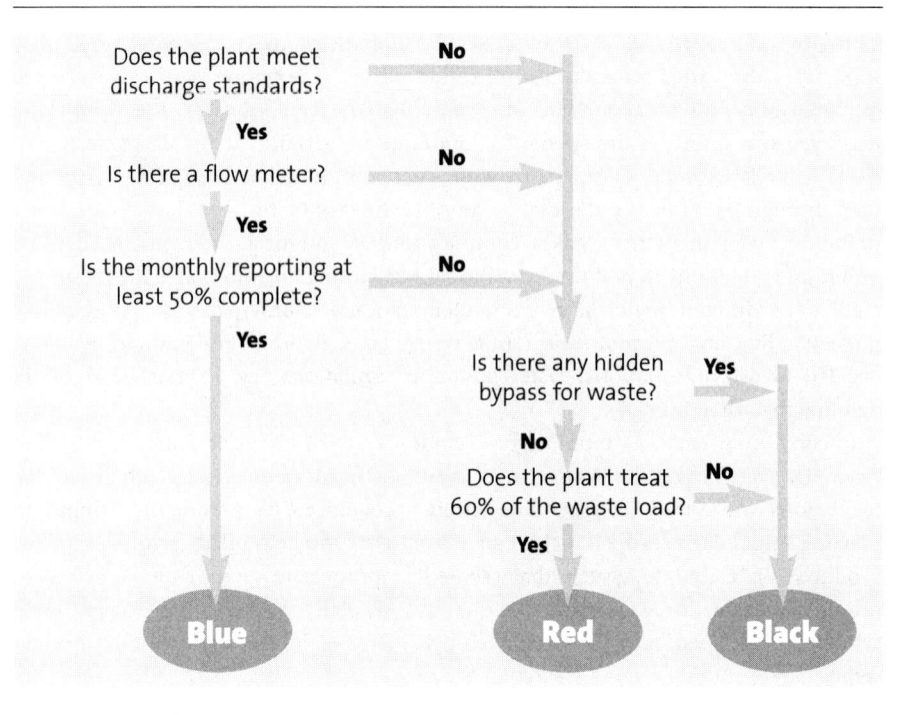

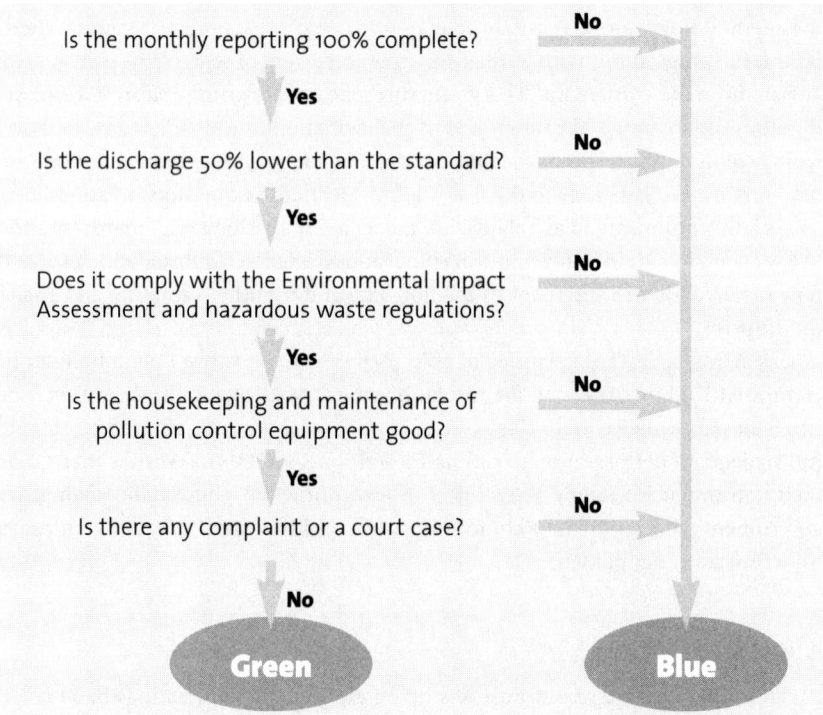

Is the monthly reporting 100% complete? — **No** →

↓ **Yes**

Is the discharge 50% lower than the standard? — **No** →

↓ **Yes**

Does it comply with the Environmental Impact Assessment and hazardous waste regulations? — **No** →

↓ **Yes**

Is the housekeeping and maintenance of pollution control equipment good? — **No** →

↓ **Yes**

Is there any complaint or a court case? — **No** →

↓ **No**

Green   **Blue**

Figure 7.1 **Examples of some of the 'yes/no' questions used to determine a facility's colour rating**
(continued opposite)

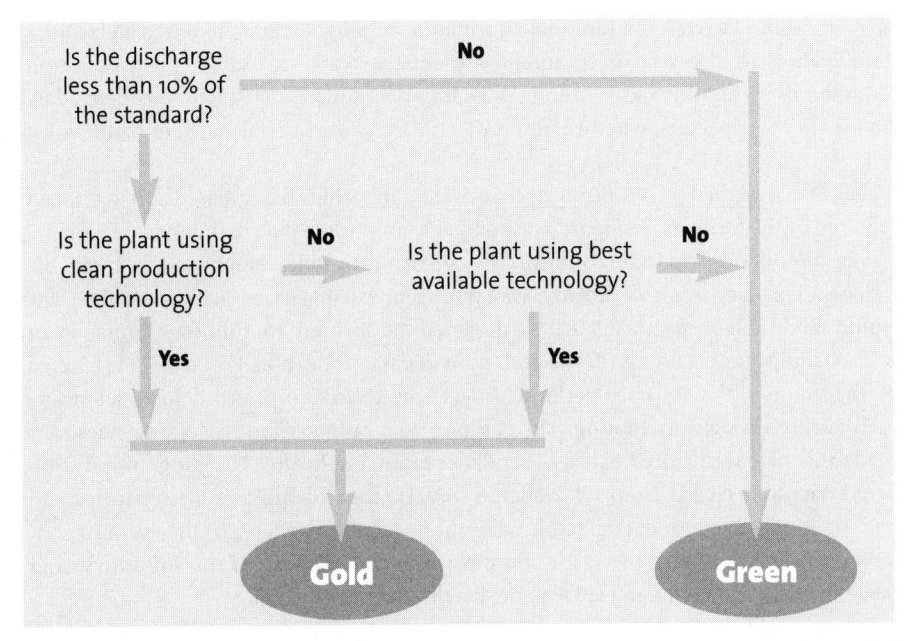

```
                                    No
Is the discharge  ─────────────────────────────────────►
less than 10% of
the standard?
      │
      ▼
Is the plant using    No    Is the plant using best    No
clean production  ────────► available technology?  ────────►
technology?
      │ Yes                        │ Yes
      ▼                            ▼
          Gold                              Green
```

Figure 7.1 (continued)

mistakes made in public could destroy the credibility of the rating system, particularly if they occurred early on. NGOs would be sure to question BAPEDAL's objectivity if they found it had assigned blue, green or gold ratings to egregious polluters, and industry organisations would question its competence if good performers were given red or black ratings.

The simplicity of the criteria for assigning colours reduced, but did not eliminate, the risk of error. The principal remaining risk came from inaccurate and incomplete data, and this had BAPEDAL worried. Privately, it doubted the accuracy of the pollution data reported by some of the PROKASIH teams. These doubts were justified: the World Bank, in its analysis of PROKASIH's performance, used data for only 155 of the 778 industrial facilities that participated in the programme in 1990 and 1991, as it judged data for the remaining facilities to be unreliable or incomplete (Afsah *et al.* 1995). BAPEDAL responded to this problem by basing the rating system on multiple sources of data, including independent inspections, by developing user-friendly computer software to analyse the data and by designing a multi-step process for reviewing proposed ratings before making them public. BAPEDAL had up to four separate sources of pollution data on individual facilities. First, it had been compiling data from the provincial PROKASIH teams since 1989. Second, its inspectors had been collecting data for JAGANUSA since 1991. JAGANUSA covered some of the same facilities as PROKASIH, generally ones whose neighbours had filed complaints with BAPEDAL. Third, BAPEDAL sent out a mail survey to prospective participants in the rating system in February 1995, and it conducted special inspections of some facilities to collect additional information. Finally, BAPEDAL required all facilities participating in the system to monitor themselves and to report their pollution discharge on a monthly basis. Among these four data sources, BAPEDAL considered data from its own inspectors, whether from JAGANUSA or from special visits, to be the most reliable.

BAPEDAL designed a data protocol that precluded ratings being based solely on data reported by the facilities. As Figure 7.2 shows, self-monitored data that indicated a facility was in compliance had to be confirmed by data from an independent source. If reliable independent data was not already available from PROKASIH or JAGANUSA, BAPEDAL would send in its inspectors. BAPEDAL designed the software to minimise human error in the handling and analysis of data and to aid in data verification. Based on data entered by the inputter, the software performed necessary calculations and helped determine each facility's provisional rating. The design of the rating criteria as yes/no questions made this task well suited to user-friendly computer software. The programme compared data and results from all available sources, thus identifying discrepancies and helping BAPEDAL conduct targeted investigations to resolve them. In the event of controversy over a rating decision, the simplicity and transparency of the software would make it easy for BAPEDAL to explain how it arrived at its decision.

Because gold, green and black ratings were considered extraordinary—exceptionally good or bad—BAPEDAL included a 'final filter' of intensive discussions about facilities

Figure 7.2 **Performance rating procedure**

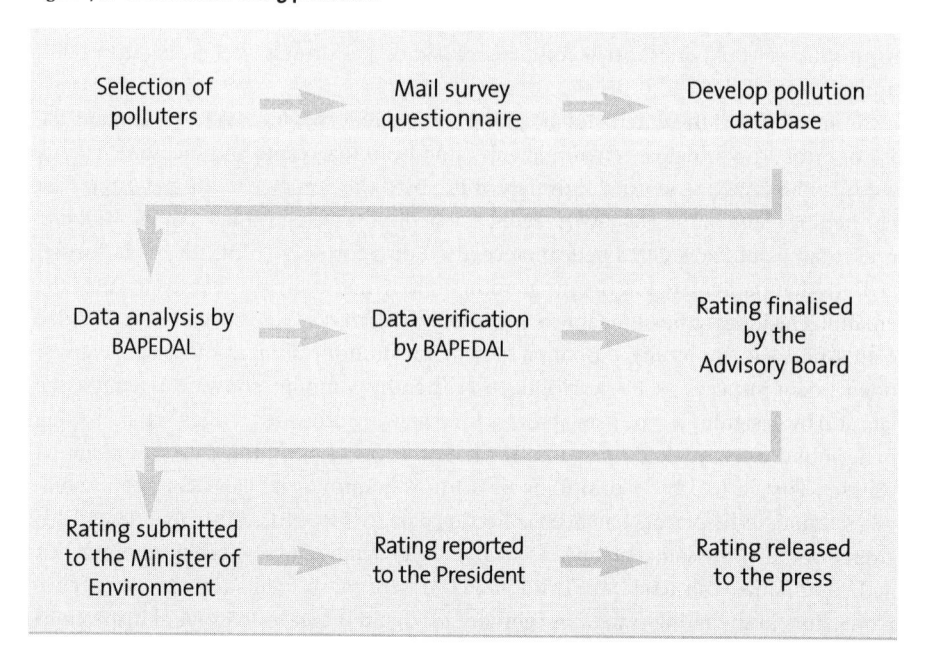

Selection of polluters → Mail survey questionnaire → Develop pollution database

Data analysis by BAPEDAL → Data verification by BAPEDAL → Rating finalised by the Advisory Board

Rating submitted to the Minister of Environment → Rating reported to the President → Rating released to the press provisionally assigned those ratings. In these discussions, staff from all divisions of BAPEDAL scrutinised ambiguities and drew on additional information to improve the accuracy of the rating decisions. BAPEDAL also included a three-step review process for the ratings: it decided to disclose the ratings only after they had been approved by: (1) a special advisory board, which included members from outside BAPEDAL (for example, the Department of Health, business associations and NGOs); (2) the State Minister of Environment; and (3) the President. Figure 7.3 summarises the steps in the rating process. To ensure that press reports on the rating system were accurate, BAPEDAL even arranged for officers involved in the project to visit the offices of major local newspapers and explain the system and demonstrate the software.

Selecting facilities to be rated

The third major challenge was to select facilities to be included in the rating system. The focus on water regulations limited the pool of potential facilities somewhat, but there were still tens of thousands of facilities discharging effluent into the country's rivers and streams. Not surprisingly, given the advantages of having as much data as possible, BAPEDAL selected most of the initial facilities from ones already participating in PROKASIH. It sent the February 1995 questionnaire to 350 facilities participating in PROKASIH. These facilities spanned the 14 industries for which KEP/MEN/03/1991 speci-

Figure 7.3 **Data verification**

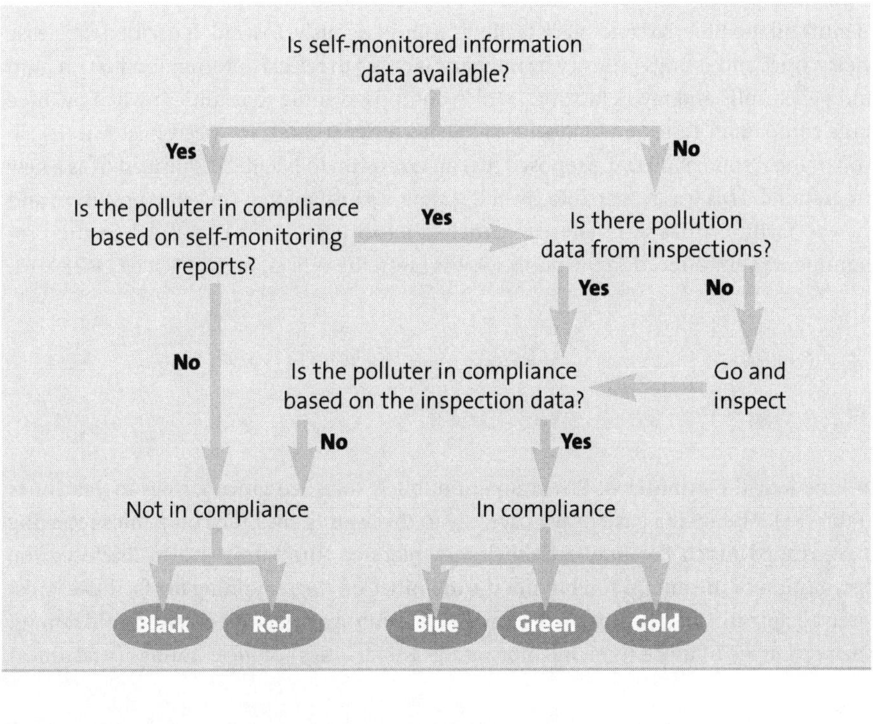

fied discharge standards. Of the facilities surveyed, BAPEDAL judged that 176, or almost exactly half, had sufficient data to be rated. BAPEDAL also invited facilities not included in PROKASIH to volunteer to be rated; 11 facilities did so. Hence, the initial number of participants in the rating system was 187.

Avoiding political repercussions

International business magazines routinely report on the business interests of senior Indonesian political and military figures. If all facilities owned by such individuals deserved blue or higher ratings, then BAPEDAL would not have a problem: the rating system stood to make powerful friends. Given the relatively recent introduction of water pollution regulations and their weak enforcement, however, this imagined scenario was too good to be true. Some of the facilities owned by well-connected figures would surely deserve red or black ratings. BAPEDAL needed to formulate a strategy to avoid turning those figures into formidable enemies. Acting on the advice of politically aware supporters, BAPEDAL decided to release initially the names of only those facilities earning green or gold ratings. This would give the system a positive image. For the remaining facilities, it decided to release initially just the number in each colour category. This would demonstrate that it was indeed serious about giving red or black ratings to facilities that were out of compliance. Finally, it decided to give the facilities whose identities were initially suppressed six months to improve their performance before it disclosed their names and ratings. This would give their owners a one-time chance to avoid public loss of face.

To sustain interest in the system and keep it in the news, BAPEDAL decided not to identify all the blue, red and black facilities simultaneously. Instead, it decided to release their names and ratings industry by industry. It chose to release information first on pulp and paper mills and rayon factories, which comprised some 30 facilities owned by three large companies, followed by textile mills and other sectors. To speed institutionalisation of the system, BAPEDAL proposed it as an extension to PROKASIH, instead of as a new programme. This was reasonable given the system's initial focus on water pollution and the 176 facilities that were already participating in PROKASIH. Hence, when the programme was announced to the public, it was given the official name PROPER PROKASIH.

The performance of PROPER

In June 1995 the Minister of Environment publicly awarded green ratings to 5 facilities (Table 7.1). The media gave heavy coverage to the awards, and the companies receiving them reaped much favourable publicity. As planned, the Minister also disclosed the distribution of the ratings for, but not the identities of, the remaining 182 facilities. Most received red ratings (115), and a few received black ratings (6). None received gold ratings. The percentage of those receiving blue ratings (61 facilities) or green ratings (5 facilities)

Rating	June 1995	December 1995	Change (%)
Gold	0	0	0%
Green	5	4	−20%
Blue	61	72	+18%
Red	115	108	−6%
Black	6	3	−50%

Table 7.1 **The short-term impact of PROPER: the number of facilities falling into each rating category**
(For a description of categories, see text.)

was more than a third of the total (35%). That this percentage was so high surprised BAPEDAL and probably most of the Indonesian public, given the prevailing weakness of enforcement. It provided additional evidence that a semi-voluntary programme such as PROKASIH could induce notable progress toward pollution reduction in a developing country such as Indonesia.

BAPEDAL met with the owners of several companies receiving red and black ratings and urged them to improve their performance during the six-month grace period. Faced with the threat of public disclosure, many companies took advantage of the opportunity and did so. In several cases, it turned out that plant managers had misinformed owners about the compliance status of their factories; once owners were better informed by the ratings, they issued stern instructions to reduce pollution. By December, when the Minister began disclosing the names and ratings of all facilities, the number of black ratings fell by half, from 6 to 3, and the number of red ratings fell by 6%, from 115 to 108 (Table 7.1). As a result, the number of blue ratings—that is, the number of facilities just meeting the basic requirements for compliance—rose by nearly a fifth, from 61 to 72. The number of green ratings fell by one, but even that was a sign of PROPER's effectiveness. One of the facilities that had been awarded a green rating in June was downgraded in response to protests by a community living in the vicinity of the facility. This incident demonstrated that PROPER could succeed in augmenting the ranks of BAPEDAL's inspectors by empowering communities to verify the accuracy of the ratings.

In just six months, PROPER raised the compliance rate from 35% to 41%. A sign of industry's confidence in the programme was that no facility, not even the one whose rating was changed from green to black, protested its rating. Another sign was that the number of facilities that contacted BAPEDAL to volunteer in the programme more than doubled, from 11 to 25. Performance varied considerably across industry groups (Table 7.2). In two cases, paper and sugar, more than half the facilities received blue ratings, although none received green ratings. In three others, however—rubber, textiles, palm oil—more than three-quarters received red or black ratings. Just over two-thirds of plywood mills received red or black ratings. From an aggregate 'compliance/non-compliance' standpoint, plywood mills would appear to be better performers than rubber, textile or

Rating	Sugar	Paper	Plywood	Textile	Palm oil	Rubber
Green	0	0	0	4	0	0
Blue	67	55	32	20	24	15
Red	33	45	52	72	76	85
Black	0	0	16	4	0	0

Table 7.2 **Percentage distribution of ratings by industry type**

palm oil mills, but nearly a quarter of those plywood mills that were out of compliance received black ratings, compared with none of the rubber and palm oil mills and only a small portion of the textile mills. PROPER gave insights into the degree of non-compliance which had previously not been available.

Performance also varied across ownership categories (Fig. 7.4). Facilities privately owned by Indonesian nationals were the worst performers: nearly 70% of their ratings were red or black. Facilities owned by multinational companies were the best performers: nearly 80% of their ratings were blue or green. The performance of state-owned facilities fell between these two extremes, with almost equal numbers being in compliance and out of compliance. The strong performance of multinationals has several possible explanations. One, which is consistent with the motivation behind PROPER, is that multinationals sell their products primarily to developed countries, where green

Figure 7.4 **Performance by ownership**

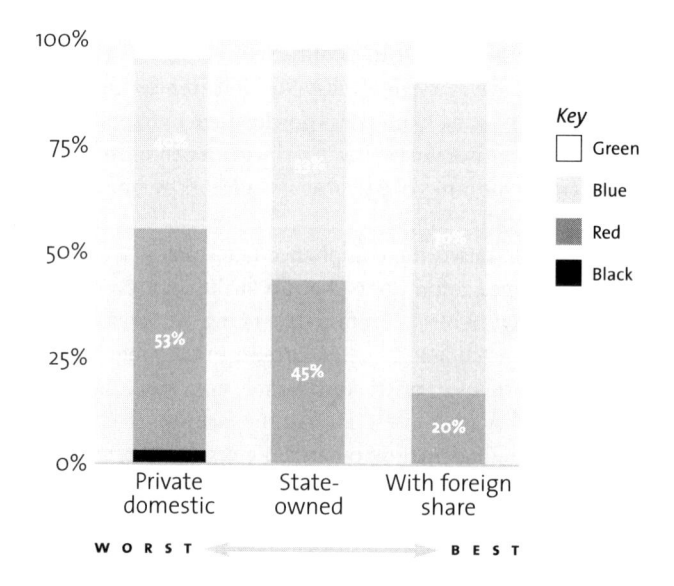

consumerism is stronger than it is in Indonesia, and are owned primarily by shareholders from those countries, who might be more environmentally inclined than the average Indonesian capitalist. On the other hand, multinationals tend to be larger than Indonesian-owned companies, and their greater size might offer economies of scale that make environmental management more affordable. Some evidence suggests that this second explanation is more likely: the World Bank study of facilities participating in PROKASIH found that the performance of multinationals did not differ from that of Indonesian-owned companies once differences in size were taken into account (Pargal and Wheeler 1995). This suggests that the Indonesian business community responds to reputational incentives just as strongly as the international business community.

The most recently available information indicates that facilities participating in PROPER are continuing to improve their performance. More than a quarter of the facilities rated red or black in December 1995 improved their ratings to blue or green by September 1996 (see Fig. 7.5).

Questioning PROPER's performance

The apparent importance of size in determining facilities' environmental performance raises a question about the ultimate impact of PROPER on Indonesian water quality. PROPER, and PROKASIH before it, include mainly larger enterprises. Yet most enterprises in Indonesia are small or medium-sized. Many are not well known to the public in terms of either their names or the products they sell. Can PROPER be extended to include the thousands of small and medium-sized enterprises in Indonesia and, if so, is it likely to be effective?

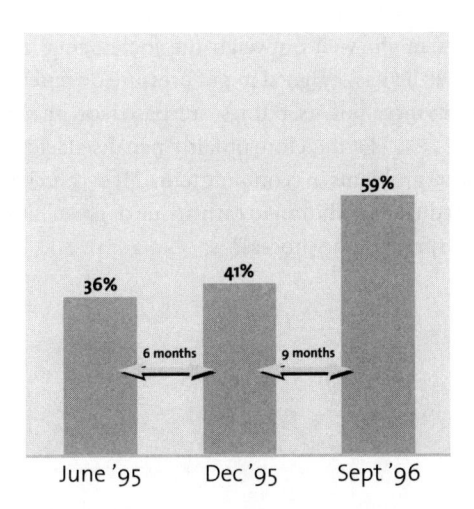

Figure 7.5 **Compliance trends**

	June '95	Dec '95	Sept '96
	36%	41%	59%
		6 months	9 months

For that matter, will BAPEDAL be able to sustain PROPER's apparent effectiveness within the group of the 187 original facilities once the programme's novelty wears off and the media moves on to other stories? How much of PROPER's initial success has been a result of the care BAPEDAL took in selecting facilities with good data, which might be the facilities that were already more capable of managing their environmental performance, and its skill at enlisting the support of political allies, whose attention and interest might wane as other pressing issues arise? Can PROPER succeed once it becomes just another, familiar government programme and must include facilities with poor data as well as good?

The World Bank study of facilities included in PROKASIH raises a third set of questions (see Pargal and Wheeler 1995). After controlling for enterprise size and other important variables, the study found that environmental performance was much worse in poorer, less-educated communities: facilities located in municipalities in locations in the bottom 25% of income and education distributions had a water pollution intensity that was *15 times as large* as the pollution intensity of facilities in municipalities in the top 25%. The study concluded that this difference was more likely to be a result of differences in the relative power of the communities than of differences in their preferences. Does this imply that PROPER is likely to work well only in the more affluent and better-educated parts of Indonesia? If so, might public pressure in such communities induce polluters to relocate to weaker communities, thus making rich communities cleaner and poor communities more polluted? This is an issue of equity.

A final set of questions has to do with efficiency. The discharge standards specified in KEP/MEN/03/1991 are uniform within the 14 industries to which they apply: that is, although standards may vary between palm oil mills and rubber mills, they do not vary between palm oil mills on rivers with downstream communities and palm oil mills on rivers without downstream communities. These uniform standards are probably not economically justified if one takes into account differences in pollution damages. Most likely, pollution from a mill on a river without downstream communities generates less economic damage than pollution from a mill on a river with downstream communities. Allowing the former mill to discharge more pollution would reduce its abatement costs, and these cost savings might well outweigh the incremental damage caused by the additional pollution. PROPER is designed to put pressure on enterprises to improve their environmental performance, but its ratings are based on the uniform standards in KEP/MEN/03/1991. Does PROPER therefore unfairly penalise facilities whose pollution is unlikely to cause serious problems in economic terms? If so, and if such facilities respond to public pressure by reducing pollution to earn a blue or green rating, how much money might they needlessly spend in the process?

WATER POLLUTION ABATEMENT IN MALAYSIA*

*Jeffrey R. Vincent and Rozali Mohamed Ali
with Khalid Abdul Rahim*

Production of crude palm oil (CPO) in Malaysia more than tripled between 1975 and 1985. This expansion solidified the industry's ranking as the largest in the world—it accounted for half of world production and three-quarters of world exports in 1980 (Ma et al. 1980)—and made it the country's second largest earner of foreign exchange by 1984. In 1975, the CPO industry held another more dubious distinction—it was the country's worst source of water pollution (Maheswaran and Singam 1977). Pollution caused by the organic waste from CPO mills was equivalent to sewage generated by a population of more than 10 million people (Fig. 8.1). This was nearly as large as the entire population of the country. At the time, no proven treatment technology for palm oil mill effluent (POME) existed. Extrapolating from 1975 pollution loads, the industrial expansion that occurred between 1975 and 1985 should have increased the population-equivalent of the industry's pollution load to 33 million people. In 1985, however, the population-equivalent pollution load actually fell to 0.08 million people (Fig. 8.1). This chapter seeks to explain how Malaysia eliminated its leading water pollution problem without blocking the growth of the palm oil industry.

The palm oil industry in Malaysia

The federation of Malaysia, formed in 1963, is a multi-ethnic nation of 18 million people in peninsular and insular South-East Asia. It was formed by the union of the former

---

* This chapter is a modified version of two previous studies on this topic. The first, 'Reducing Effluent while Raising Affluence: Water Pollution Abatement in Malaysia' appeared as a Harvard Institute for International Development (HIID) case study in 1993. The second appeared as Chapter 10 in Vincent et al. (1997) and is reproduced with permission from HIID. It was adapted for this volume by Michael Rock.

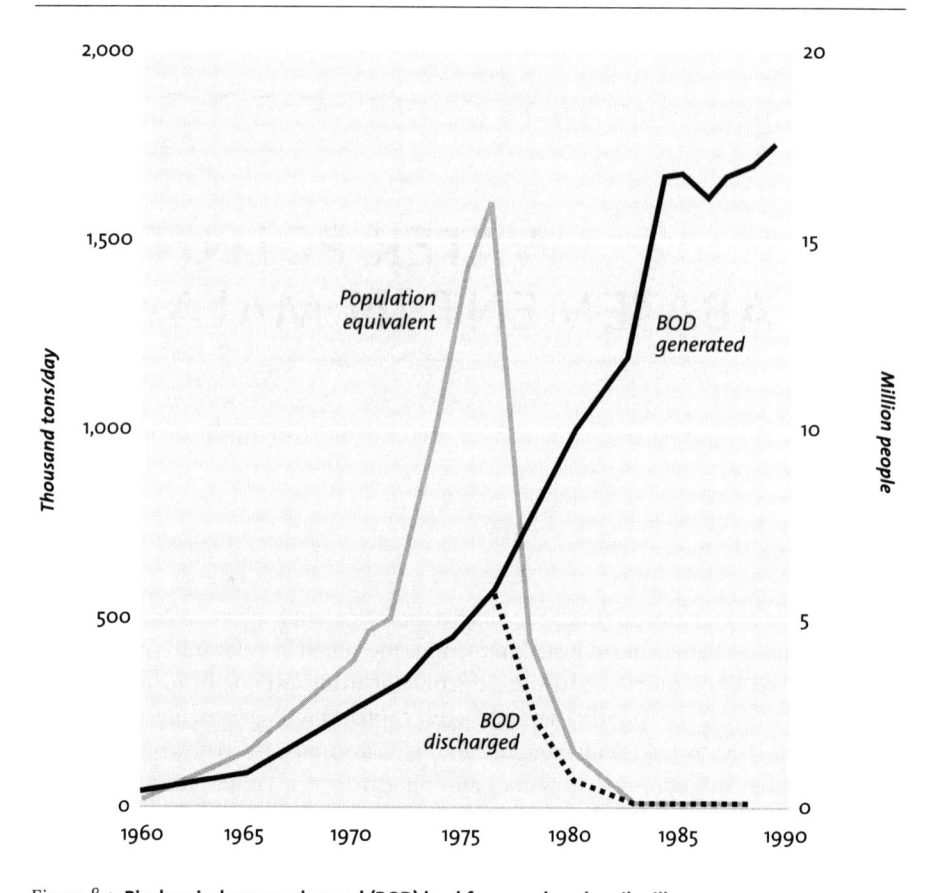

Figure 8.1 **Biochemical oxygen demand (BOD) load from crude palm oil mills**

British colonies of Malaya (peninsular Malaysia), British North Borneo (now Sabah) and Sarawak. At independence in 1957 Malaysia was a typical primary commodity exporter. It was the world's leading producer of natural rubber and tin. Together these two commodities accounted for 62% of export earnings in 1965 (IMF 1991). Following a recommendation of a World Bank team to diversify its economy, the government began promoting palm oil production, which came to be concentrated in peninsular Malaysia. In 1962 the government announced that it would permit estates and smallholders to use rubber replanting grants to establish oil palm (Yusof, in Ma et al. 1980). The response was dramatic: the area devoted to palm oil production increased more than sixfold between 1960 and 1975, and by 1975 it was two-thirds as large as the area devoted to rubber production.

The government became directly involved in oil palm plantations. In 1956, to deal with a persistent gap in yields of rubber between estates and smallholders and with the problem of rural poverty, the government established the Federal Land Development Authority (FELDA). FELDA's mission was to develop new land and to provide settlers with the infrastructure, land-holdings and technical assistance needed to increase crop yields and rural incomes. FELDA's initial land development schemes focused on rubber, but in 1961 it began focusing on palm oil. Between 1965 and 1975 three-quarters of the area developed by FELDA was planted with oil palm.

Most of the settlers in FELDA schemes were ethnic Malays, who constituted half of peninsular Malaysia's population in 1970. Historically, ethnic Malays had a lower economic status than the ethnic Chinese Malaysians, the second largest population group. Many in the Malay community felt that the economic gulf separating ethnic Malays, who lived primarily in rural areas, and the ethnic Chinese widened following independence. To make matters worse, even though ethnic Malays did migrate to cities, they frequently found it difficult to find jobs. Squatter areas, lacking in public services and with squalid conditions atypical of the bucolic *kampungs* that the rural Malays left behind, sprouted around the country's urban areas. In May 1969, discontent with these conditions, kindled by fears that the Chinese were adding political clout to their economic dominance, exploded into racial riots. This experience fortified the government's resolve to attack rural and urban poverty and to reduce income disparities between ethnic groups. These two objectives formed the twin prongs of the government's New Economic Policy (NEP). The NEP guided economic development between 1971 and 1990. It granted various preferences—government hiring, scholarships in higher education, and shareholding requirements for new investments—to the *bumiputera* (Malays and other indigenous ethnic groups).

Land development schemes by FELDA and other agencies became a chief tool for fighting rural Malay poverty. The schemes increased Malays' economic assets, as settlers received title to land after repaying a portion of the development costs at favourable interest rates. These agencies and schemes became prominent because of their role in social engineering and because they were well run. Benefit–cost analyses of the schemes show that they earned positive rates of return and succeeded in raising smallholder incomes.

Palm oil processing and pollution from palm oil effluent

Like most of Malaysia's leading crops, the oil palm, *Elaeis guineensis*, is an exotic plant. It is native to Africa and was first planted commercially in Malaya in 1917 (Khera 1976). Only minor expansion of the crop occurred during the following five decades because of higher returns for the principal competing crop, rubber. As noted earlier, the plantation sector's shift toward oil palm began in the 1960s, when rubber prices began a prolonged decline and the government enacted various policies to promote agricultural diversification. Most expansion occurred in peninsular Malaysia.

Oil palm fruits produce two commercially valuable oils. The mesocarp (middle layer) of the fruit produces palm oil, an all-purpose vegetable oil. The nut or kernel yields a smaller amount of palm kernel oil. Fruit bunches must be harvested as soon as they are ripe and processed within 24 hours. CPO mills are therefore usually located in or adjacent to plantations. By the early 1970s, the processing needs of the plantations had led to the establishment of a large number of moderately sized mills scattered around the countryside.

CPO mills utilise a primarily mechanical processing technology (Aiken *et al.* 1982; Ma *et al.* 1982). They tend to be located on watercourses because they require about one ton of water to process one ton of fresh fruit bunches (Aiken *et al.* 1982). Palm oil effluent is a mixture of wastes generated at three points: when the bunches are sterilised with steam to loosen the fruits and to arrest the formation of free fatty acids; when residual oil is separated from other components of the sludge from the clarification tank, where palm oil is decanted; and when palm kernels are processed. Mills generate 2.5 tons of effluent for every ton of CPO produced (Ma *et al.* 1980, 1982).

CPO mills in Malaysia in the early 1970s disposed of effluent by discharging it, untreated, into the nearest body of water (Ma *et al.* 1980). When there were only a handful of mills in the country, the environmental impacts of this disposal method were minor. Effluent consists mainly of dissolved organic compounds. At low concentrations, these compounds do not pose a serious environmental threat. However, between 1965 and 1975 the output of CPO grew at an exponential rate. The environmental impact of rising palm oil mill effluent (POME) discharge was worsened by the fact that most of peninsular Malaysia is drained by a relatively small number of major rivers. Although CPO mills were dispersed, many were on one of these rivers or a tributary. Virtually overnight, CPO mills became the major source of water pollution in almost every major river basin in the peninsula.

Pollution problems caused by POME relate mainly to its oxygen-depleting effects. Ton for ton, the oxygen-depleting potential of POME is 100 times as great as domestic sewage. When dumped into watercourses, POME is initially decomposed by aerobic (oxygen-requiring) bacteria. As decomposition proceeds, oxygen is depleted. If POME discharge continues, the population of aerobic bacteria eventually crashes and so do the populations of other organisms that require oxygen, such as fish, prawns, and other aquatic animals that provided a significant share of the protein of Malay villagers. When the oxygen concentration in fresh-water falls below 2 parts per million (ppm), most aerobic life cannot survive. Decomposition of POME continues but now by anaerobic bacteria that release hydrogen sulphide, ammonia and other malodorous compounds. These compounds are toxic to fish, and their odour is more than a nuisance. In the 1970s, some villagers were forced to leave their homes and set up new villages because the stench from CPO mills was so overpowering. Between 1974 and 1978, CPO mills were cited in the majority of water pollution complaints to the government. At the time, it seemed inevitable that the problem would only deepen. Worse still, no known technology for treating POME existed.

The government's response

The POME pollution problem put the government between a rock and a hard place. On the one hand, the government could no longer afford to ignore the problem. POME discharge and the problems associated with it were large and growing. On the other hand, the government could not shut down the palm oil industry: it realised that it could not alleviate poverty without economic growth, and the palm oil industry was a principal engine of that growth.

The government's first step was to pass the Environmental Quality Act (EQA) in 1974. Pollution from palm oil mills was not the sole factor motivating the passage of the EQA—for example, rubber mills were also a major source of water pollution—but it was arguably the most important factor. The EQA authorised the Department of the Environment (DOE) to 'prescribe' certain industrial premises, that is to require the operators at those premises to obtain a licence before they can operate, and to attach licence conditions related to pollution control. In determining the conditions, the EQA directed the DOE to consider factors related to the economic costs of pollution abatement (EQA 1974: part III, section 11):

1. Whether it would be practicable to adopt the existing equipment, control equipment or industrial plant to conform with the varied or new condition
2. The economic life of the existing equipment, control equipment or industrial plant
3. The estimated cost to be incurred by the licensee to comply with the varied or new condition

The EQA did not similarly require the DOE to consider the economic benefits of pollution abatement, but it did direct the DOE to consider 'the quantity or degree of cut-back of emission . . . to be achieved by the varied or new condition'. The DOE began the process of formulating licence conditions by forming an expert committee with representatives from both industry and government (Maheswaran and Singam 1977). The committee's assignment was to investigate possible treatment technologies and to advise the DOE on regulations that were 'not only environmentally sound but also sensible within the framework of economic feasibility and available technology' (Ong et al. 1987: 37).

The passage of the EQA, the establishment of the DOE and the formation of the committee apparently convinced the palm oil industry that the government was intent on reducing pollution from palm oil effluent. The industry initiated research on effluent treatment technologies. No treatment system for the effluent existed, but systems for treating similar types of organic waste had been developed in industrialised countries (Wood 1977). These systems typically involve a series of treatment ponds that provide proper conditions for growing bacteria that can decompose the waste. Water containing the waste becomes cleaner and cleaner as it proceeds from one pond to the next. The industry's research strategy was to adapt one of these systems and thereby avoid the higher costs involved in designing a treatment system from first principles.

Within two years, the industry's research and development (R&D) efforts were sufficiently promising to convince the DOE that it did not need to wait much longer to invoke its licensing authority. On 7 July 1977 the DOE announced the Environmental Quality (Prescribed Premises) (Crude Palm Oil) regulations (Maheswaran 1984). The regulations imposed standards on eight parameters of palm oil effluent, with biochemical oxygen demand (BOD) being the key parameter. The regulations required CPO mills to apply for an operating licence every year. In their application, mills needed to describe their systems for treating and discharging effluent. The DOE could reject a mill's application if it disapproved of the proposed treatment system. Every three months mills also needed to file 'quarterly returns' in which they reported the amount of effluent discharged and its BOD concentration, based on tests by independent laboratories.

The DOE announced that the standards would be made increasingly stringent over the four-year period 1978-81 (Table 8.1). Standard A would take effect on 1 July 1978. Existing mills were therefore given a one-year grace period to construct treatment systems. From standard A to standard D, mills would be required to reduce the BOD concentration from 5,000 ppm—one-fifth the level in untreated effluent (PORIM 1986)—to 500 ppm. The DOE informed the industry that standard D would not necessarily represent the final level (Maheswaran and Singam 1977).

Some industry representatives complained that the regulations were too stringent (Thillaimuthu 1978). An engineer with one of the largest plantation companies argued:

> What is quite clear at this stage is that there is no single generally applicable solution [i.e. treatment system] that is proven on a large scale, over an acceptable period of time ... There can be little doubt that the limits for discharge will be very difficult to meet, except at [a] cost incompatible with the economics of operating a mill, unless there are unforeseen advances in technology (Wood 1977).

Table 8.1 **BOD (biochemical oxygen demand) standards on palm oil effluent**
Source: DOE 1985

Standard*	Date effective	Level (ppm)†
A	1 July 1978	5,000
B	1 July 1979	2,000
C	1 July 1980	1,000
D	1 July 1981	500
E	1 July 1982	250
F	1 July 1984	100

* Standards A–D were announced in the 1977 Crude Palm Oil Regulations.
† ppm: parts per million (mg / l).

The DOE responded to these complaints by pointing to three provisions that eased the burden of the regulations. First, in granting the one-year grace period and phasing in the standards over four years it recognised that the industry needed time to construct treatment facilities, to gain experience operating them and, in other ways, to move from experiment to practice. The DOE might have wanted to reduce the industry's pollution discharge more rapidly, but it recognised that the only way many mills could comply with more stringent standards would be to shut down. Given the importance of the industry, this was not a politically feasible option.

Second, the DOE did not make the BOD standard mandatory in the first year. It recognised that many mills did not have treatment systems in place, and that others could substantially reduce their pollution discharge even if they fell short of achieving the standard. For example, a mill that reduced its BOD concentration from the 25,000 ppm typical of untreated effluent to 'only' 15,000 ppm would nevertheless have reduced the BOD load it discharged by 40%. Given the time needed to construct a treatment system, such a mill could well be on its way to full compliance with the standard.

Instead, the DOE used a novel feature of the EQA to relate the size of the annual licence fee to a mill's effluent discharge. The licensing provisions in the EQA authorised the DOE to vary the size of the licence fee for prescribed premises according to (EQA 1974: part III, section 17): '(a) the class of premises; (b) the location of such premises; (c) the quantity of wastes discharged; (d) the pollutant or class of pollutants discharged; (e) the existing level of pollution'. The regulations invoked provision (c), which gave the DOE the latitude to make the licence fee equivalent to an effluent charge. Mills could choose the least-cost option: either treating effluent to reduce the BOD load in their effluent or paying a higher licence fee.

The regulations made the licence fee the sum of two parts, a flat processing fee of RM100 and a variable, effluent-related, fee. If the mill discharged effluent onto land, the effluent-related fee was linked simply to the quantity of effluent, with the mill required to pay RM0.05 per ton. If it discharged effluent into a watercourse, the effluent-related fee was more complicated and was linked to the BOD load in the effluent. Mills were required to pay RM10 per ton of BOD for concentrations up to the standard, and RM100 per ton of BOD above the standard. For example, an average-sized mill that discharged 90,000 tons of effluent a year with a BOD concentration of 15,000 ppm would owe a basic fee of RM4,500 ([90,000 tons] × [5,000 ppm] × [RM10/ton]) and an excess fee of RM90,000 ([90,000 tons] × [the difference between 15,000 ppm and 5,000 ppm] × [RM100/ton]). In effect, the regulations imposed a dual effluent charge system. The charge was higher above the standard than below to induce compliance.[1]

An interesting feature of the system is that it involved an element of self-enforcement. The DOE calculated the fees on the basis of information on projected effluent volume

---

1   Czechoslovakia and Italy have used similar dual effluent charge systems (Tietenberg 1988, 1990). In the case of Italy, the charge was, as in Malaysia, devised mainly 'to encourage polluters to achieve provisional effluent standards as soon as possible' (Tietenberg 1990: 20). The ratio of the excess charge to the basic charge in Italy was nine, very similar to the ratio in Malaysia.

and BOD levels for the forthcoming year reported by mills in their licence applications.[2] The quarterly returns and unannounced site visits provided means for the DOE to check that the estimates were reasonably accurate. A minimum payment applied: mills had to pay an effluent-related licence fee of at least RM 150, even if the calculated value was less.

The DOE did not set the charges by determining the marginal benefits of effluent abatement at the optimal level. The DOE's officers, none of whom were economists, were more uncertain about the level of the charges than any other part of the regulations. In the end, the DOE set the charges at levels that it judged would provide some financial inducement, without being onerous, for the industry to reduce its pollution discharge.

Third, the DOE recognised that ongoing R&D would be necessary for addressing the problem of effluent disposal. Therefore, it included in the regulations a provision authorising it to reduce the licence fee for mills conducting research on effluent treatment. Section 17 of the regulations stated:

> If the Director-General [of the DOE] is satisfied that research on effluent disposal or treatment of a kind or scale that is likely to benefit the cause of environmental protection is being or to be carried out at any prescribed premises, he may, with the approval of the Minister, completely or partially waive any effluent-related amount payable by virtue of [the licence fee] regulation.

In effect, the DOE allowed the industry to retain some of the revenue that it otherwise would have had to pay out in effluent charges.

Two features of the regulations reveal the DOE's reasons for imposing effluent charges. The first is the minimum payment. Given that the effluent charge for watercourse disposal was RM 10 per ton of BOD, the minimum payment removed the incentive for mills to reduce watercourse discharge below 15 tons of BOD: they had to pay RM 150 even if their actual abatement level was lower. The second is the mandatory standard after the first year. Once the standard was mandatory, the DOE could suspend or cancel the licence of any mill that violated the standard. It retained the effluent charges, however, with the excess charge being applied to mills whose violations were not considered serious enough to merit licence suspension. For example, the DOE might apply the excess charge to a mill that was in the process of constructing a treatment system.

These features indicate that the DOE viewed the effluent charges less as a means of promoting cost-effective abatement than as a means of reinforcing a system of uniform standards. The minimum payment made the regulations less than a pure effluent charge system in the first year, by reducing the incentive effect of the basic charge. The mandatory standard made the regulations even less like a pure system in following years, by reducing mills' flexibility in making abatement decisions. These features were attractive from the DOE's point of view because the former guaranteed that licensing would generate a minimum amount of revenue, while the latter guaranteed that mills would reduce their BOD discharge by a minimum amount.

---

2   Effluent charges in France are also based on expected levels of discharge (Hahn 1989).

Performance of the regulations

The DOE was somewhat disappointed with the performance of the regulations during the first year (Maheswaran et al. 1980). On the positive side, it collected a substantial amount of revenue from licence fees (RM 2.8 million) as many mills chose to pay the excess charge rather than meet the standard. Most important to the DOE, the average mill reduced its discharge of BOD by about two-thirds. Although this was an impressive reduction, the DOE had expected the average mill to meet the standard and thereby reduce its discharge by about 80%. The DOE's high expectations were not met because so many mills opted to pay the excess charge.

The DOE could have responded by raising the basic and excess charges, since the charges were apparently too low to induce, on average, compliance with the BOD standard.[3] Instead, as discussed above, the regulations committed it to making standard B and subsequent standards mandatory. The DOE signalled its willingness to take strong measures to enforce the standard almost immediately. In November 1979 it suspended the licence of a mill on the Sungai Langat in the state of Selangor. During the period 1981–84 it took similar action against 27 mills (DOE 1985).

After the first year, the motivation to comply with the standard was therefore not the excess charge but rather the risk of being shut down for violating the mandatory standard. It follows that the standard, not the excess charge, deserves most of the credit for the rapid reduction in the aggregate BOD load discharged during the initial years of the regulations (Fig. 8.1). One might argue that the excess charge deserves two-thirds of the credit because the BOD load dropped by about two-thirds during the first year, before the standard was mandatory. This argument is not very convincing, however. The substantial decrease in load probably occurred because mills were building systems capable of meeting the more stringent, mandatory standards that the regulations had announced would immediately follow standard A.

It is nevertheless possible that the basic charge had some residual impact on BOD levels after the first year. An average-sized mill discharging 90,000 tons of effluent annually would discharge a BOD load of 180 tons if it complied with standard B, 90 tons if it complied with standard C and 45 tons if it complied with standard D. Given the basic charge of RM10 per ton, its effluent-related licence fee in each case would be RM1800, RM900 and RM450, respectively. These amounts were larger than the minimum licence fee, RM150. Depending on its marginal abatement cost curve, an average-sized mill could save up to RM1650, RM750 and RM300, respectively, by voluntarily decreasing its BOD concentration to 167 ppm, at which concentration it would owe the minimum fee (90,000 tons × 167 ppm × RM10).

During the second year of the regulations, the average mill reduced its daily discharge of BOD to about half the level it discharged during the first year. Although this still did not match the DOE's expectation (Maheswaran et al. 1980), the discrepancy between performance and expectation was narrowing. The ponding treatment system was prov-

---

3 The impact of the charges was probably not as weak as it appeared; many mills had simply not yet completed their treatment systems.

ing to be effective at an operational scale. In just two years, the regulations had reduced the population-equivalent of the pollution load from 15.9 million people to 2.6 million people (Fig. 8.1), despite a 50% increase in CPO production. The BOD load continued declining in succeeding years. The industry's efforts to improve treatment technologies were given a boost in 1980 when the government established the Palm Oil Research Institute of Malaysia (PORIM). A survey conducted by PORIM and the Rubber Research Institute of Malaysia (RRIM) during the period 1980–81 found that 90% of the 40 mills surveyed were discharging effluent with a BOD concentration at or below standard D (500 ppm) and that 40% were achieving a BOD concentration below 100 ppm (Ma et al. 1982). Most of the mills were using the ponding system, although seven were using tank digestion systems that they had developed to overcome land constraints.

Although the basic charge is one possible explanation for the mills beating the standard, it probably does not explain why so many achieved concentrations below 100 ppm. As mentioned above, the basic charge gave an average-sized mill an incentive to reduce its concentration down to, but not below, 167 ppm. An alternative explanation is that mills expected the DOE to impose even more stringent standards, as it had indicated it might when it announced the regulations, and that they had therefore constructed systems to meet those standards, not standards A–D. If so, they guessed correctly: the PORIM/RRIM survey and other evidence of ongoing improvements in treatment technology led the DOE to announce more stringent fifth- and sixth-generation standards, standards E and F, in the early 1980s (Table 8.1). In fact, under standard F the DOE included a provision allowing it to impose a special standard of 20 ppm on mills upstream from intakes for water supply systems.

By the mid-1980s the population-equivalent of the industry's BOD load dropped to the nearly insignificant level of less than 100,000 people (Fig. 8.1). In 1991, the DOE randomly monitored 112 of the 265 CPO mills in Malaysia and found that 75% were complying with standard F (DOE 1992). The industry's ability to reduce its BOD discharge to such low levels was facilitated not only by improvements in treatment technology but also by the development, by it and by PORIM, of various commercial by-products made from palm oil effluent (Khalid and Wan Mustafa 1992; Ong et al. 1987). As early as 1977, a Danish company perceived a market opportunity in the forthcoming regulations and began marketing a process to convert separator sludge into animal feed (Jorgensen 1977). By 1982, ten large pig and poultry farms were using palm oil meal in their feed mixes (Jorgensen 1982). Mills that discharged effluent onto land found that it had a fertilising effect (Ma et al. 1980; Maheswaran and Singam 1977), and this discovery enabled many plantations to eliminate their purchase of fertilisers (Ma et al. 1980; Yusof and Ma 1992). One company saved an estimated RM390,000 per year on its fertiliser purchases (Tam et al. 1982). In 1982, three mills with tank digesters were recovering methane, which comprises 60%–70% of the gases generated during anaerobic digestion, and using it to generate electricity for mill use.[4] The industry has discussed selling electricity generated from the 'bio-gas' to the National Electricity Board (Lim et al. 1984). One analysis found that the payback period for the investment required to build an integrated fertiliser/bio-

---

4 Personal communication with A.N. Ma, 27 November 1992; Ong et al. 1987.

gas recovery system was only 3.1 years (Quah et al. 1982). In 1984, four mills found uses for all their effluent and consequently had zero discharge (Maheswaran 1984).

Although the industry expanded after the regulations were implemented and was able to turn some of the effluent into commercially viable products, it did incur costs as a result of the regulations. Capital costs accounted for most of the costs associated with treatment systems. Excluding the cost of land, estimates of capital costs ranged from RM330,000 for a ponding system to RM600,000–RM950,000 for systems involving tanks (Khalid and Wan Mustafa 1992, and various sources cited therein). By 1984 mills other than those owned by FELDA had spent an estimated RM100 million to construct and operate treatment systems (Lim 1984). Relative to the industry's total production costs, however, treatment costs were low: only 0.2% in 1983, according to one estimate (Chooi 1984).

The regulations might not have impaired overall competitiveness of the processing sector of the palm oil industry, but they did affect the distribution of economic returns within the industry (Khalid 1991; Khalid and Braden 1993). Most of Malaysia's output of crude and refined palm oils was, and is, sold in an extremely competitive world market for fats and oils.[5] This prevented mills from passing much of the costs of effluent treatment onto consumers. Instead, they shifted the costs upstream to oil palm growers, who had no sales outlet aside from CPO mills, by reducing the price they paid for fresh fruit bunches. Ultimately, oil palm growers bore 66%–75% of the costs. Cumulative losses in producer surplus[6] during the period 1982–86, relative to the value of output during the same period, were only 1.2% for CPO mills and 4.7% for refined palm oil mills, but 44% for growers.

Cost-effectiveness of the effluent charges

Although the effluent charges probably had only a minor impact on the industry's aggregate BOD load after the first year, they still might have generated some cost savings. The basic charge gave mills flexibility to adjust their BOD concentrations between the floor created by the minimum effluent-related licence fee and the ceiling imposed by the mandatory standard, and the excess charge gave the DOE an alternative to shutting down mills that violated the standard only temporarily. We estimated the potential magnitude of these cost savings for the case of mills on the Sungai Johor. In 1991, this river had more mills (16) than any other in the country, and it received the largest aggregate BOD load. We restricted our analysis to the ten mills that discharged effluent into the river instead of onto land. All used the standard ponding system to treat their effluent.

Most of the mills were established several years after the regulations came into force. To control for the standard they faced, we analysed their abatement behaviour from 1984 to 1991, when only standard F applied. Half of the mills were established during this period. Treatment technologies were well established by this time, so mills' behaviour was unlikely to have been affected by uncertainties related to effluent treatment.

5 By 1985 Malaysia exported mainly refined palm oil.
6 Roughly speaking, producer surplus is analogous to short-term profits.

The Appendix provides details of the analysis. The general approach was: (1) to estimate a system of marginal cost curves (one curve for each mill) and (2) to use these curves to simulate abatement behaviour in response to both a uniform standard and an effluent charge that yielded the same aggregate reduction in the BOD load as the standard. From the simulation results we calculated aggregate abatement costs under the two pollution control options. The simulated effluent charge was a pure effluent charge: unlike the actual charges, it did not vary according to the BOD level and there was no minimum licence fee. Hence, the simulation results indicated the aggregate abatement cost savings that the ten mills *would have enjoyed* if they had faced a pure effluent charge instead of a uniform standard. The simulated savings therefore provide an upper bound on the magnitude of the cost savings that the actual charges provided.

The DOE provided data for the analysis from the mills' annual licence applications and quarterly returns. The quarterly returns report tons of effluent discharged each month and BOD concentrations during the first week of each month. We converted these to quarterly estimates by summing the effluent estimates and weighting the BOD estimates by the quantity of effluent. We calculated quarterly BOD abatement by multiplying tons of effluent by the difference between 25,000 ppm, the mean concentration of BOD in untreated effluent, and the actual concentration.

We treated the quantity of effluent as an exogenous variable. This is a reasonable assumption, since CPO production, and therefore effluent generation, is largely determined by the ripening of oil palm fruits, which is subject to climatic and biological factors beyond mills' control. Hence, we assumed that mills minimised abatement costs purely by selecting the BOD concentration of the fixed amount of effluent they discharged. In practice, mills do not have precise control over the BOD concentration, but they can influence it by the amount of residual oil and solids they allow in untreated effluent entering the treatment system, by the length of time they retain effluent in the various treatment ponds, and by the effort they put into maintaining the ponds (primarily to prevent build-up of suspended solids and sludge).

To estimate the marginal cost curves we also needed data on the marginal costs at actual abatement levels, as well as on other variables that might be expected to affect abatement costs. Regarding the latter, we obtained data on processing capacity utilisation from the licence applications and quarterly returns. One might expect abatement costs to rise with capacity utilisation, assuming that the capacity of a mill's treatment system is proportional to its processing capacity.

Unfortunately, no direct data on treatment costs was available. We assumed that mills were rational cost-minimisers that abated up to the point where the marginal abatement cost equalled the marginal effluent charge they actually paid. If they complied with the standard, we set the marginal cost equal to the basic charge; if they violated it, we set it equal to the excess charge. Out of the 172 quarterly observations in the dataset, mills complied with the standard in about 50% of the quarters. We deflated the charges by using the gross domestic product (GDP) deflator (1978 price levels). Although the charges were constant in nominal terms, because of inflation they were about 10% lower in 1991 than in 1984. At 1978 price levels, the values in 1991 were RM6.66 for the basic charge and RM66.62 for the excess charge.

	Standard (ppm)			
	5,000	1,000	500	100
Aggregate discharge *(tons of BOD)*	1,029	206	103	21
Minimum marginal cost *(RM)*	6.51	7.03	7.10	7.15
Maximum marginal cost *(RM)*	37.62	40.67	41.03	41.31
Aggregate abatement costs *(RM)**	67,117	87,065	89,666	91,763

Note: ppm = parts per million; BOD = biochemical oxygen demand.

* Sum of areas under marginal cost curves

Table 8.2 **Simulated abatement costs for uniform standards: crude palm oil mills on Sungai Johor, average quarter in 1991**

The estimation results confirmed that marginal abatement costs were positively and significantly related to both abatement level and capacity utilisation. On average, a 10% increase in abatement resulted in a 4.3% increase in the marginal cost. It was therefore meaningful to proceed with the simulation of cost savings.

We simulated the cost savings at 1991 values for effluent discharge and capacity utilisation.[7] The difference in aggregate abatement costs between effluent charges and uniform standards depends on the magnitude of differences between mills' marginal cost curves. In the simulation model, mills' marginal cost curves differed for three reasons: the amount of effluent they treated (which affected their abatement level), their capacity utilisation, and the intercepts of their marginal cost curves, which varied across mills. All three factors varied substantially across mills. Average quarterly tons of effluent discharged and capacity utilisation varied by factors of nearly ten. Differences in the intercepts implied that, at given levels of effluent discharge and capacity utilisation, the marginal abatement cost for the lowest-cost abater was 49% below the average for the mills, whereas that for the highest-cost abater was 95% above. These differences suggest that cost savings from effluent charges could potentially be large.

Table 8.2 shows results of simulations of the uniform standard. We simulated four standards, ranging from 5,000 ppm (standard A) to 100 ppm (standard F). The aggregate BOD load dropped from 1,029 tons under the 5,000 ppm standard to only 21 tons under the 100 ppm standard. Aggregate abatement costs rose sharply when the standard increased from 5,000 ppm to 1,000 ppm, but more modestly as the standard increased even further. This reflected not only diminishing *differences* between the four standards analysed (4000 ppm, 500 ppm, and 400 ppm), but also the inelasticity of the marginal cost curve: marginal costs increased less and less rapidly as abatement rose. For each standard, the difference between the marginal cost for the mill with the lowest marginal

---

7 We used average values for 1989 for one mill, because data was not available for 1991.

	Effluent charge (RM)			
	29.03	37.57	39.00	40.15
Aggregate discharge *(tons of BOD)*	1,029	206	103	21
Mills exceeding 100 ppm	4	2	2	1
Minimum concentration *(ppm)*	0	0	0	0
Maximum concentration *(ppm)*	14,098	5,064	3,243	1,711
Aggregate abatement costs *(RM)**	57,632	84,250	88,184	91,451
Cost savings *(RM)*: absolute terms**	9,485	2,815	1,482	312
Cost savings: relative terms†	1.165	1.033	1.017	1.003
Aggregate effluent fees *(RM)*	29,848	7,735	4,022	824
Total cost to mills *(RM)*‡	87,480	91,985	92,206	92,275

Note: ppm = parts per million; BOD = biochemical oxygen demand.

* Sum of areas under marginal cost curves

** Aggregate abatement costs for the uniform standard minus the aggregate abatement costs for the effluent charge

† Aggregate abatement costs for the uniform standard divided by the aggregate abatement costs for the effluent charge

‡ Sum of aggregate abatement costs and aggregate effluent fees

Table 8.3 **Simulated abatement costs for effluent charges: crude palm oil mills on Sungai Johor, average quarter in 1991**

cost curve and the marginal cost for the mill with the highest curve was more than RM30. The ratio of highest to lowest was in the order of five to six, which is comparable to what O'Neil (1980, cited in Tietenberg 1988) reported in a simulation study of BOD abatement on the Fox River in the United States. These differences suggest that effluent charges might generate sizeable cost savings.

Table 8.3 shows the effluent charges corresponding to each standard and the abatement costs associated with them.[8] All the charges were much greater than the 1991 basic charge and much lower than the 1991 excess charge. This implies that, in 1991, the basic charge alone would not have achieved the aggregate abatement level of even standard A, and the excess charge alone would have achieved an abatement level even higher than standard F. The DOE probably did not need to make the standards mandatory after the first year: the excess charge would have provided sufficient incentive for the industry to meet the standards, *on average*.

8   The effluent charge figures also indicate the trading price for tradable permits if that system were employed instead of effluent charges.

In the presence of only effluent charges, however, abatement levels would have varied greatly across mills. Under the lowest charge, RM29, only four mills discharged effluent with a BOD concentration exceeding 100 ppm, but those that did exceeded it by a great amount. One had a BOD concentration of 14,098 ppm. At higher effluent charges, only two mills exceeded 100 ppm. These results suggest that a pure effluent charge system would induce most mills to reduce their BOD discharge to very low levels, but it might create local pollution problems in the vicinity of the few mills that continued to discharge effluent with high BOD concentrations.

As expected, aggregate abatement costs under effluent charges were lower in all cases compared with the uniform standards. The savings were greater in both absolute and relative terms at lower abatement levels (less stringent standards), but they were modest even there: about RM1,000 per mill at the RM29 effluent charge. Differences were still detectable between the 100 ppm standard and the RM40 effluent charge, but they were tiny: about RM30 per mill. The effluent charges thus generated very low cost savings when abatement levels were high.

From the standpoint of total costs—abatement costs plus effluent charges—the industry would actually be better off under a uniform standard. Aggregate abatement costs under this approach were less than the sum of aggregate abatement costs and aggregate effluent fees under an effluent charge. Effluent charge payments would more than offset the industry's savings on abatement costs.

Conclusions

Malaysia's crude palm oil regulations were not equivalent to a pure effluent charge system. They most closely resembled a pure system in the first year, when the BOD standard was not mandatory. They were more complicated even then, however, in that the level of the charge depended on whether or not the BOD concentration of a mill's effluent exceeded the standard. Mills' abatement decisions were also affected by the provision for payment of a minimum fee. In being linked to a standard and being motivated by objectives other than cost-effective pollution control, the palm oil effluent charges were similar to effluent charge systems implemented in other countries.

Malaysia's experience with the regulations offers several lessons for pollution control efforts in the country. The most general, and most important, is that pollution abatement and industrial expansion can occur simultaneously. The government need not be reluctant on the grounds of competitiveness to address the pollution problems of leading industries, as indeed it was not in the case of the palm oil industry. Output of the industry expanded as rapidly after 1978 as before.

A facilitating factor was that CPO mills' concerns about excessive abatement costs proved unfounded. One reason was the mills' ability to shift the costs of the regulations onto oil palm growers. Another, which reduced instead of simply shifted costs, was technology. R&D efforts softened the potential trade-off between pollution abatement and industrial expansion. The ponding system was relatively inexpensive, especially in terms of operating costs, and mills were able to develop numerous by-products from the effluent. The credit for R&D provided an incentive to investigate these technologies, and the low value of the basic charge made it easier for mills to finance their implementation from internal resources.

The research credit and the low value of the basic charge were critical in gaining industry support of the regulations. They reflected the industry's influence on the expert committee that proposed draft regulations to the DOE. Industry representation on this committee helped ensure that the structure and timing of the regulations took into account the costs the regulations imposed on the industry and offered the industry flexibility in responding to them (and therefore opportunities for cost savings). The risk in this arrangement, of course, was that industry could have 'captured' the DOE and weakened the regulations to the point of ineffectiveness. The government's commitment to solving the palm oil effluent problem prevented this from happening.

The government's commitment was illustrated most strikingly by its well-publicised actions to suspend the operating licences of mills that violated the regulations. Such actions, coupled with the minimum fee requirement and, especially, the imposition of mandatory standards after the first year, strongly suggest that the risk to a mill of having its operating licence suspended, not the financial incentive created by the effluent charges, was primarily responsible for the reduction in BOD discharge. This was probably true even in the first year, as the mandatory standards were pre-announced, and mills knew they had just a one-year grace period.

Mandatory standards reduced the ability of the charges to promote cost-effective abatement. Simulation results for the Sungai Johor suggest that the increased cost to the industry was small, because the difference in marginal costs across mills was low in absolute terms. The simulation results understated the cost impacts, however, because they measured only short-run marginal costs (operating costs). The estimated marginal cost curves were for mills' existing treatment systems, which were constructed to meet standard F. If in reality mills had faced an effluent charge rather than a mandatory standard, mills that chose a concentration higher than the standard might have been able to save on capital costs as well as operating costs by altering the design of their treatment systems.

The simulation results also illustrate a potential environmental risk associated with effluent charges: charges can result in local pollution problems when pollutants do not disperse readily. The simulation results indicated that some mills would discharge effluent with a BOD concentration well above historical standards if faced with only an effluent charge. Although it is possible to design a cost-effective system of 'ambient charges' whose levels vary across pollution sources and are linked to local ambient conditions (Tietenberg 1988) such a system is information-intensive and difficult to administer. In essence, it is not much different from a system of mill-specific effluent standards,[9] which offers an environmental agency greater certainty about pollution abatement.

---

9 Standard F can be viewed as a crude system of mill-specific standards, given the special standard for mills upstream from a water intake.

In sum, Malaysia's experience with the crude palm oil regulations is by no means a straightforward testimony to the advantages of economic instruments over command-and-control regulations. The regulations had many aspects of the latter. These aspects undermined the advantages of the charges related to cost-effectiveness, but they minimised the risk of continued pollution problems downstream from high-cost mills. The economics of this trade-off are not clear, as neither the effluent charges nor the standards were chosen by comparing marginal abatement costs and marginal abatement benefits. The regulations undeniably led to a massive reduction in the BOD load in the country's rivers, but it is possible that the reduction was either too big or too little or that it occurred too quickly or too slowly. Resolving these issues would require incorporating estimates of environmental benefits into the analysis.

Appendix: estimation and solution of the simulation model for CPO mills on the Sungai Johor

As discussed above, we assumed that marginal abatement costs, $f(A)$, for mills on the Sungai Johor equalled the marginal effluent charges, F^m, they paid:

$$F^m = f(A). \qquad [A8.1]$$

where the marginal cost curve f is a function of the BOD load abated, A. F^m equals the basic charge for concentrations less than or equal to the standard and the excess charge for concentrations above.

Estimation of the marginal cost curve was complicated by characteristics of the data on effluent charges. When a mill's BOD concentration exactly equals the standard, its marginal cost does not necessarily equal the basic charge. The mill would pay a marginal effluent charge of RM10 per tonne, but its marginal cost could be as high as RM100 per tonne. The marginal cost is bounded by the basic and excess charges, but its precise value cannot be determined.

We dealt with this problem by defining the dataset in two ways. The first, full, panel included all the data. As it turned out, in no case did the quarterly BOD concentration exactly equal the standard. The second, partial, panel excluded data for which the standard fell within a range defined as the quarterly BOD concentration plus or minus two standard deviations. We calculated the standard deviation by using the three monthly estimates for each quarter, and we calculated it separately for each mill and each quarter.[10] The full panel included a total of 172 quarterly observations, and the partial

10   We also considered excluding data on mills whose quarterly BOD discharge was less than 3.75 tonnes. This is the quarterly equivalent of the annual 15 tonnes associated with the minimum effluent-related licence fee. Such mills' marginal costs could be below the basic charge. Excluding this data, however, would have reduced the number of observations in the partial panel to only 28 and would have entirely eliminated half the mills. We judged that the potential increase in accuracy in measuring the marginal costs was not worth this massive loss of data.

one included 86 observations. Both panels were unbalanced: there were differing numbers of observations for the mills, because some mills were older than others. The partial panel included data on only nine of the mills.

A second complication relates to the possible lack of comparability between the marginal cost curves for BOD concentrations above and below the standard. An unusually large harvest is perhaps the most common reason for temporary violations of the standard. The size of the harvest affects the output of crude palm oil, which in turn affects the amount of effluent to be treated. When the amount is large, a mill might be forced to move the effluent through the treatment system more rapidly than usual, and this results in a higher BOD level. This implies that the mill's marginal cost curve has shifted temporarily upward. We added capacity utilisation, the ratio of CPO output to processing capacity, to equation [A8.1] to reflect this shift.

A third complication relates to the amount of variation in the effluent charges. If the charges vary little over time then estimation of equation [A8.1] will not yield meaningful results. The basic and excess charges were set at RM10 and RM100 per tonne in the regulations, and they have remained the same since. Despite this, there was substantial variation in the data on the charges, for two reasons. First, the basic and excess charges differ by a factor of ten. Second, although each charge has remained constant in nominal terms, its real value has decreased as a result of inflation. Malaysia has an excellent record in controlling inflation, but nevertheless the real value of the charges, determined by dividing them by the GDP deflator (base year of 1978), fell by more than 30% between 1978 and 1991.

We converted equation [A8.1] into a form that could be econometrically estimated through the following steps. We assumed that the quarterly total variable cost (C^v) of operating a treatment system was a Cobb–Douglas function of the BOD load abated (A) and capacity utilisation (U):

$$C^v_{it} = \exp(\alpha_i)(A_{it})^{\beta+1}(U_{it})^{\gamma}/(\beta+1). \qquad [A8.2]$$

where exp(.) represents the exponential function (the mathematical constant e [approximately equal to 2.7] raised to the power indicated in parentheses); α, β and γ are parameters to be estimated; and i and t are subscripts distinguishing data by mill, i, and time-period, t. Note that α was allowed to vary across mills. We will say more about this below. Equation [A8.2] indicates that total variable costs are zero when there is no abatement, which makes sense. A is defined here as:

$$A_{it} = E_{it}(0.025 - c_{it}). \qquad [A8.3]$$

where E is the amount of effluent discharged and c is the BOD concentration of the effluent. The number 0.025 represents 25,000 ppm (25,000/1,000,000), the BOD concentration of untreated effluent.

A mill selects the value of c that minimises the sum of total variable costs related to effluent treatment, given by equation [A8.2], and effluent discharge, which equals the total effluent charge (F^t, for total fees):

$$F^t_{it} = F^m_t E_{it} c_{it}. \qquad [A8.4]$$

where F^m is the effluent charge per ton of BOD load. If we sum the expressions on the right-hand sides of equations [A8.2] and [A8.4], substitute equation [A8.3] into the expression to eliminate A, take the derivative with respect to c, and set the result to zero, we obtain:

$$F^m{}_t = \exp(\alpha_i) A_{it}{}^\beta U_{it}{}^\gamma. \qquad [A8.5]$$

The mill abates up to the point where the effluent charge (the left-hand side) equals the marginal treatment cost (the right-hand side). This demonstrates that the general form of equation [A8.1], which ignores the effect of capacity utilisation, is valid.

Taking the natural logarithm of each side of equation [A8.5] converts it into a linear expression that is more readily estimated:

$$\ln(F^m{}_t) = \alpha_i + \beta \ln(A_{it}) + \gamma \ln(U_{it}) + \varepsilon_{it}. \qquad [A8.6]$$

This equation includes an error term, ε, which one normally assumes is independently and identically distributed (i.i.d.) across mills and time-periods.

The proper procedure for estimating equation [A8.6] depends on the characteristics of the intercept α_i (Judge et al. 1985). If α_i reflects purely fixed differences between mills—for example, permanent differences related to the scale or design of their treatment systems—then equation [A8.6] can be estimated by ordinary least squares (OLS), with a different intercept included for each mill. The estimated intercepts equal the α_is. This is the *fixed-effects*, or *dummy-variable*, model.

On the other hand, α_i might reflect random differences between mills. The values of α_i might be drawn from a common distribution with a mean of α^*, with differences among them being due to a random error term, μ_i:

$$\alpha_i = \alpha^* + \mu_i. \qquad [A8.7]$$

The mean of μ_i is zero. This implies that equation [A8.6] can be rewritten as:

$$\ln(F^m{}_t) = \alpha^* + \beta \ln(A_{it}) + \gamma \ln(U_{it}) + (\varepsilon_{it} + \mu_i). \qquad [A8.8]$$

The error term now has two components, ε_{it} and μ_i. Hence, this model is called the *error-components*, or *random-effects*, model.

Using OLS to estimate equation [A8.8] yields inefficient estimates of α^*, β and γ: the standard errors of the OLS coefficient estimators are larger than necessary because they ignore information on the non-i.i.d. nature of the error term. Moreover, estimates of the standard errors of the coefficients would be biased, because OLS estimators of the standard errors of coefficients assume that error terms are i.i.d. Efficient estimation of equation [A8.8] requires generalised least squares.

Following estimation, the random effect μ_i for each mill can be estimated by the formula:

$$\mu_i = \{\Sigma[\ln(F^m{}_t) - \alpha^* - \beta \ln(A_{it}) - \gamma \ln(U_{it})]\} \sigma_\mu{}^2 / (T_i \sigma_\mu{}^2 + \sigma_\varepsilon{}^2). \qquad [A8.9]$$

The term in curly brackets is the sum of the residuals for the mill in question (here, α^*, β and γ are estimates of the coefficients), $\sigma_\mu{}^2$ and $\sigma_\varepsilon{}^2$ are the estimated variances of μ_i and ε_{it}, respectively, and T_i is the number of observations on the mill.

A chi-squared (χ^2) test can assist in choosing between the models (Hausman 1978). The null hypothesis is that the random effects are uncorrelated with the explanatory variables on the right-hand side of equation [A8.8]. If the test fails to reject this hypothesis, then the random-effects model is preferred: it provides more efficient coefficient estimates than does the fixed-effects model. If one rejects the hypothesis, however, then the fixed-effects model is preferred: its coefficient estimates are inefficient but unbiased, whereas those of the random-effects model are biased.

Table A8.1 presents the results of estimating the fixed-effects model (equation [A8.6]) and the random-effects model (equation [A8.8]) for both panels. The table also includes results for a third model, identical to equation [A8.6] except that the α_is are constrained to have the same value. These results are for OLS with a common intercept (ordinary OLS).

Results for each model differed relatively little between the samples. This is a satisfying result: it suggests that effluent charges did indeed equal marginal costs for observations in the larger sample. Since the larger sample includes not just more data but data on all the mills, we regarded estimates based on it as the preferred estimates. Ordinary OLS, fixed-effects, and random-effects estimates of β and γ were fairly similar, indicating that the estimates of the coefficients were fairly robust. The estimates of β and γ were significantly different from zero at the 1% level in the OLS and random-effects models; only γ was significant, at the 5% level, in the fixed-effects model.

We used an F test to test the null hypothesis that the mill-specific intercepts in the fixed-effects model were identical. Failure to reject this hypothesis would indicate that the fixed-effects model is not statistically different from the ordinary OLS model. The value of the test statistic was 4.94 (10 and 160 degrees of freedom), which rejected the null

Table A8.1 **Estimates of parameters in the marginal cost curve for abatement by mills on the Sungai Johor (standard errors are in parentheses)**

Parameter	Full panel OLS[a]	Full panel FE[b]	Full panel RE[c]	Partial panel OLS[a]	Partial panel FE[b]	Partial panel RE[c]
β	0.488* (0.092)	0.316 (0.211)	0.427* (0.150)	0.571* (0.116)	0.295 (0.196)	0.413† (0.159)
γ	0.694* (0.128)	0.412† (0.168)	0.487* (0.149)	0.844* (0.131)	0.276 (0.178)	0.388† (0.157)

Notes
* Significant at the 5% level
† Significant at the 1% level
a OLS: ordinary least squares estimates, with the intercept (not shown) constrained to be the same across mills
b FE: fixed-effects estimates
c RE: random-effects estimates

hypothesis at the 1% level (critical value of approximately 2.47). Hence, the values of the α_is did differ between the mills.

We used the χ^2 test to determine whether it was more appropriate to treat the differences in the α_is as resulting from fixed or random effects. The value of the χ^2 statistic (2 degrees of freedom) was 1.86, which meant that we failed to reject the null hypothesis of no correlation between the random effects and the explanatory variables at even a 10% level (critical value of 4.61). Hence, the random-effects estimates were preferred.

The random effects estimates of β and γ for the full panel were both positive, as expected: marginal costs rose with abatement level and capacity utilisation. The value of β was less than one, which, given the Cobb–Douglas form, indicates that the marginal cost curve was inelastic: a 10% increase in abatement raised the marginal cost by 4.3%. The estimate of γ was also less than one.

To simulate a uniform standard, we used equation [A8.3] to predict the amount of abatement for a given value of c, and then we substituted the result into equation [A8.2] to predict total variable costs of abatement. We knew the parameter values in equation [A8.2] from estimating equation [A8.8]. We set the values of E and U equal to their average quarterly values in 1991 (we used 1989 values for mill 7 because data was not available for that mill for 1991). The exponential value of each random effect—$\exp(\mu_i)$, which is part of $\exp(\alpha_i)$ in equation [A8.2]—was calculated by using equation [A8.9].

To simulate an effluent charge, we first solved equation [A8.5] for c and constrained it to predict only non-negative values:

$$c_{it} = \text{pos}\{0.025 - [F^m_t/\exp(\alpha_i)/U_{it}^\gamma]^{1/\beta}/E_{it}\}. \qquad [A8.10]$$

where pos is a mathematical operator which sets c equal to zero if the expression in curly brackets is negative. The calculated value of c was then substituted into equation [A8.2] and simulation proceeded as in the case of the uniform standard.

9

TOWARD MORE SUSTAINABLE DEVELOPMENT
The environment and industrial policy in Taiwan

Michael T. Rock

The stark contrast between economic development in the high-performing economies (HPEs)[1] of East Asia (World Bank 1993) and rapidly deteriorating urban environments[2] raises several interesting questions. Why have governments in the HPEs been so slow to respond to accumulating industrial pollution? Once governments responded to industrial pollution, how did they do so? Did these economies, including Japan, simply follow practices established elsewhere or did they draw on their own institutions of growth to devise unique strategies for controlling and reducing industrial pollution? Finally, how effective have environmental policies been in reversing the decline in urban environmental quality?

Because so little is known about governmental responses to industrial pollution in the HPEs,[3] these questions are tackled by examining the government's response to industrial

1   The high-performing economies of East Asia are South Korea, Japan, Hong Kong, Singapore, Taiwan, Malaysia, Indonesia and Thailand.
2   Growth in the HPEs proved to be very intensive in energy and electricity use (Brandon and Ramankutty 1993; World Bank 1992). Because of heavy reliance on coal (Brandon and Ramankutty 1993), this growth is highly polluting. Three of the HPEs' largest cities are among those cities with the highest concentrations of particulate matter (Brandon and Ramankutty 1993). Emissions of carbon dioxide (CO_2) per capita are also among the highest in the developing world (WRI 1994). A rapid rate of growth of manufacturing has been accompanied by an equally rapid rate of growth in the toxic intensity of production. Between 1974 and 1987 the toxic intensity of production increased 5.4 times in Indonesia, 3.2 times in Malaysia, 3 times in Korea and 2.5 times in Thailand (Brandon and Ramankutty 1993). This has contributed to serious water pollution and hazardous waste problems (Steering Committee 1989).
3   This does not mean that nothing is known. O'Connor (1994) reviews the establishment of environmental agencies, promulgation of environmental laws and regulations, the setting of emissions and effluent standards and the monitoring of performance.

pollution in one of the HPEs—Taiwan. What follows is decidedly exploratory. It is meant to whet appetites for more and better research on this important topic. The argument proceeds in four steps. In the next section, I describe Taiwan's well-known institutions of growth. I then examine how those institutions have been adapted to deal with the pressing problems of industrial pollution. In the penultimate section, I consider the impact of Taiwan's unique approach to industrial pollution control, abatement and prevention on industrial pollution. In the final section, I offer a few concluding remarks.

◢ Economic (and industrial) policy and the institutions of growth with equity in Taiwan

Shortly after moving to Taiwan, the Kuomintang (KMT) used its pre-eminent political position there to restructure both the Taiwanese state and state–society relationships. What élite state actors did and how they did it had a profound effect on the institutions of economic and industrial policy. To begin with, the KMT reformed and reconstituted itself as a Leninist-like party. This was followed by tight government–party control over youth groups and labour, destruction of the indigenous landlord class and a restructuring of relationships between the state and agriculture, and government control over the commanding heights of the economy, including banks (Haggard 1990; Wade 1990). The net effect of these actions was the creation of a strong autonomous state and weak organisations in civil society (Wade 1990).

Within this strong autonomous state, economic policy was limited to a small number of agencies and individuals.[4] Ideologically, this small policy-making group was wedded to the economic philosophy of Dr Sun Yat-Sen. Sun's three principles and KMT ideology were an odd mixture of private enterprise, central planning and socialism (Haggard 1990). These principles are enshrined in the constitution of Taiwan and have been repeated in the country's five-year plans (Wade 1990). As one of those plans states,

> For private enterprise . . . [the state will protect] . . . reasonable personal income . . . and freedom of economic enterprise . . . However, . . . manipulation of society's economic lifeline in the hands of a few and over concentration of wealth will not be allowed. Consequently, the government must take part in all economic activities and such participation cannot be opposed on the ground of any free economic theory (Wade 1990: 261).

Thus the KMT, like its counterparts in Japan and Korea, recognises a pre-eminent role for the state in guiding market development to meet avowedly political ends.

4 At the top of the system is the president and an informal inner group of the cabinet known as the Economic and Financial Special Group (EFSG). The EFSG consists of the minister of economic affairs, the governor of the central bank, the minister of finance, the director-general of the budget and several ministers without portfolio. This group is advised by the Council for Economic Planning and Development (CEPD), the Industrial Development Bureau (IDB) of the Ministry of Economic Affairs (MOEA) and the Council for Agricultural Planning and Development (CAPD) (Wade 1990).

Successful state guidance of market development hinged on maintenance of macroeconomic stability, significant public investment in human and physical infrastructure and the mobilisation of an investable surplus which could be channelled into efficient and productive industry. The Taiwanese state did each of these things exceedingly well. An investable surplus was obtained from agriculture by government reorganisation of agricultural markets. Barter of rice for fertiliser at prices unfavourable to farmers, compulsory sales of rice at low prices and land taxes proved to be an effective way of extracting a surplus out of agriculture (Johnston and Kilby 1975). Substantial investments in rural infrastructure, which contributed to increases in agricultural productivity, made it possible to extract this surplus without impoverishing the peasantry (Amsden 1979).

Although the Council for Economic Planning and Development (CEPD) planned the country's industrial development, the Industrial Development Bureau (IDB) of the Ministry of Economic Affairs (MOEA) had extensive influence over where this surplus was invested. It used its control over trade policy and a large array of fiscal incentives that accompany the 1960 Statute for Encouragement of Investment to affect which industries private firms invested in. It occasionally relied on subsidised credit from state-owned commercial banks to particular industries and firms. It relied heavily on state-owned enterprises in the commanding heights of the economy—petrochemicals, steel and basic metals and shipbuilding (Wade 1990). Also, it played a leading role in public-sector research and development (R&D) policy, including strong ties to the country's premier science and technology institute, the Industrial Technology Research Institute (ITRI).

Progression up a product ladder pioneered by others provided the IDB with the blueprint for industrial development. Studies of trends in income elasticity of demand, technical change and of the current composition of imports helped to identify industries that should be developed next (Wade 1990). Each new stage of industrial development resulted in amendments to the Statute for Encouragement of Investment (between 1960 and 1982 the statute was amended 11 times [Wade 1990]). It also led to changes in the list of industries eligible for administratively allocated and subsidised credit. When government feared that the private sector might be too slow to respond to these fiscal and financial incentives, it turned to state-owned enterprises as in plastics and steel or to the quasi-public ITRI for development of a high-technology computer-chip industry (Wade 1990). As is now known, this approach to economic and industrial development has been profoundly successful and it more or less characterises the rest of the East Asian HPEs (Johnson 1987).

Industrial policy and environmental protection

Although there is general agreement that this particular political economy did make an 'economic growth with equity' difference, less is known about its influence on the environment. Some have suggested that it has created an environmental disaster (Bello and Rosenfeld 1990). Others have speculated that the relative strength and political insulation of the so-called economic ministries contributed to a delayed response to environmental degradation (Chun-Chieh 1994). Unfortunately, there is virtually no rigorous research on either of these important topics. Even more surprising, though, is that little, perhaps nothing, is known about whether this particular political economy has been positively engaged in environmental protection.

There are several reasons to suspect that this political economy might make a significant positive difference in the environment. First, one of the HPEs—Japan—has already done so. At least some of the comparative environmental data suggests that it has outperformed its counterparts in the Organisation for Economic Co-operation and Development (OECD).[5] But this is not the only reason to suspect that the other HPEs of East Asia might also do so. Systematic and quick movement up the product ladder in industry shows that governments in the HPEs have been particularly adept at capitalising on the real advantages of being late industrialisers. Perhaps they could do the same in terms of the environment.

Late industrialisation and openness to foreign capital and technology create several opportunities for the HPEs to leapfrog less costly and more effective urban industrial environmental outcomes.[6] Because each of the HPEs is dependent on the OECD countries for capital and technology there is some possibility that at least some of that capital and technology will be less polluting and/or use less raw materials than the current capital stock in the HPEs. Because industrial growth rates in the HPEs are significantly higher than the developing-country average, this creates more opportunities for accessing this newer and less-polluting capital and technology. Taken together, higher industrial growth rates and dependence on the OECD for capital and technology alone could make a substantial environmental difference.

Two examples make this clear. At projected growth rates, more than three-quarters of the industrial capital stock in Indonesia will be new by 2010 (Brandon and Ramankutty 1993). If this new capital stock could be made cleaner or less materials-intensive than the existing stock, it could put Indonesia on a less pollution-intensive industrial growth path.[7] Since much of this stock will come from OECD countries, some of this should happen. There is evidence that this has already happened elsewhere in at least one industry—wood pulp-making—where diffusion of a cleaner industrial process has been shown to be correlated with a country's openness to foreign trade (Wheeler and Martin 1992).

But these are not the only forces that might speed emergence of a less pollution-intensive industrial growth path in the HPEs. Because the HPEs tend to be more open to foreign direct investment (FDI), there may be greater opportunities for environmental 'demonstration effects' than exist in the more closed economies.[8] Finally, because envi-

---

5 Total emissions of sulphur oxides, particulates and nitrogen oxides in Japan as a share of gross domestic product are less than a quarter of the OECD averages (World Bank 1992).
6 What follows is drawn from O'Connor 1994.
7 Whether it will ultimately do so depends on shifts in the structural composition of industry away from pollution-intensive activities. Work by the World Bank (1994a) suggests that this may well happen.
8 This can happen when multinationals extend their home-country practices in developing countries where they do business. There is growing evidence that multinationals carry their home-country environmental practices with them when they establish production facilities in developing countries (Brown et al. 1993). Some multinationals have even extended these practices to their developing-country supply firms (Mapes 1994).

ronmental concerns are making their way into international trade agreements and into international industrial standards, including voluntary environmental management standards, the greater openness of the HPEs means that those responsible for industrial and trade policy are likely to be keenly aware of these trends.[9] Awareness is at least the first step toward integrating environmental considerations into industrial policy.

Is there any evidence of this in Taiwan? If so, what are or have been the implications for urban environmental quality? Taiwan, like the rest of the HPEs, followed a grow-first clean-up-the-environment-later export-led industrial development strategy (Steering Committee 1989). The country's first environmental laws date from the mid-1970s. Although those laws covered specific media (air, water and solid waste) they had little effect. They were not part of a comprehensive environmental law; no clear emission or effluent standards accompanied them; and jurisdictional authority for standard setting, monitoring and enforcement were unclear (Tsong-Juh 1994). By 1982 some of the jurisdictional problems were eased by creation of an Environmental Protection Bureau in the Department of Public Health. Belatedly, domestic public pressure and international developments led the government to be more forceful. Between 1980 and 1992 the media-specific acts of 1974-75 were amended and strengthened. The Waste Disposal Act was amended in 1980, 1985 and again in 1988. The Water Pollution Control Act was amended in 1983 and 1991.[10] The Air Pollution Control Act was amended in 1982 and 1992.[11] In 1987, the government created a cabinet-level Taiwan Environmental Protection Agency (TEPA) and modelled it on the US Environmental Protection Agency (EPA).[12] By 1993 Article 18 of the Constitution stated, 'Environmental and ecological protection should be given the same priority as economic and technological development' (OSTA 1995).

Despite the continuing lack of a basic environmental law, the TEPA was given responsibility for standard setting, environmental monitoring and enforcement. In a few short years, the TEPA evolved into a respectable environmental agency. Ambient air and water quality standards generally follow US standards.[13] The TEPA developed a rigorous

9   New voluntary environmental management standards (the ISO 14000 series) are being developed by the International Organization for Standardization in Geneva, Switzerland (House 1995). This is just one of a host of voluntary environmental standards being developed that might impact on either competitive advantage or a developing-country firm's ability to export to developed-country markets. There is a large and growing literature in this area (e.g. Feldman and Shiavo 1995; Freeman, et al. 1995; Porter and van der Linde 1995a).
10 Amendments tightened biochemical and chemical oxygen demand (BOD and COD) emission standards, granted the TEPA authority to levy a water pollution fee on water users and mandated that the government's water quality-monitoring results be made part of the public record (O'Connor 1994).
11 Amendments to the Air Pollution Control Act tightened emissions standards for sulphur oxides (SO_x) and carbon monoxide (CO) and granted local governments authority to set standards that were stricter than the national standards (O'Connor 1994).
12 The government also engaged the US Environmental Protection Agency to provide technical assistance to the TEPA (author's interviews, United States and Taiwan, November 1995).
13 For air, this is true for sulphur dioxide (SO_2), carbon monoxide (CO), nitrogen dioxide (NO_2), lead and photochemical oxidants. Taiwan's ambient standard for TSP is lower than that for the United States. For water, Taiwan's standards for chemical concentrations for cadmium, cyanide, chromium, arsenic and mercury are equal to or more stringent than US standards (O'Connor 1994).

emissions/effluent standard-setting process and an equally rigorous monitoring and enforcement programme.[14]

Standard setting depends heavily on expert committees. Until recently, those committees have not included representatives from industry. The process is described by officials at the IDB as closed and done without consultation. As a result, firms often complain to the IDB that the standards are too tough and, in at least one instance, an entrepreneur went directly to the president to complain about a particular standard.[15] In response to this, the IDB has begun to play a mediating role between the TEPA and industry. It now regularly interacts with the TEPA and it is considering developing its own technical capability to assess the TEPA's standards. Senior government officials and prominent academics advising the TEPA openly applaud its tough emissions and effluent standards and see them as needed and effective 'sticks' to force compliance.[16]

Monitoring and enforcement appear to be equally tough.[17] By 1991 the TEPA had 184 stations for monitoring ambient air quality for the pollutants PM10, SO_2, NO_2, CO and O_3. Results from these stations are aggregated into individual indices which are then aggregated into an overall pollution standards index (PSI). The TEPA now routinely reports the percentage of days in a year or month when the PSI is below 100 (considered healthy), between 101 and 199 (considered unhealthy) and above 200 (considered very unhealthy). It also closely monitors the reliability of monitoring stations.

In 1989, two years after the TEPA was formed, the TEPA committed itself to meeting the country's standards for ambient air quality by 2002. To facilitate movement in this direction, it screened over 9,000 factories, identified 24,343 different sources of air pollution in those factories, 10,484 pieces of pollution control equipment and 74,343 smokestacks (TEPA 1993). Of the 9,000 factories screened, 6,959 were identified as major polluters (O'Connor 1994). The TEPA required each of these factories to draw up and submit environmental improvement plans. Following this, the TEPA carried out random inspections of 1,543 smokestacks and ascertained that 23% failed to meet emissions standards. This was followed by the Flying Eagle Project and the Rambo Project. Flying Eagle used police helicopters to respond to citizen complaints about factory emissions whereas the Rambo Project was a 'get tough' project that randomly rechecked state-owned and private factories that had failed earlier inspections.

14 The setting of emissions standards begins with an expert committee composed of academics, government officials and knowledgeable others who advise on standards. The process permits informal consultations with industry, but this is not required (O'Connor 1994).
15 The process also seems to be changing. Representatives of the National Federation of Industries (NFI), the peak industry association in Taiwan, state that members from their Pollution Control and Industrial Safety Committee go over proposed TEPA regulations line by line. This is followed by attendance at TEPA public hearings for all new regulations. In one recent instance, the NFI got the TEPA to water down an emission standard for biochemical oxygen demand because firms in the porcelain industry said they could not meet the tougher standard (author's interviews, Taiwan, November 1995).
16 In the words of one academic interviewed, tough standards are necessary to get the large number of small and medium-sized enterprises to understand that pollution control is now part of the cost of doing business (author's interview, Taiwan, November 1995).
17 Unless otherwise noted, comments in this paragraph and the next are taken from TEPA 1993.

These actions have resulted in fines, suspensions of operating permits (government-mandated factory closings) and voluntary factory closings. The TEPA carried out intensive inspections of the factories that failed to respond to its request for environmental improvement plans. These inspections led to fines for some and to suspension of operating permits for others. In some instances, inspected plants, such as the Kaoshiung factory of the Taiwan VCM Corporation, which failed to improve its environmental performance within a specified time-period, were fined on a daily basis. In the case of the Taiwan VCM Corporation this led the company to voluntarily suspend operations until it was able to reduce its emissions (TEPA 1993). In other instances, repeating violators of emission standards, such as the Hualien plant of the Chung Hwa Pulp Corporation, voluntarily suspended operations until they could meet standards (TEPA 1993).

In still other instances, factories have been closed[18] and/or the government has banned certain heavily polluting activities completely or has restricted them from locating in heavily populated areas.[19] Alkali/chlorine factories have been prohibited from using a mercury electrolysis process and by 1993 all such processes were replaced by an ion electrolytic membrane process. New factories producing pesticides based on pentachlorophenol (PCP) have been prohibited and all pre-existing factories are being eliminated. Production of certain carcinogenic organic dyes has been banned. New factories working with paper pulp have been banned from central areas and they have been required to install equipment to recycle the black liquor.

The aggregate evidence on the use of negative sanctions reinforces these case examples.[20] The number of inspections has grown dramatically, from about 338,000 in 1988 to about 729,000 in 1994. Even though only about 6% of the inspections are for stationary-source water and air pollution, this amounted to approximately 40,000 site-specific inspections annually between 1990 and 1994. Over this same period, fines for violations of standards have increased from NT$2.66 billion to NT$7.34 billion, and average fines have increased from NT$7,869 in 1988 to NT$10,068 in 1994. This represents a 28% increase in the average fine.

The TEPA has gone one step further. It has developed an impressive database on the output (metric tons) of various pollutants by industry for 11 industry groups and three specific state-owned enterprises—Taipower, China Petroleum Company and China Steel. These estimates have been used to extrapolate output of pollutants under a no-enforcement, or natural growth, scenario. One measure of the effectiveness of enforcement involves comparison of this natural growth scenario to output of pollutants after enforcement. For example, the TEPA estimated the natural growth of total suspended particulates (TSP) in 1991 at 1,137,746 tonnes. Actual emissions after pollution control was estimated to be only 1,002,012 tonnes, or a reduction of 11.4% (TEPA 1993). The TEPA has developed a similar database for solid and hazardous industrial waste. The production of such waste is subdivided by industry for 20 different industries. It is further subdivided by type, into generalised waste, toxic waste, corrosive waste and infectious waste.

---

18   Between 1990 and 1994 an average of 37 factories were closed annually (OSTA 1995).
19   Examples in the following are from IDB n.d.b: 4-10.
20   What follows draws from data in OSTA 1995.

But environmental protection neither starts nor stops with the TEPA. The government of Taiwan has begun to integrate environmental considerations into industrial policy. This is being done in three principle ways. The government is following an import-substitution industrial development strategy for the creation of an indigenous environmental goods and services industry. It is heavily subsidising industry purchases of pollution control and abatement equipment. In addition, the government is financing research into pollution prevention and providing industry with subsidised technical assistance in waste reduction and minimisation.

All of this is part of the latest industrial development strategy. Appreciation of the exchange rate, rising wage rates, emerging labour shortages and increased demands for a cleaner environment contributed to an export of industry that some in the IDB feared was leading to a 'hollowing out of industry' (IDB 1995). To prevent this, the government promulgated a six-year national development plan to upgrade industry and replaced the 1960 Statute for Encouragement of Investment with the Statute for Upgrading Industry. This statute provides selective incentives to firms to purchase automated production equipment and technology; increase R&D expenditures; improve product quality, increase productivity, reduce energy use and promote waste reclamation; and to purchase pollution control and abatement equipment (IDB 1995).

The government also began to promote 24 key high-technology high-value-added items in ten emerging industries. Promoted industries included those involved in communications, semiconductors, precision machinery, aerospace and, most notably, environmental goods and services. These industries were selected because they cause little pollution, have strong market potential, are technologically demanding, are not energy-intensive and have high value-added (IDB 1995).

The government is relying on several promotional privileges to facilitate the growth of a domestic environmental goods and services industry. Firms in this nascent industry have, by law, been organised into industry-specific associations.[21] Government environmental contracts, such as for the building of public-sector waste incinerators or providing technical assistance to private-sector firms for waste minimisation or reduction, are reserved for firms in these highly industry-specific associations. Since the government has adopted an explicitly private-sector approach to environmental clean-up, these benefits are likely to be substantial.[22] Domestic environmental goods and services providers are also favoured by tax, commercial bank lending and land-use policies. Firms in all of the ten emerging industries, including the environmental goods and services industry, are eligible for either a 20% investment tax credit or a five-year tax exemption plus a double retaining of surplus earnings (IDB 1995). They are eligible for loans from commercial banks and the executive Yuan's development fund at preferential rates and they are given

---

21 Government law requires firms to organise into an industry association whenever there are more than five firms in an industry (author's interviews, Taiwan, November 1995.).

22 For example, the government is planning to build 22 large waste incinerators and to contract construction out to private-sector firms. Private-sector engineering firms in the Taiwan Environmental Engineering Association expect to get most, if not all, of this business (author's interviews with members of this industry association, Taiwan, November 1995).

priority consideration in the acquisition of industrial land.[23] Local environmental hardware providers benefit from a 20% tax credit that accrues to firms purchasing pollution control and abatement equipment (the tax credit for imported equipment is only 10%). Firms in this industry are also eligible for export assistance, but little is known about how this programme works. If it follows practice elsewhere in East Asia, access to assistance may be conditioned on export performance (Rhee *et al.* 1984). Since the government has established explicit export targets for the environmental goods and services industry through 2002, something like this may be happening.[24]

In addition to this broad array of support for development of a domestic environmental goods and services industry, the government offers a wide range of programmes to assist firms trying to reduce industrial pollution. Some of these programmes offer fiscal and financial incentives; others provide technical assistance, particularly for pollution prevention and for waste treatment. The purchase of pollution control and abatement equipment entitles purchasers to tax credits (of either 20% or 10%), and 5%–20% of the costs of expenditures on energy conservation and on recycling equipment or technologies can be credited against profits (IDB 1995). A joint IDB/TEPA Waste Reduction Task Force provides free technical assistance to firms on waste reduction and minimisation.[25] The IDB also runs an information service for exchange of industrial wastes; it sponsors demonstration projects; and it has a programme to congregate small and medium-sized enterprises (SMEs) in industrial parks. SMEs who move to these parks are provided with good infrastructure, including common waste-water treatment facilities. In return, the SMEs are required to elect an SME committee to enforce emission and effluent standards within the park.[26]

Finally, the IDB finances a growing research programme on clean technologies. It has created a clean technologies unit in Division 7 of the IDB and contracted its research on clean technologies out to the UCL laboratory of ITRI.[27] Most remarkably, ITRI's clean-technology researchers are going well beyond plant-by-plant pollution prevention. They are watching closely what others (such as the 3M company and the US EPA Toxic Release Inventory) are doing and they are exploring several cost-effective alternatives for developing policy-relevant estimates of the pollution intensity of output by industry sub-sectors. One of these measures compares the weight of materials used to produce a product with the weight of the final product. Another looks at waste (the difference between the weight

23 Because of rampant NIMBY ('not in my back yard') problems, preferential access to new industrial land is extremely important.
24 In 1992, domestic production of the environmental goods and services industry equalled NT$1.3 billion. Of this 6% (NT$75 million) was exported. Production is expected to grow at an annual average rate of 11% through 2002. In that year domestic production will equal NT$3.75 billion, of which NT$464 million is expected to be exports (14%) (IDB n.d.a: table titled 'Status and target of environmental protection industry').
25 A representative of a US firm trying to break into this business stated his firm had a difficult time doing this because the Taiwan government's assistance was both very good and free (author's interviews, Taiwan, November 1995).
26 Author's interviews, Taiwan, November 1995.
27 This is part of UCL's Environmental Sciences and Technology Division. This division works on the ISO 14000 series and life-cycle analysis. It houses the new National Centre on Cleaner Production (author's interview, Taiwan, November 1995).

of inputs and the weight of final product) per NT$ of sales. A third disaggregates waste into four categories (raw materials, industrial water, energy and toxic chemicals) per NT$ of value-added.

Although calculating the pollution intensity per NT$ of value-added by highly disaggregated industry sub-sectors is likely to be viewed by outsiders as either prohibitively expensive or too difficult, in Taiwan it appears to be little more than an extension of what the IDB already does to administer its duty drawback system for exporters.[28] Technical staff in the new National Cleaner Production Centre at ITRI are doing this for several reasons.[29] Most importantly, they hope that better understanding of the pollution intensity of production processes will enable them to redesign production processes to reduce the pollution intensity of production. Because they see this as too risky for the private sector, scientists at ITRI see this as an important role for government. Second, they see this metric, pollution intensity per NT$ of value-added, as a way to assess industry-specific performance in Taiwan against international best practice and to track industry-specific performance over time. This metric is expected by them to become the metric by which the government judges the environmental behaviour of individual firms and industries and its own environmental performance. If this metric were tethered to government performance monitoring which linked rewards, such as preferential access to subsidised credit or to increasingly scarce new land for industrial development, it could spur firms to search for ways to reduce the pollution intensity of output much the same way it encouraged them to increase exports in an earlier time. If this happened, it could well serve as the basis for more cost-effective environmental outcomes (for discussion of this in another context, see Porter and van der Linde 1995a).

Impact on environmental quality

What tangible evidence is there that this myriad of plans, policies and programmes actually works? Unfortunately, there is no easy answer to this question. Lack of a consistent set of environmental data makes it difficult, if not impossible, to track long-run movements in virtually all environmental indicators. The best that is possible is to patch together piecemeal data to create at least a partial picture. What does this data show?

If firm expenditures on pollution control and abatement are any indication, Taiwanese industry seems to have turned an environmental corner.[30] By 1991, four years after the

---

28 The IDB's administration of the duty drawback system requires about 20 technicians working full-time to calculate the input–output coefficients for large numbers of export items and the imported inputs used to produce those items (Wade 1990). Since many of the inputs in production are imported, it appears that it might be relatively easy to adapt this scheme to calculate the pollution intensity of value-added by industry sub-sector.
29 What follows is based on interviews conducted in November 1995 at the new National Cleaner Production Centre in the Environmental Sciences and Technology Division of the UCL laboratory of ITRI.
30 What follows is from O'Connor 1994.

TEPA was created, private-sector investment in pollution control and abatement equipment equalled almost 6% of total private-sector manufacturing investment. In 1992, it equalled 4.3% of total investment. These figures are dwarfed by the pollution control expenditures of state-owned enterprises. In 1992, Taipower invested 8.4% of its fixed investment in pollution control. Pollution control investments for the China Petroleum Company rose from almost nothing in 1988 to 18.9% of fixed investment in 1990 and then to 30% of fixed investment in 1992. By 1992 total investments in pollution control in state-owned enterprises was nearly 350% greater than in the private sector. Given the large role of state-owned enterprises in industry in Taiwan,[31] it would appear that industry in Taiwan in 1992 expended a larger share of its investment budget in pollution control than Japanese firms did at the height of Japan's pollution control effort.[32]

In addition, there are many examples that suggest that the joint IDB/TEPA Waste Reduction Task Force is providing high-quality technical assistance to firms trying to reduce the pollution intensity of production.[33] Several notable examples include the following. Factories using stearic acid cadmium, a compound-decomposition method composed of stearic acid and cadmium, have been assisted in a shift to cadmium oxide as a raw material and in replacing this highly polluting compound-decomposition method with a fusion method production process. The metal-finishing industry has been assisted in replacing a highly polluting acid wash with airtight sandblasting. The scrape metal industry has been assisted in replacing a highly polluting acid wash with a copper(II)chloride ($CuCl_2$) etching solution. The electroplating industry has been assisted in a shift to continuous automatic electroplating and to cyanide-free electroplating. It has also been assisted with installation of equipment for recovering chromic acid. The recovery rate is now 98% or more. This kind of assistance has been extended to state-owned enterprises. The China Steel Corporation has been assisted in a programme to re-utilise water-quenched clinkers, normally a waste product, from steel-making. As a result of this programme, the annual re-use of clinkers reached 77.5% of output in 1988. In 1993 alone, this reduced dumping of clinkers into the sea by 1.36 million tonnes.[34]

Although there is no easy way to characterise the impact of these developments on overall pollution loads, industry-specific progress in several industries suggest that pollution intensities are falling. Two examples should suffice (IDB n.d.c: 15). Pollution loads for the major pollutants in the printed circuit board industry fell dramatically after emission controls. The pre-emission pollution load of lead was 120.6 kg per day for firms in this industry. After emissions controls, the pollution load for lead fell to 67.8 kg per day. This met 1993 emissions standards for this industry. The pollution load of chemical oxygen demand (COD) prior to emissions standards was 32,844 kg per day. This fell to 11,922 kg per day after pollution control devices were installed in firms in this industry. Similar success has been recorded in the heavily polluting electroplating industry. Prior

---

31 Taiwan's relatively large state-owned enterprise sector accounted for 14% of gross domestic product and 33% of gross domestic capital formation in the period 1978–80 (Wade 1990).
32 This occurred in 1975 in Japan. In that year, Japan allocated 7.5% of the fixed investment budget of industry to pollution control (O'Connor 1994).
33 The examples that follow are from IDB n.d.b: 4-10.
34 In that year only 22.5% of clinker wastes (540,000 tons) were dumped in the sea (TEPA 1993).

Pollutant	Emissions (kg/day) Before control	Emissions (kg/day) After control
Lead*	120.6	67.8
Chemical oxygen demand*	32,844	11,922
Suspended solids†	31,504	9,924
Nickel†	5,650	633
Chromium†	5,996	495
Zinc†	5,403	1,253
Copper†	4,382	758
Iron†	2,746	195

* Printed circuit board industry
† Electroplating industry

Table 9.1 **Pollutants before and after emission control**

Source: IDB n.d.c: 15

to emissions control, the pollution load of suspended solids equalled 31,504 kg per day, after emissions control emissions fell to 9,924 kg per day. This was only 13% above the emission standard for this industry (the effects of emission control on these and other pollutants are shown in Table 9.1).

Ultimately, if all of these pollution control, abatement and prevention activities work, there should be measurable improvements in ambient environmental indicators. Is there any evidence of this? Several ambient indicators reveal significant, perhaps substantial, environmental progress. Following the first round of environmental regulation in the early 1980s, ambient concentrations of sulphur dioxide (SO_2) and nitrogen dioxide (NO_2) in the air fell steeply (O'Connor 1994). Implementation of a new policy requiring significant reductions in the sulphur content of heavy oil and diesel fuel in 1993 led to even further reductions in SO_2 concentrations.[35] There has been an equally precipitous decline in the percentage of days with a PSI index greater than 100.[36] Chlorofluorocarbon (CFC) consumption has fallen sharply (by 86%), from 16,255 tons in 1988 to 2,304 tons in 1994 (OSTA 1995). Although data on lead concentrations in air is not available, the sale of unleaded gasoline has risen from about 13% of total sales in 1988 to 65% of total sales in 1994 (OSTA 1995).

Less progress has been made elsewhere. Between 1984 and 1993, PM10 concentrations have hovered between 90 and 100 $\mu g/m^3$, but at least concentrations have not risen.[37] Data

---

35 Following this, sulphur dioxide (SO_2) concentration in the air fell from 26 parts per billion (ppb) in 1993 to 8 ppb in 1994 (OSTA 1995).
36 In 1991, 16.25% of days had a PSI greater than 100; by 1994 this fell to 6.99% of days (OSTA 1995).
37 Since only 34% of PM10 emissions come from industrial processes, constant ambient levels of PM10 may suggest that emissions from industry are falling (TEPA 1993).

on raw river water quality tells an even less encouraging story. The unpolluted portion of 21 major rivers fell from about 75% to about 66% between 1983 and 1990, and the heavily polluted portion of rivers increased from 5.7% to 11.3% (O'Connor 1994). Since the pollution load of industrial waste-water has been falling, at least some of the deterioration is a result of the extremely low rates of treatment of domestic sewage (less than 3%) and the growth of livestock waste-water.[38]

Concluding remarks

Why has the Taiwanese government done all these things? There are several answers to this question. Democratisation, growing public concern over the environment and an almost intractable NIMBY ('not in my back yard') problem has meant that governing élites can no longer ignore industrial pollution.[39] Criticism from the influential overseas Chinese community, particularly that from the United States, has added to domestic pressures.[40] Loss of international recognition has increased élite sensitivity to international criticism. Some of this has extended to the environment.[41] In another instance, the government has gone out of its way to demonstrate exemplary environmental behaviour.[42] Growing domestic and international political pressures have been reinforced by basic economic considerations. Four different types of economic development predominate.

First, appreciation of the exchange rate, rising wage rates and emerging labour shortages alongside increased demands for a cleaner environment contributed to an export of industry during the 1980s. Many of the firms that migrated to China and Vietnam, among other places, were in industries considered by the Taiwanese to be excessively dirty or polluting (e.g. textiles, leather goods, and metal and electroplating industries). In contrast, the new high-technology industries promoted under the Statute for Upgrading Industries are considered relatively clean. Put another way, the latest

---

38 The biochemical oxygen demand (BOD) of industrial waste-water has fallen from 54% to 24.6% of total BOD from all pollution sources (IDB n.d.c: 6).
39 Recent public opinion polls put industrial pollution as one of the top three problems facing the country. The NIMBY problem has made industrial siting almost impossible. One response to this has been to dredge the ocean to create new offshore islands on which the most polluting industries will be located (author's interviews, Taiwan, November 1995).
40 Each year the government holds a national reconstruction conference where scholars, government officials and overseas Chinese are invited to review accomplishments and assess the major issues facing the country. At least since the mid-1970s the overseas Chinese, particularly those from the United States, have been critical of Taiwan's growing pollution problems (author's interviews, Taiwan, November 1995).
41 Recent condemnation of Taiwan by the United States for violation of CITES (Convention on International Trade in Endangered Species) stung the country's political élite (author's interviews, Taiwan, November 1995).
42 Despite the fact that the country was not permitted to be a signatory to the Montreal Protocol, it met the Protocol's stipulations (OSTA 1995).

competitive shift in industrial structure is away from industries with high pollution intensities toward industries with low pollution, or at least lower, intensities. Industrial policy merely speeds up this process.

Second, there is acute recognition in government that international competitiveness demands better environmental behaviour. Taiwan's exporters are learning from direct experience that they have to certify to importers in some developed countries that their products meet the environmental regulations of the importing country.[43] Officials in the TEPA, the CEPD and the IDB are also aware of the competitive value of 'green' labelling. For this reason, Taiwan has created its own green labelling programme—the Green Mark. By 1995, 197 products qualified for Green Mark status (OSTA 1995). These same officials are all too acutely aware of the possible impact of the ISO 14000 series on Taiwan's international competitiveness. The government is reputed to be engaged in bilateral discussions with the European Union to assure Taiwanese access to the European market following implementation of the ISO 14000 series there. It has also established an inter-ministerial committee to consider how Taiwan should respond to this series of standards.[44]

Third, the Taiwanese expect the demand for environmental goods and services in South-East Asia to grow rapidly over the next several years. They are preparing to capture a substantial share of this market. They see this as part of the next step up the industrial ladder and an important way Taiwan can distinguish itself in the region. Some see it as a part of the way Taiwan can establish itself as an Asia–Pacific regional manufacturing centre in the region's vertical division of labour system (IDB 1995). The National Cleaner Production Centre in ITRI is contributing to this by establishing an Asian network for clean production.[45]

Last, the Taiwanese view their approach to industrial pollution control and reduction as more cost-effective than the alternatives. They believe that tough emission and effluent standards and equally tough enforcement are necessary to get firms to accept protecting the environment as part of the cost of doing business in a highly competitive global economy, but they believe equally strongly that the IDB's fiscal and financial incentives can quicken the shift to a less-polluting industrial growth strategy, and they know from direct experience that pollution prevention sometimes pays.

For all these reasons, then, the Taiwanese state has decided to act to reduce industrial pollution. That action appears to have mobilised not only a relatively new environmental agency but also the pre-existing industrial policy machinery, particularly the IDB and the CEPD. The CEPD's latest development plan, which promotes ten emerging industries,

---

43 Manufacturers of wood furniture who export to the Nordic countries must certify that the wood in furniture does not come from virgin tropical forests (author's interviews, Taiwan, November 1995).
44 In addition, the private sector is looking to the government to advise it on what the ISO 14000 series will mean for them (author's interview with representative of the National Federation of Industries, Taipei, November 1995).
45 This is being done in co-operation with the Federation of Asian Chemical Societies, the Hong Kong Productivity Council, the Philippine Business for the Environment, the Korean Institute for Chemical Technology and the Asian Institute for Technology in Thailand (author's interviews, Taiwan, November 1995).

presupposes a structural shift in industry to less-polluting industries.[46] If this happens, it alone should contribute to a lowering of the pollution intensity of industry. The IDB is complementing this with technical assistance in waste reduction and minimisation and with research on indicators of pollution intensity per NT$ of value added by industry sub-sector. Taken together, this work is quite remarkable. Because it is being done with real Taiwanese data, it puts Taiwan ahead of others who are exploring these issues in developing countries,[47] but how all this will play out in the end is not clear. Some in both the public sector and the private sector decry the focus on the environment. Since much of the current environmental effort is predicated on international developments, such as the ISO 14000 series, an international turn away from concern over the environment could slow or impede further progress. A heightening of political and military difficulties with the People's Republic of China might also slow environmental efforts by drawing attention and resources to more pressing security considerations. Despite these possibilities, it does appear that the Taiwanese have found a way to turn the attention of some of its 'institutions of growth with equity' to industrial pollution. Because of this, Taiwan may be in the midst of leapfrogging to more cost-effective environmental solutions. If this proves to be the case, others, particularly in South-East Asia, might have something to gain by paying more attention to what is happening in Taiwan.

---

46 This mirrors work being done at the World Bank under the Industrial Pollution Projection system (Hettige *et al.* 1995).
47 The World Bank is extrapolating US pollution intensities of specific industries to developing countries. By projecting changes in industrial structure they are able to project changes in the pollution intensity of industry (World Bank 1994a).

10
MEASURING UP
Toward a common framework for tracking corporate environmental performance

Daryl Ditz and Janet Ranganathan

◢ Introduction

The need for environmental performance indicators

'You can't manage what you don't measure', states a well-known adage. Business environmental performance is a case in point. Many forward-thinking firms recognise the need to better manage the environmental impacts of their activity. At the same time, others outside of firms—communities, regulators, investors and customers—are demanding more meaningful information on corporate environmental performance. While these diverse audiences share a common interest in environmental performance, no agreement has emerged on how to measure it. Environmental performance indicators (EPIs)—information used to measure and to motivate progress toward environmental goals—can serve the information needs of business managers as well as those of others outside corporate walls. But for EPIs really to be effective, a common set of metrics must emerge that are universally adopted and understood by all.

This demand for meaningful information comes at a time of explosive change in communication and other information technologies. As a senior Clinton administration official noted, 'Today, a high school student in Pittsburgh, Pennsylvania has better access to federal statistics than a top government official five years ago' (Katzen 1997: A27). But is it possible to find a common set of environmental metrics to serve such diverse audiences? Does not every firm desire different, custom-tailored measures? Would universal environmental reporting compromise confidential business information?

The idea of tracking corporate environmental performance is nothing new. Driven by regulations and self-interest, most firms keep tabs on environmental compliance, spills and accidents, and pollutant releases. Unfortunately, other than staying on the right side of the law, such EPIs are not much use in most business decisions. Compliance metrics, such as the number of violations, tend to ignore superior performance and rarely do they lower regulatory burden or create positive public recognition. Since these indicators are predicated on today's rules, they do not anticipate future requirements. Fundamentally, compliance does not guarantee that countries or regions will live within their means environmentally.

The topic of EPIs has attracted increasing attention in the United States and many other countries (Atkinson and Hamilton 1996; UNCSD 1995; EPA 1995). Yet most EPI efforts focus on either the internal world of corporate management (Epstein 1996) or the external world of public right-to-know (Unison 1995; UNEP/SustainAbility 1996). This chapter shows that the full potential of corporate EPIs is realised only when they serve decision-makers both inside and outside company walls.

Building a common framework for environmental accountability

A universal framework governing the measurement and communication of corporate environmental performance is coming. It will resemble today's system of financial reporting and for much the same reasons—comparability, transparency and completeness are prerequisites for independent evaluation.

This movement is already under way in some parts of the world but will reach across national borders, bolstered by trends in public access to information and the globalisation of business. Its leaders include the multinational Nortel, activists such as Friends of the Earth United Kingdom, Dutch government planners and progressive businesses in Colombia.

In less than a generation, society will look back at the availability of environmental performance information today and be amazed at the gaps, clutter and noise. We will wonder how environmentalists, regulators and firms could have claimed so much progress with so little credible information. With the gradual adoption of standard EPIs, comparability—across firms, sectors and countries—will permeate internal management systems and external reporting. This framework will facilitate the comprehensive analysis of resource consumption and pollution, extending far beyond manufacturers to capture the environmental dimensions of the energy sector, agriculture, transportation, housing and other economic activities. Transparency will encourage greater competition in environmental performance.

For many segments of society, the implications of this transformation will be profound.

- **Business** will view standardised reporting on environmental performance as a routine business output and a potential source of customer and shareholder value.
- **Financial institutions** will connect environmental performance with economic value and incorporate EPIs into investment, insurance and lending decisions.
- **Consumers** will be able to distinguish between products and providers based on environmental performance during production, use and disposal.

- **Communities** will have ready access to information on the environmental performance of local facilities and be able to benchmark them against operations, firms and industries around the world.
- **Government agencies**, including environmental authorities, will foster this change by revamping their own management of information and drawing on environmental performance information to evaluate policy reforms.

Is this vision unrealistic? Is it overly optimistic? This framework is not only plausible but also ultimately inevitable. Consider the evolution of financial accounting. A century ago a firm's accounts were a private matter. The suggestion of standardised reporting on earnings, assets and liabilities was nearly unimaginable, but, after the collapse of the stock market in 1929, firms and investors saw the value of systematic disclosure. As financial and managerial accounting has matured, firms have accommodated this requirement while maintaining detailed internal records for the sake of internal management, and yet drawing from this information and the body of 'generally accepted accounting principles' (GAAP) to report publicly.

This new global framework for measuring and reporting corporate environmental performance is being assembled from pieces available today—voluntary corporate disclosure, increasing transparency in regulatory policy and the shift in mind-set from compliance to performance. New information technologies, especially the nexus of computing and communications, will provide the infrastructure and tools to acquire, analyse and disseminate information worldwide. Anyone with Internet access will find credible, comparable information on the environmental performance of businesses across the globe. Corporate environmental leaders will be recognised, whereas corporate laggards and those who shy away from public accountability will feel the pressure from environmentally aware communities, consumers and investors.

In the next section, we place corporate EPIs into a larger context and argue for four key categories of EPI to satisfy managerial and public policy needs; we also present the results of a survey of environmental managers, highlighting some commonly used EPIs and the characteristics that make them useful. We then examine how such EPIs are being applied within companies. Applications of EPIs outside of companies are explored in the penultimate section. In the final section we provide recommendations for actions to lay the foundation of a new framework for measuring corporate environmental performance.

◢ In search of common yardsticks

Although environmental performance indicators (EPIs) are attracting attention from all corners, opinions on what to measure vary wildly. Many business managers are drawn to EPIs by the rising interest in environmental management systems, particularly as these systems relate to the new international standards of the ISO 14000 series and Europe's Eco-Management and Audit Scheme (EMAS). Policy-makers and others with an eye on the national environmental picture tend to view EPIs in terms of a country's, not a company's, environmental performance. At the same time, some communities and environmentalists approach EPIs from the standpoint of their 'right to know', emphasising public disclosure at the facility level.

The result is a jumble of indicator jargon that mixes similar terms with different, even contradictory, interpretations. Although controversial in many respects, the ISO series has helped distinguish three types of EPI: operational indicators, management indicators and environmental condition or 'state-of-the-environment' indicators (ISO 1996). This three-way taxonomy mirrors the 'pressure–state–response' model developed by the Organisation for Economic Co-operation and Development (OECD) to assess the environmental performance of countries.

- ☐ **Operational indicators** measure potential stresses on the environment, such as burning fossil fuels or converting forest resources at a paper mill.

- ☐ **Management indicators** measure efforts to reduce or mitigate environmental effects, such as company spending on energy efficiency or environmental training programmes.

- ☐ **Environmental condition indicators** measure environmental quality, such as ambient air pollution concentrations or global climate change.

As illustrated in Figure 10.1, these three classes of environmental indicator are related, at least in principle. For example, management efforts, such as environmental training programmes, should reduce spills and pollutant releases, resulting in higher environmental quality. In practice, the links are often tentative, unclear and disputed. Burning fossil fuels releases carbon dioxide (CO_2) to the atmosphere, but any single company's contribution to the aggregate is practically undetectable, and, even with the accumulating weight of scientific opinion, the effects of CO_2 emissions on global climate remain highly uncertain.

All three types of indicator can be useful to business decision-makers. Management indicators are the most familiar metrics for internal use. Tracking spending and time devoted to environmental objectives is fundamental to good cost accounting (Ditz et al. 1995). Such measures also demonstrate an investment in putting environmental policies into practice.

As for state-of-the-environment indicators, they too can be directly relevant to business decisions. The quality of local water can be a decisive factor in whether a facility is allowed to expand production, even if there are many other sources of pollution. Similarly, firms with large timberland holdings have a natural stake in forest health. Even global environmental conditions can alter the political and economic landscape of business, as our experience with stratospheric ozone protection reminds us (Cook 1996). Clearly, it would be absurd for every firm to monitor global CO_2 concentrations or sea level rise, but regulations and consumer preferences are shaped by local and global environmental conditions, altering a company's costs and markets.

Although management indicators are necessary and although tracking environmental conditions is important, operational indicators are special, because operational perfor-

Operational indicators: e.g. fossil fuel use

Environmental condition indicators: e.g. global average temperature

Management indicators: e.g. hours of environmental training

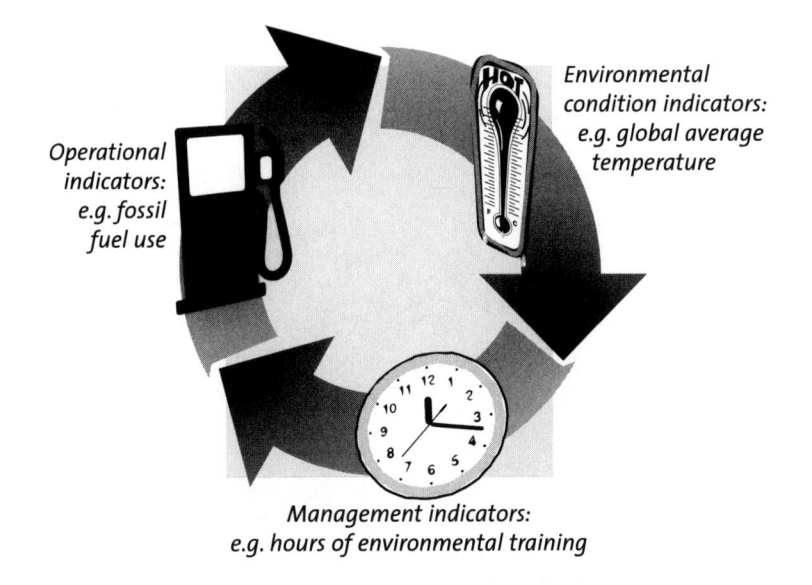

Figure 10.1 **Three types of environmental performance indicator under the ISO 14000 series**

mance links internal management practices with the external environment. To company managers, operational EPIs should reflect what they are accomplishing through their environmental expenditures—the results of their efforts. To the public, operational EPIs should help to describe a company's 'ecological footprint', their explicit contribution to environmental problems and solutions (Wackernagel and Rees 1996).

The remainder of this publication concentrates on these operational indicators (or 'pressure' indicators, to use OECD's terminology) of corporate environmental performance. Unfortunately, even after narrowing the universe of EPIs, a profusion of potential metrics remains. The draft guidance on environmental performance evaluation, ISO 14031, lists over 100 illustrative EPIs (ISO 1996). The problem is that too many EPIs can cloud, rather than clarify, the situation.

Unlike measures of financial performance, EPIs do not lend themselves to a common unit. Rather, EPIs are recorded in such disparate units as pounds of waste generated, litres of water used, British thermal units (BTUs) of energy consumed, or hectares of forest harvested. These multiple dimensions of environmental performance quash any hope for a single, universal, EPI to suit all firms. In the absence of common ground, companies are adopting customised, but fundamentally incompatible, metrics. This situation frustrates the public and private desires to combine EPIs generated at the facility or community level into information that is meaningful at larger scales. As suggested in Figure 10.2, the desire to aggregate EPIs arises both in corporate management and in public policy, giving another meaning to 'measuring up'.

There are compelling reasons why a power plant would adopt EPIs different to those of a paper mill. The two companies use different materials, with different environmental

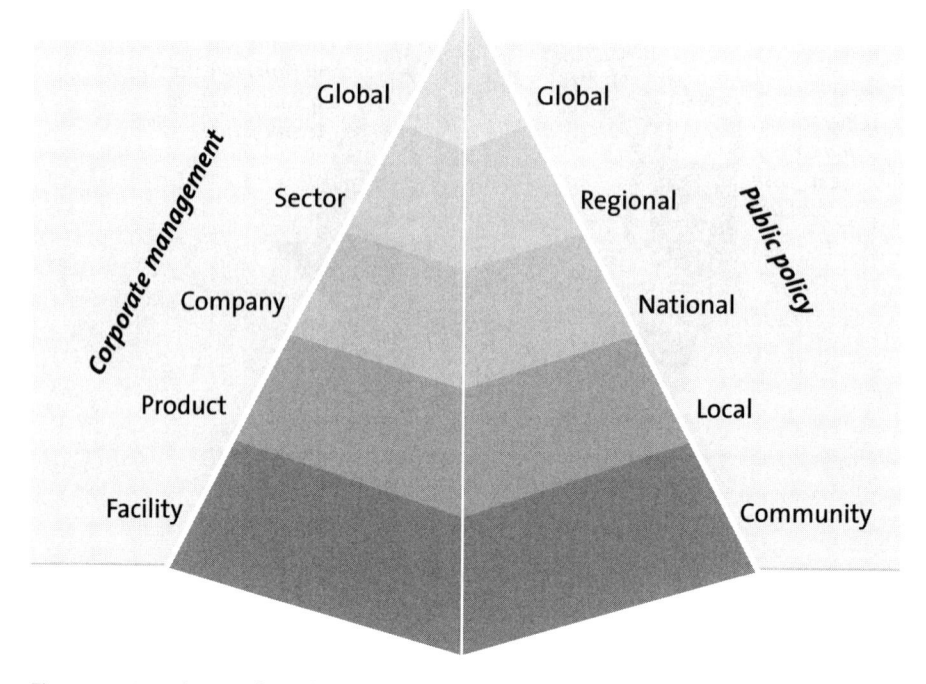

Figure 10.2 **Two views on the scale of environmental performance indicators**

impacts, regulated under different laws. But some agreement on the fundamentals of measuring performance is necessary to draw meaningful comparisons across facilities and industries, to set sensible public priorities and to track the performance of countries and regions. Even individual businesses benefit from the use of some common EPIs that can be applied internally across operations and over time.

A common set of EPIs

Which EPIs should be part of this common set? Four key categories of environmental performance can be derived from the fundamental resource inputs and outputs of a firm. In contrast to most compliance-oriented EPIs, these performance measures bear little similarity to most regulatory requirements. Instead, they focus manufacturers, customers and others on products, processes and services that prevent pollution and boost resource efficiency.

- **Materials use** (quantities and types of materials used): this EPI tracks resource inputs, distinguishing their composition and source.
- **Energy consumption** (quantities and types of energy used or generated): this EPI, the energy analogue to materials use, also differentiates fuel types.

❑ **Non-product output** (quantities and types of waste created before recycling, treatment or disposal): this EPI distinguishes production efficiency from end-of-pipe pollution control.

❑ **Pollutant releases** (quantities and types of pollutant released to air, water and land): this EPI includes toxic chemicals, as well as greenhouse gases, solid wastes and other pollutants.

To highlight environmental performance along the product chain, Figure 10.3 illustrates how these four EPIs might be applied to a hypothetical computer manufacturer. Materials use—the throughput of raw materials and sub-components purchased, stored and processed—is intimately connected with production efficiency or yield and therefore with cost. Energy consumption—whether by suppliers, in manufacturing, or in the use and disposition of computers—reflects another cost item with environmental implications. Non-product output—considering all inputs not contained in products—represents under-utilised inputs plus the extra costs of their subsequent management. Pollutant release captures the introduction of materials to the environment. A computer maker has no legal mandate to worry about emissions of hazardous air pollutants from upstream facilities that supply semiconductors or printed circuit boards, but it should take an interest. Changing regulations or permit limits could disrupt supply, slow the time to market or even shut down a facility.

Unlike classical life-cycle analysis, this framework does not require a full-blown analysis of every supplier's suppliers. Companies along the product chain must each assume responsibility for quantifying their contribution in terms of materials use, energy consumption, non-product output and pollutant releases. This framework presumes a cascading flow of information along the product chain, information that has practical relevance to decision-makers at several links in the chain. Thus, the computer manufacturer would rely on information provided by upstream chip suppliers, energy providers and others to assess its environmental performance. This example illustrates the practical complexity of evaluating environmental performance, especially in global manufacturing. The advantage of this framework is that it could be applied to a chemical company or an electric utility, even though their activities have different environmental profiles.

Is this framework realistic? Eventually, it will be. To date, the most visible progress has been made on pollutant releases. In some countries and companies information is beginning to emerge on non-product output and energy use. In very few countries and companies information on inputs is appearing. Many companies are sceptical. Some information, particularly quantitative data on inputs, is closely guarded by firms as confidential business information. In some cases, firms are willing to share aggregated environmental performance indicators for the whole corporation but resist more detailed disclosure at the facility, process or product levels. Reaching agreement on the specifics of an environmental performance reporting scheme will be difficult and will take time, but it is possible.

Who will define GAAP for environmental performance? How would these relate to the existing regulatory requirements, both financial and environmental? As yet, these

Figure 10.3 **Four key environmental performance indicators illustrated for a computer manufacturer**

questions cannot be fully answered. The four key EPIs are core elements of such a framework and present evidence of how this system is beginning to emerge.

An important start is the identification of this set of EPIs. Putting them into practice will take some painstaking work. For example, tracking pollutant releases (to choose one of the better-developed EPIs) requires designating which chemicals must be tracked, the units of measurement, the time-period, the level of aggregation and other data elements (Irwin *et al.* 1995). Similarly, issue of materials use begs the question of which inputs to track: all inputs, hazardous materials, water? Even something as seemingly straightforward as energy would need to be disaggregated by fuel type, an increasingly difficult matter for consumers of electricity. Yet it is necessary to work out the fine structure of EPIs whether they are publicly divulged or used only internally by companies.

These four EPIs draw from previous efforts to characterise corporate environmental performance; for example, the work of the Wuppertal Institute in Germany on the 'ecological rucksack', the hidden flows of materials associated with upstream activities. Emissions inventories, such as the US Toxic Release Inventory (TRI), already capture pollutant releases and non-product output, at least for some chemicals and industries. The public 'environmental statement' required under Europe's new EMAS system requires site-specific reporting on energy and materials use (including water), waste generation, pollutant release, noise and other significant impacts. Other efforts aimed at 'closing the loop' on industrial processes are also arriving at a similar conclusion: key EPIs must describe inputs, outputs and the transformations in between (Steger 1996).

There have been several notable efforts to standardise corporate environmental reporting. The Coalition for Environmentally Responsible Economies (CERES), a non-profit membership organisation of social investors, public pension trustees, environmental groups and others, has convinced an impressive array of firms to subscribe to a broad set of principles. Companies such as General Motors and Polaroid that have endorsed the CERES principles commit themselves to taking steps toward energy conservation, sustainable use of natural resources, waste reduction and other social and environmental goals. CERES has also convened the Global Reporting Initiative (GRI), whose mission is to develop globally applicable guidelines for companies reporting on the environmental, social and economic aspects of their activities. What sets the four EPIs apart from these other efforts is their collective application as part of an overarching framework for measuring and communicating environmental performance. Ultimately, this framework will guide internal corporate management and public policy.

How do these four key EPIs compare with current business practices? For some clues, the World Resources Institute (WRI) teamed with the Tellus Institute, a Boston-based non-profit research and consulting organisation, to survey two groups of environmental managers in the United States and Canada. As detailed in Box 10.1, the respondents, mostly corporate environmental managers from major manufacturers and electric utilities, rely heavily on compliance-oriented EPIs and mandatory reporting of pollutant releases. However, a majority of respondents 'regularly used' other unregulated metrics, including greenhouse gas emissions, water and energy use, and chemical inputs.

The results of the WRI–Tellus survey, although clearly showing the traditional appeal of compliance measures to a small sample of environmental professionals, also reveals

THE WRI–TELLUS SURVEY, ALTHOUGH NOT A SCIENTIFIC SAMPLING OF ALL corporate EPI users, offers a glimpse into the use of EPIs by 33 environmental managers. In September 1996, participants at two conferences on environmental accounting and pollution prevention were invited to complete a survey.* The majority of respondents were corporate environmental professionals, most representing companies with annual revenues over US$1 billion. Approximately 33% of the respondents work in the chemical industry, 28% for electric utilities, and the remainder representing diverse industrial sectors. Full details of the methodology and results are available elsewhere (White and Zinkl 1997).

The survey defined EPIs as information used on a continuing basis to measure environmental performance. Respondents were asked which of a set of EPIs they currently use, which business decisions were most affected and how valuable they considered characteristics such as comparability and public availability. Several respondents offered examples of how EPIs had influenced decisions inside their firms.

When asked which EPIs they regularly used, over 90% of respondents selected environmental compliance and chemical releases—two categories of EPI in which mandatory reporting is well established. As Figure 10.4 illustrates, other EPIs such as water and energy use were cited less frequently, with chemical-use efficiency (perhaps the better term is 'yield') selected by fewer than 43% of respondents.

* The conferences were: 'Putting Pollution Prevention into Action', held in Washington, DC, and sponsored by the Hampshire Research Institute and the US Environmental Protection Agency (EPA); and the 'Whistler Environmental Accounting Seminar', held in Vancouver, BC, and sponsored by the Whistler Centre.

Box 10.1 **WRI–Tellus survey of EPI users and uses**

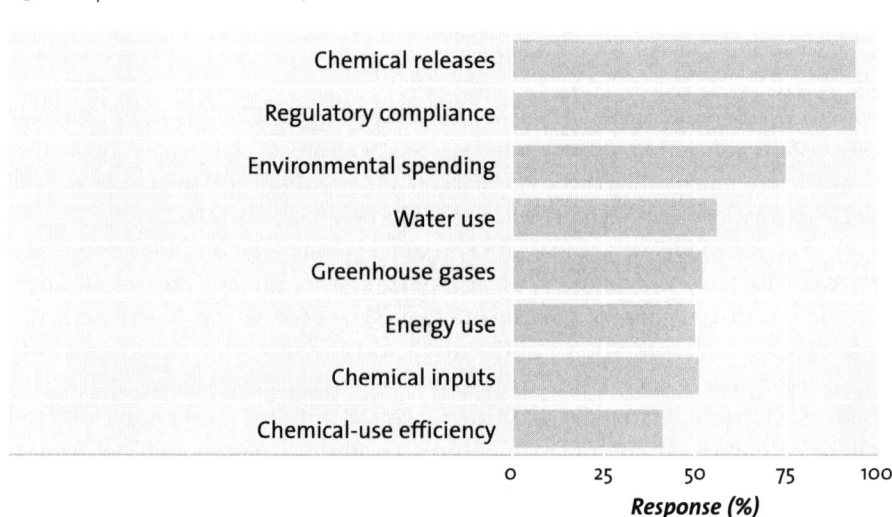

Figure 10.4 **Use of environmental performance indicators**

a reliance on input and efficiency metrics. This may signal the expanding recognition of environmental issues as opportunities to simultaneously reduce costs, enhance economic value and satisfy customer expectations, as others are pointing out (Hayes 1996). The survey also underscores that many different metrics are in use, an issue explored in greater detail in the following sections of this chapter.

The four EPIs proposed here will ultimately drive firms to make changes in products, processes and practices that prevent pollution and wring higher efficiency from the use of chemicals, water, energy and other resources. In the hands of business leaders with a genuine commitment to 'eco-efficiency' and product stewardship, these EPIs will help track progress and motivate improvement. If applied to measuring macro-level environmental performance—for example, of an entire sector—these EPIs will serve as building blocks of national, even global, environmental management systems. The challenge for companies, as well as countries, lies in integrating EPIs into practical decisions.

◢ Integrating EPIs into business decisions

To guide their decisions, businesses routinely use such indicators as production rates, inventories, return on investment and earnings per share. These indicators are used both inside and outside of companies to track performance. The power of such indicators is that they are universally understood, comparable and, as such, create a shared vocabulary for tracking progress. Unfortunately, there is no agreement on how to measure environmental performance. Given that environmental performance is now high on the agenda of many firms the lack of comparable and commonly accepted indicators is a major obstacle to continued progress.

Firms need EPIs for the very same reason that others outside firms want them—to introduce accountability for environmental performance, particularly crucial in companies moving to decentralise responsibility to business units. Corporate environmental managers, especially those under mounting pressure to justify their budgets, have a major stake in measuring environmental performance. For firms operating internationally, across jurisdictions with multiple and conflicting standards, consistent EPIs are a practical necessity to manage environmental performance in an integrated fashion.

By drawing on recent business experience, supported by results from the WRI–Tellus survey, this section shows how business is using EPIs. Although the following examples represent only a small sample, they do help answer three basic questions: Which EPIs are being used? Where do EPIs make a difference? What makes EPIs most useful?

Tracking progress toward corporate goals

Many companies have adopted environmental policies with lofty goals. In fact 94% of manufacturing respondents to an Investor Responsibility Research Center (IRRC) survey of Standard & Poor's 500 companies indicated that they have articulated broad environmental principles or commitments as part of their overall policy (IRRC 1996). Goals carry no weight, however, unless supported by concrete EPIs. As one respondent to the WRI-Tellus survey stated, 'By examining our indicators on a regular basis, we have been able to see where we need improvements and make plans to carry us there' (White and Zinkl 1997). In short, measuring environmental performance is basic to internal accountability and communication, especially in companies that span the globe.

Nortel, a US$12 billion developer of communication products, networks and systems, has demonstrated leadership by adopting progressive environmental goals and creating an interesting set of metrics to measure their success. In the words of John Roth, Nortel's chief operating officer: 'We believe our approach to environmental management is a sound business strategy that gives Nortel a competitive edge.' Between 1993 and 2000 Nortel set goals to reduce pollutant releases and solid waste disposal by 50%, cut paper purchases by 30% and improve overall energy efficiency by 10%. Since publishing these targets, Nortel has taken bold steps to measure progress toward these and other environmental goals with a set of EPIs (see Box 10.2). For internal purposes, Nortel has devised a composite index of individual EPIs and a complex set of scores and weighting factors. This index can serve as a single barometer of environmental performance for senior management, although it must be amenable to disaggregation to track progress on the component parts or compare its environmental performance against that of another firm. To its credit, Nortel has made a habit of disclosing information on the component EPIs.

Box 10.2 **Tracking Nortel's progress toward corporate goals**

IN PURSUIT OF ITS OVERALL ENVIRONMENTAL COMMITMENT, NORTEL TRACKS 25 indicators, including releases of air toxics, ozone precursors and greenhouse gases, consumption of water, electricity and paper, and regulatory violations. Some EPIs are normalised against the annual cost of sales to offset fluctuations in business volume. These individual environmental metrics are entered in a complex scoring scheme, matching actual performance against targets and the previous year's level. The targets decrease each year, which promotes continued improvement. The result is a single aggregate environmental performance index (EPIndex) for Nortel. A baseline of 100 was set for 1993, with higher annual scores indicating improved environmental performance. Under this custom scoring scheme, the maximum score in any year is 175.

Nortel's performance in three categories—paper purchases, pollutant releases and energy use—is summarised in Figure 10.5, along with the composite EPIndex for 1993 to 1996. In constructing the EPIndex, Nortel weights the individual EPIs (e.g. pollutant releases count for up to half the total, compliance for a quarter, and resource consumption and remediation each for an eighth). These weighting factors, admittedly subjective, reflect the relative importance Nortel attaches to each performance category, considering risk and potential impact. The EPIndex enables managers—environmental or others—to grasp the company's overall environmental performance at a glance. Although the EPIndex was developed primarily to measure progress for internal consumption, Nortel has made the information publicly available in their annual environmental report (Nortel 1997) and on their website (www.nortel.com/cool/environ).

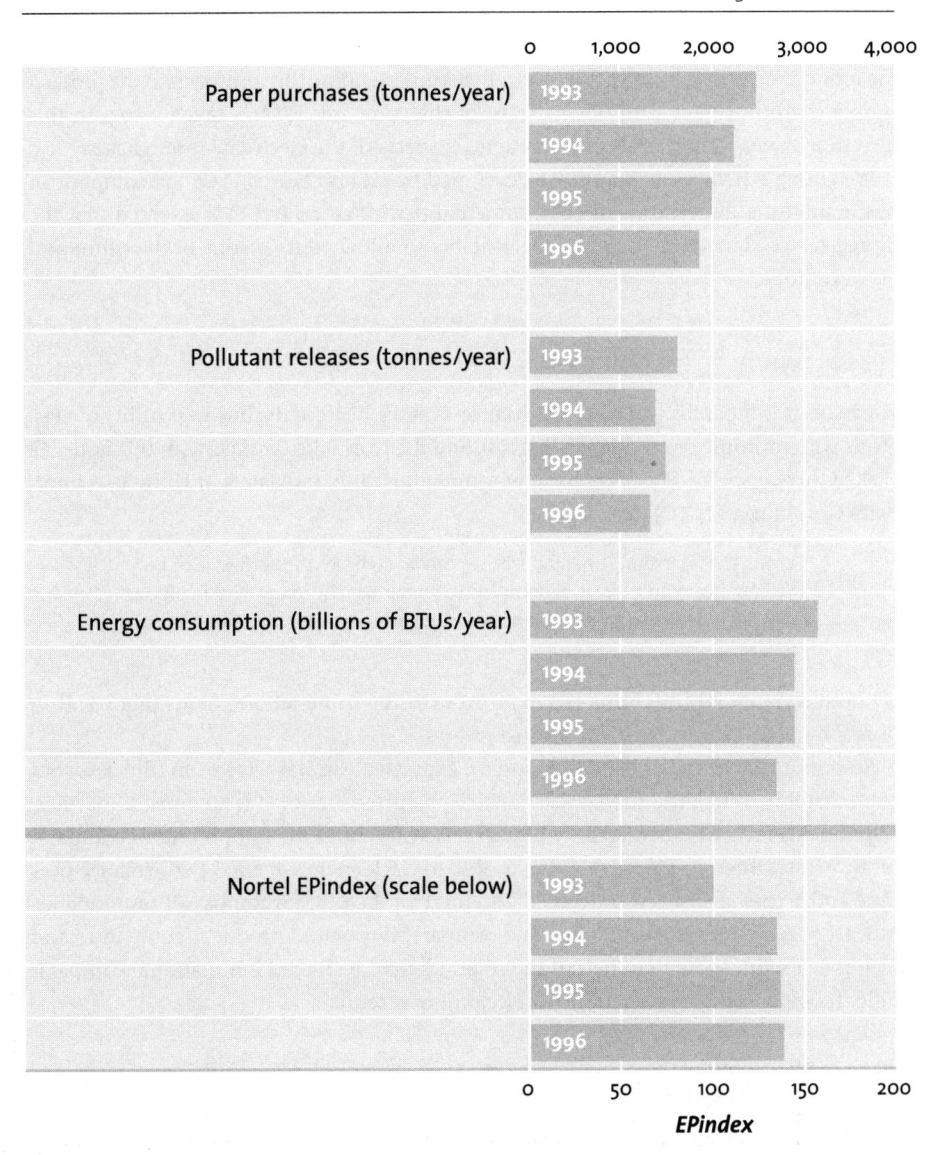

Figure 10.5 **Nortel's environmental performance index (EPIndex) and elements of performance, 1993–96**

ICI, the UK-based chemical giant, has a similar approach to accounting and reporting waste. Rather than reporting total emissions in its annual environmental report, ICI classifies pollutants by type of impact. Within such categories as global warming, human health effects and ozone depletion, individual emissions are multiplied by a potency factor to define a quantitative measure of 'environmental burden'. Given the lack of data on the relative environmental or health risk of all emissions, this requires some use of judgement. ICI intends to use this approach to frame its emissions reduction strategy for the future. Future developments will also focus on extending this approach to the product supply chain upstream and downstream of the company. ICI has taken a step in this direction by evaluating some environmental aspects of the electricity it purchases.

Weighting schemes, such as those developed by ICI and Nortel, ease the comparison of the multiple dimensions of environmental performance, but it is essential that the disaggregated data be made publicly available to allow comparisons both within and between firms.

Comparability

The issue of comparability or standardisation is at the heart of the business utility of EPIs. Firms with multiple operating facilities around the world are familiar with this issue. As Leslie Carothers, Vice-President for Environment, Health and Safety at United Technologies Corporation (UTC), notes,

> Getting UTC's far-flung operations to report data is by far the hardest job I have. This is going to be a very long-term effort, undertaken by many of us in phases. Still, having common measures would greatly enhance the value of this exercise.[1]

Without standardisation, firms face a plethora of reporting requirements that result in non-comparable information of limited utility.

Respondents to the WRI–Tellus survey expressed similar views on the issue of comparability. When the survey asked which characteristics of EPIs are most important, respondents revealed a remarkably broad consensus about the importance of comparability. As presented in Figure 10.6, being able to track environmental performance over time and across facilities was deemed 'essential' or 'helpful' by almost all respondents. At least 75% of those surveyed also saw comparability across products, companies and sectors as useful. More than half of the respondents considered international comparability (comparability across countries) helpful or essential.

Energy use in production and products

Energy use can be a major economic consideration in manufacturing and product use. General Motors (GM) is driving down the energy intensity of its manufacturing operations by setting such annual goals as a 5% reduction in energy use between 1995 and 1996. In the 1995 reporting year, GM's North American operations included more than 170 facilities in the United States, Mexico and Canada. GM tracks energy consumption and costs at each, distinguishing among electricity, natural gas and other fuels. As a significant component of manufacturing costs, better management of energy translates into cost savings for GM.

---

1 Personal communication with Leslie Carothers, Vice-President, Environment, Health and Safety, United Technologies Corporation, 1997.

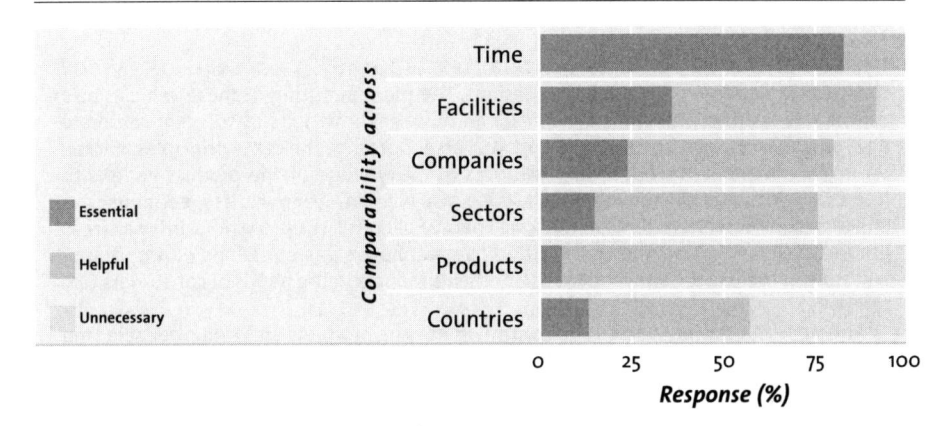

Figure 10.6 **Which environmental performance indicator characteristics are most important?**

As part of its energy conservation programme and voluntary reporting of greenhouse gases, GM tracks energy use from stationary sources (notably their manufacturing plants) and from the use of GM vehicles. To facilitate comparison, GM also estimates and reports greenhouse gas emissions from mobile and stationary sources (see Box 10.3). By examining energy used in both manufacturing and product use, GM can compare environmental performance at different points in its product chain. Clearly, energy use and greenhouse gas emissions are more significant factors in the use of cars than in the manufacture of cars.

Greening product chains

Many firms are redefining environmental performance to reach beyond the point of sale. The emergence of product 'take-back' legislation in Europe, together with consumer pressure and voluntary initiatives, such as the US Energy Star programme, are extending corporate environmental responsibility to encompass product use and disposal (Dillon 1996). In the electronics industry, international product specifications and market expectations are forcing firms to develop products that incorporate energy-saving features and recyclability. Hewlett-Packard, for example, has developed a set of metrics to improve product stewardship and to equip management with the tools to measure progress. These focus on energy consumption during product use and materials used in manufacturing (e.g. product mass, recycled material content, disassembly time, and number and variety of parts). By using these metrics, product lines may have assigned to them set specific goals for improvement. Over the past five years, Hewlett Packard's Vectra personal computer line has reduced product mass by 46%, reduced the number of parts from 1650 to 350 and improved the ease of disassembly and recycling.

Not only do EPIs allow firms to evaluate their own performance but also EPIs can also be applied upstream to select and manage business relations with suppliers. Some corporations, such as B&Q, Sun Microsystems and The Body Shop, have already begun to develop environmental performance evaluation schemes for their suppliers. B&Q, the

THE USE OF MOTOR VEHICLES CONTRIBUTES ABOUT 16% TO OVERALL US ENERGY consumption and greenhouse gas emissions. The manufacturing of these vehicles also contributes to energy consumption and greenhouse gas emissions. In its recent corporate environmental, health and safety report, GM, an endorser of the CERES principles, backed up its commitment to conserving resources at every stage of the product cycle with interesting information about energy use. GM also voluntarily reports its greenhouse gas emissions under section 1605b of the 1992 Energy Policy Act. These GM reports break fresh ground by reporting on the contribution to greenhouse gas emissions during manufacturing as well as the subsequent use of their products in the hands of consumers (GM 1996).

Although the company has little control on how its products are used after sale, and although there is no legal responsibility to track the environmental performance of its products, GM projected energy use and greenhouse gas emissions from GM cars and light trucks in the United States. These estimates, presented in Figure 10.7, were based on assumptions about fleet fuel efficiency, the number of vehicles on the road and the miles driven. The results indicate that the use of GM vehicles accounts for significantly more greenhouse gases than does their manufacture. As part of its longer-term commitment to improve the energy efficiency of vehicles, GM has joined the US government, Ford and Chrysler in the 'Partnership for a New Generation of Vehicles' to help promote the development of affordable, fuel-efficient, low-emission automobiles.

Box 10.3 **Energy use at General Motors**

Figure 10.7 **General Motors greenhouse gas emissions from US sources**
(in millions of tonnes carbon dioxide equivalent per year)

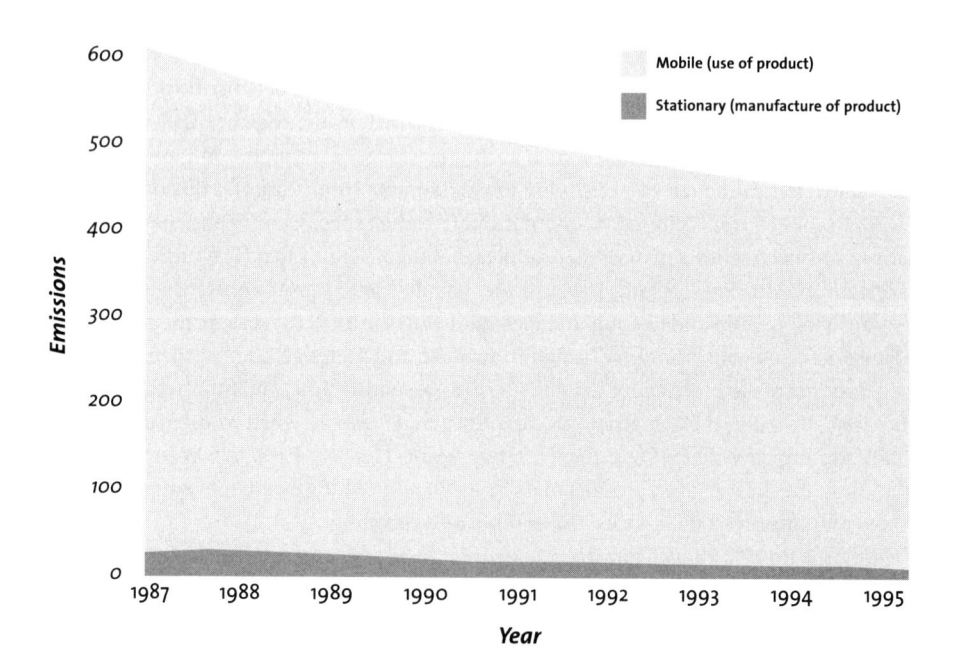

UK's largest home-improvement and garden centre, grades some 500 suppliers in 60 countries. Suppliers are evaluated both on their understanding of environmental issues and on their performance. Faced with a growing number of customer queries regarding environmental issues, other companies are following suit. Harmonising what should be measured and how the measurements should be taken will reduce the burden of firms responding to multiple queries.

With many governments around the world in the process of developing 'green' procurement guidelines both for suppliers and for their products, and with firms coming under increasing pressure from customers, the time is ripe for applying these EPIs. In the United States, Executive Order 12873 requires federal agencies to buy and use 'environmentally preferable' products. In a first pilot project focusing on cleaning products, the US Environmental Protection Agency (EPA) discovered strong interest in information on ingredients and their characteristics, information basic to the environmental performance of the product, not just the eco-label alone (EPA 1997a).

Materials efficiency and pollution prevention

In an influential *Harvard Business Review* article, Michael Porter and Claas van der Linde (1995b) argued that firms 'must learn to frame environmental improvement in terms of resource productivity'. Polaroid, guided by the premise that decreased use and reduced generation of toxics minimises the subsequent management of waste, has developed metrics to track materials use and waste generation (Epstein 1996). Information on chemical use and generation of wastes is used to set priorities for changes in manufacturing processes and the design of new products, processes and facilities. The company's centralised environmental accounting and reporting system produces corporate and divisional reports that monitor existing production and predict the impact of new chemicals on these key EPIs.

Many firms in the states of New Jersey and Massachusetts have also found that information on toxic chemical use helps identify opportunities to reduce waste, increase productivity and save money (see Box 10.4). In both states, firms who use a large quantity of chemicals report a simple method of materials accounting for toxic chemical use at their facilities, which includes the amount of toxic chemical used, produced, brought on site and shipped off-site as waste or product. The requirement to track chemical use is mandatory for some facilities in these states, which can create economic and environmental benefits for the firms.

Progressive businesses in other countries are embracing the importance of resource efficiency. The Colombian Business Council for Sustainable Development (CECODES), which includes 30 large firms contributing 4% to the national economy, has developed tools for measuring and improving eco-efficiency.[2] Since 1995, participants have reported on a variety of quantitative indicators. For example, overall these firms estimate US$1.6 in value-added per cubic metre of water and US$4.5 in value-added per kilowatt-hour of energy.

2 Personal communication with María Emilia Correa, Executive Director, Colombian Business Council for Sustainable Development, 11 February 1997.

IN THE UNITED STATES SOME STATES REQUIRE INDUSTRIAL FACILITIES TO GO beyond TRI to provide information on chemical use. New Jersey has required chemical-use reporting since 1987. Both New Jersey and Massachusetts have set overall goals for reducing wastes and have asked facilities to report their goals and their progress on pollution prevention. A recent analysis of data from Massachusetts revealed a 17% reduction in toxic chemical use and a 25% reduction in generation of hazardous waste between 1990 and 1994 (TURI 1997). A separate report from the New Jersey Department of Environmental Protection estimates a 50% decline in non-product output (NPO) in New Jersey between 1987 and 1994 (Aucott et al. 1996).

These encouraging results from Massachusetts and New Jersey stand in sharp contrast to national trends where the generation of toxic chemical wastes prior to treatment has steadily grown, as illustrated in Figure 10.8. In addition to these environmental benefits, in New Jersey for every dollar spent by firms and the regulators on the additional reporting and planning requirements, companies reported economic savings between US$5 and US$8 (NJDEP 1996).

Box 10.4 **Firms in New Jersey and Massachusetts benefit from tracking chemical use**

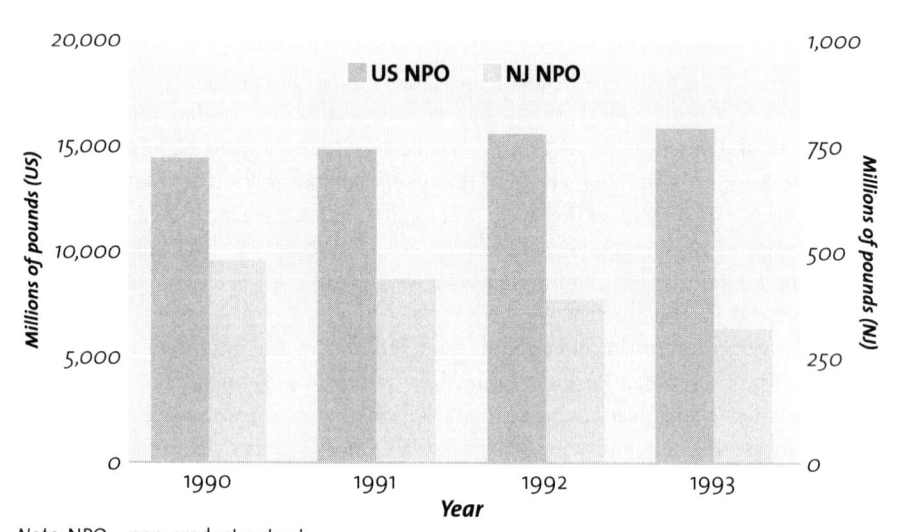

Figure 10.8 **New Jersey and US NPO (using TRI section 8 data)**

Note: NPO = non-product output

EPIs in other business decisions

The WRI–Tellus survey revealed that EPIs are being used by respondents inside firms to make business decisions, not just environmental decisions. Respondents were asked to choose the top three business decisions most affected by the use of EPIs (see Fig. 10.9). Regulatory compliance was the most frequently cited use of EPIs; the majority of these environmental managers scored compliance decisions as those decisions most affected by EPIs. This comes as no surprise; however, EPIs are also evident in other business decisions. Strategic planning was rated in the top three by almost two-thirds of respondents. Benchmarking, investment and other core business decisions—such as purchasing, product design and research and development (R&D)—were influenced by EPIs, albeit to a lesser extent. One respondent noted that EPIs were used in evaluating the prospective acquisition of a new business. These results are encouraging in that they suggest that the respondents use EPIs to inform mainstream business decisions, far beyond the traditional boundaries of the environmental, health and safety department.

The fact that employee compensation failed to make the top three decisions of any of these environmental managers is particularly noteworthy. This reinforces the findings of a 1996 Conference Board survey that revealed most firms do not include metrics for pollution prevention, materials efficiency or product stewardship in employee evaluation and incentive schemes. This is an unfortunate omission, since internal incentives are a powerful means for driving the integration of environmental considerations into routine decisions.

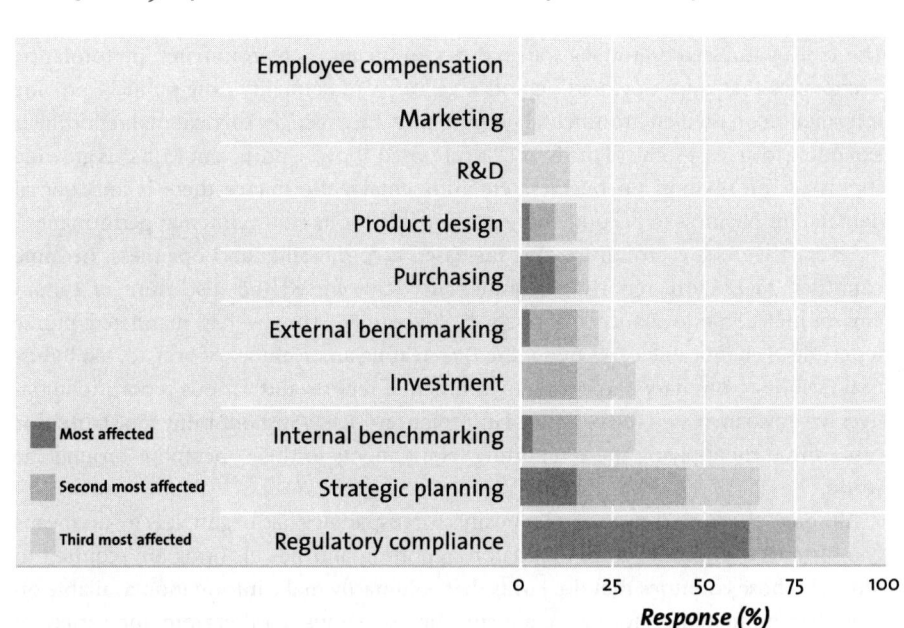

Figure 10.9 **Top three business decisions most affected by environmental performance indicators**

Lessons from internal uses of corporate EPIs

What does this brief foray into the world of corporate performance measurement indicate about current business practice? First, the four key EPIs are already being used inside firms to support a range of decisions, not just environmental decisions. There is a clear business case for these EPIs. By creating environmental accountability within an organisation, EPIs are driving improvements in resource efficiency and increasing profitability. Perhaps EPIs are already being used in many other ways, a question that should be explored through surveys and case studies of other business decision-makers and industries. Second, comparability is a critical component in the business utility of EPIs. Businesses cannot benchmark their own performance over time or against others without this. This is especially important for corporations with multiple facilities operating across international boundaries.

Although the business examples and the results of the survey offer promising signs, further progress is needed. More firms need to measure environmental performance. Firms that have started need to adopt a broader view of performance, upstream and downstream of their manufacturing activities. Ultimately, firms that are genuinely committed to improving their environmental performance will benefit from the widespread adoption of a common set of standardised, transparent metrics. In so doing, firms will be able to track their progress, take credit for achievements and distinguish themselves from their competitors.

Integrating EPIs in decisions outside the company

The 1990s witnessed a growing transparency in business environmental performance. Companies that once operated behind a veil of confidentiality now publicly provide information on their environmental performance. Increasingly, this external reporting is extending to at least some of the four EPIs proposed above. Concurrent with this growing supply of information on business environmental performance there is an external demand for business to become more accountable for its environmental performance.

Several factors are prompting this increased accountability and openness. In some countries, mandatory reporting requirements have forced the disclosure of certain environmental performance information. Since 1986, US law has mandated public reporting of chemical releases under the TRI, creating an inventory of over 20,000 industrial facilities. Voluntary corporate environmental reports and various reporting initiatives have also increased the volume of information (UNEP/SustainAbility 1996). In some cases, firms are applying these reporting requirements to their operations around the world.

Although the transition to greater business transparency has begun, it is by no means complete. Few countries have mandatory reporting and not all firms are required to report in those countries that do. Firms that voluntarily make information available on their environmental performance are still the exceptions. Furthermore, the variety of approaches to reporting environmental performance information makes it difficult, if not impossible, to compare products, facilities, firms, sectors and countries.

Using examples drawn from around the globe, this section illustrates how the same EPIs that are useful inside firms can also serve users and uses outside firms, including governments, communities, financial institutions and others. This section also highlights the need for a unified reporting framework that embraces transparency, comparability and completeness. In the absence of such a framework, governments, communities and companies will be awash in a sea of contradictory, disconnected and incomparable measures of performance.

◢ Using EPIs to track national goals

Just as firms need EPIs to track progress toward corporate goals, countries also need macro-level EPIs. Unfortunately, efforts to track progress toward national goals are typically limited by the quality and completeness of information and the receptiveness of policy-makers. National commitments under international conventions on climate change, biodiversity and forestry could propel demand for better indicators of national environmental performance and shift greater responsibility to companies for measurement and reporting progress.

The Netherlands offers one of the best examples of a comprehensive national approach to environmental performance goals and indicators, an approach similar to Nortel's, described above. The Dutch have established a set of national environmental goals covering climate change, ozone depletion, acidification, eutrophication, dispersion of toxic substances and solid waste disposal (VROM 1997). Each is measured by a composite EPI that serves to signal progress, as shown in Figure 10.10. The broad scope of these indicators allows for the tracking of national progress and for comparisons across economic sectors. Specific targets to reduce toxic substances, for example, also enter into sectoral and facility negotiations.

The Dutch indicators are evolving in interesting ways. Although Dutch environmental authorities have maintained a broad inventory of emissions for more than 20 years, they have allowed very limited public disclosure. Beginning in 1999, the largest industrial facilities will be required to submit annual pollutant reports, which will be made public. The contribution of smaller industrial sources and non-point sources, including transportation and agriculture, will be estimated separately. This merging of the macro-level indicators with a micro-level reporting (more akin to the TRI) represents a powerful combination of goal-directed policy and disclosure-based pressure for improvement.

Although the United States lacks such a comprehensive set of national environmental goals, some aspects of environmental performance have given rise to quantitative milestones. EPA's 33/50 programme, a voluntary initiative to reduce emissions of 17 high-priority TRI chemicals, established measurable goals and urged firms to commit to meeting them. Using the 1988 TRI reporting year as a baseline, the 33/50 programme sought emission reductions of 33% by 1992, and of 50% by 1995. What happened was that

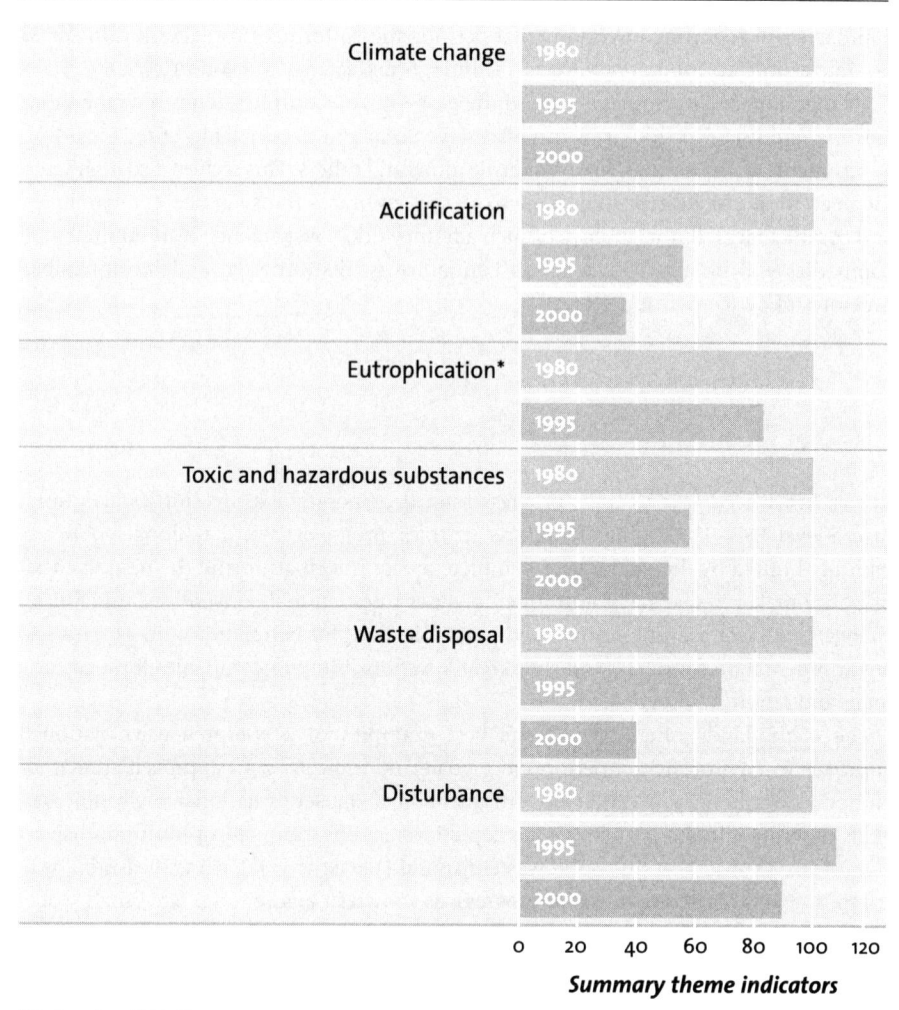

[Bar chart showing summary theme indicators for Climate change, Acidification, Eutrophication*, Toxic and hazardous substances, Waste disposal, and Disturbance, with bars for 1980, 1995, and 2000]

* No target available for 2000

Figure 10.10 **Tracking progress toward national goals: the Dutch experience**

emissions of these chemicals were halved a year ahead of schedule. Although the success of the 33/50 programme contributed to the overall reduction in TRI emissions, the actual generation (i.e. non-product output) of toxic chemicals has risen by 7% since 1991 (EPA 1997b). This suggests that improvements in pollutant emissions have been gained by pollution control rather than pollution prevention. Clearly, if pollution prevention is a national priority, then goals and indicators must reflect these concerns.

External comparability and transparency

The US TRI, a mandatory information tool, provides the prototypical example of how public disclosure of standardised environmental performance information not only measures progress but also drives improvement. TRI requires annual reporting on specific amounts and types of several hundred chemicals released or otherwise managed by industrial facilities. Essentially, all of the information contained in the TRI database is publicly available through public libraries, electronic channels,[3] and other means. The original legislative language specified that the data be made publicly available in a computerised format. Between 1988 and 1995, total releases of these toxic chemicals fell by approximately 46% from the reporting facilities, a testimony to the power of public disclosure to drive improvement (EPA 1997b).

TRI clearly illustrates the utility of data that is transparent, accessible and comparable, but it has some serious shortcomings. It covers only toxic chemicals, the largest manufacturers, firms in the United States and two of the four EPIs proposed earlier in this chapter—non-product output and pollutant releases. In a move to fill some of these gaps, the US EPA published final regulations in May 1997 extending TRI reporting to electric utilities, mining, waste incinerators and four other industries (EPA 1997c). The EPA has also announced its intention to propose an expansion of TRI data elements to include information on chemical use, which would begin to address another of the four key categories of EPI proposed earlier in this chapter. This proposal has triggered strong objections from industry associations and their member firms. In particular, there is concern that making chemical-use data available will reveal trade secrets (see Box 10.5). Although evidence from New Jersey and Massachusetts suggests that these concerns may be overstated, the architects of this new framework must be sensitive to the protection of legitimate commercial confidentiality.

National pollutant release inventories

Outside the United States, several countries have operational TRI-style national pollutant inventories; these include Australia, Canada, the Netherlands, Ireland and the United Kingdom. In addition, Japan and Sweden will introduce national pollutant inventories within the next few years. National environmental authorities, firms, researchers and non-governmental organisations (NGOs) have shown keen interest in learning from each others' experiences. Through a series of international workshops, the OECD developed a guidance document to aid governments in developing pollutant release and transfer registers (PRTRs) (OECD 1996b).

In a unique move toward truly international comparability, the national environmental authorities of the United States, Canada and Mexico agreed to create a regional pollutant inventory for North America featuring consistent data (see Box 10.6). Canada had modelled its National Pollutant Release Inventory (NPRI) on US experience, and Mexico is benefiting from the joint collaboration. Soon North America will be the largest region in the world reporting on industrial pollution in a comparable way. In Europe the

3   rtk.net

MANY FIRMS REMAIN WARY OF DISCLOSING PERFORMANCE INFORMATION, particularly materials use, for fear of disclosing trade secrets to competitors. Respondents to the WRI–Tellus survey also indicated little willingness to publicly report information on energy and chemical use. Only about a third of respondents were in favour of publicly reporting these two indicators compared with almost 75% for pollution emissions. In general, respondents opposed reporting at the product or process level but were more favourably inclined toward disclosure at the facility and corporate level.

Yet experience from New Jersey and Massachusetts, two states that require materials accounting, shows that few companies ever claim chemical use data as a trade secret. In New Jersey, which automatically accepts trade secret claims as valid unless contested, only three out of the nearly 600 firms claimed trade secrets in the 1996 and 1997 reporting years (USPIRGEF 1997). Similar results were found in Massachusetts. One possible explanation is that the time-lags between actual performance and public disclosure are long enough that the information is not very valuable to competitors. Nonetheless, public reporting remains a cultural barrier within industry and must be addressed while fulfilling the needs of stakeholders for information on corporate environmental performance. Perhaps Harris Gleckman, director of Benchmark Consulting, was only half right when he stated that 'Business efficiency is not a public issue. Business environmental impact is' (BEC 1995). To the extent that business efficiency means more effective resource utilisation and lower environmental impacts, it is very much a public issue.

Box 10.5 **Confidential business information**

Box 10.6 **Comparing environmental performance in North America**

THE COMMISSION FOR ENVIRONMENTAL CO-OPERATION (CEC), THE TRINATIONAL organisation set up under the North American Agreement on Environmental Co-operation, is developing a series of reports that assemble existing public information on pollutant release and transfer registers in Canada, the United States and Mexico. The CEC believes that tracking environmental pollution is essential to enhance environmental quality, to increase public and industry understanding of the types and quantities of toxic chemical released into the environment, to encourage progress in emission reduction and to assist governments in setting priorities for action.

In an important step forward the CEC published a report with the first rigorous comparison of reporting systems on industrial pollution in three countries (CEC 1996).

There are some differences between the US and Canadian inventories: in 1994, the TRI reported on 368 chemicals, the NPRI on 178. Differences also exist in the sizes and types of industry that must report. Mexico is currently in the process of a developing a similar system, the Registro de Emisiones y Transferencia de Contaminantes. A second CEC report compares 1994 data from Canada with data from the United States (CEC 1997). Despite the differences, there are remarkable similarities in terms of which industries and chemicals top each country's lists. One surprising, but as yet unexplained, finding was that Canadian facilities reported sometimes more than twice as many emissions as US facilities, as indicated in Figure 10.11.

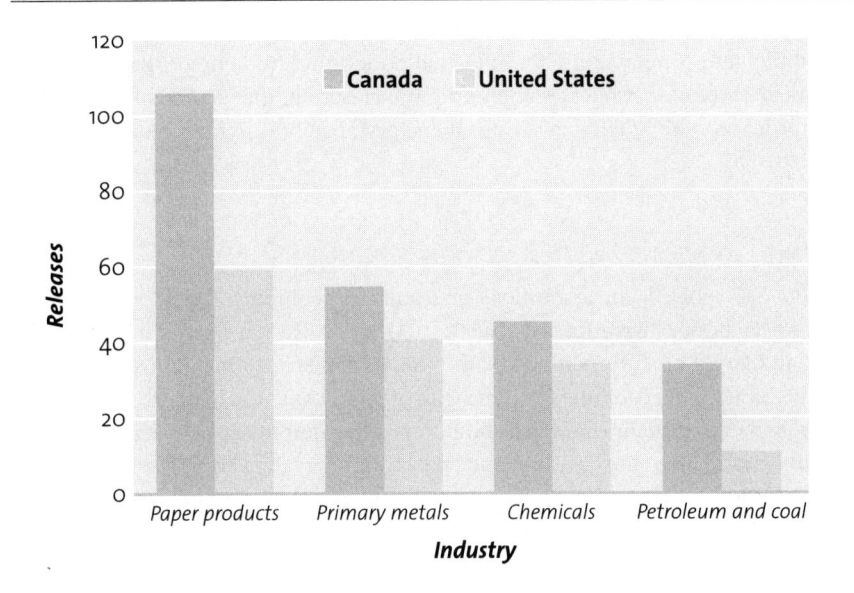

Figure 10.11 **Comparison of chemical releases between the United States and Canada (in kg per facility), 1994**

UN Economic Commission for Europe is taking the lead under the Aarhus Convention on Access to Information, Public Participation in Decision-Making and Access to Justice in Environmental Matters, which was signed in 1998. A Task Force on PRTRs has been established to implement Article 5.9, which calls for parties to develop a nationwide 'structure, computerized and publicly accessible database compiled through standardized reporting'. At the same time, the UN Institute for Training and Research is working with countries in Central Asia and Africa on PRTRs, and non-governmental groups are encouraging adoption of Aarhus principles in other regions.

Some governments and NGOs are making detailed information available via the Internet on pollution from facilities in Canada, the United States and the United Kingdom. The Internet has emerged as a particularly quick, powerful and cost-effective tool for disseminating and accessing environmental performance information around the world. For example, when Friends of the Earth UK posted data from the country's Chemical Release Inventory as a map-linked database on the Internet, they clocked some 25,000 visits in the first year of operation. In contrast, local offices recorded only 750 visits to review similar information on paper—a striking example of the distinction between public availability and public accessibility.

The trend toward greater public disclosure of industrial pollution information is by no means limited to TRI-style pollutant inventories. Denmark's environmental reporting law, which took effect in 1996, requires nearly 2,000 companies to publish environmental reports, known as 'green accounts', for each production site (Rikhardsson 1996, 1996). These reports, modelled on a financial reporting approach, will help standardise performance information on use of raw materials and release of pollutants.

The actions described above all signal a major step in the direction of greater public accountability and comparability for industrial pollution. Although most actions have concentrated on pollutant releases and non-product output, they represent key opportunities for incorporating materials use and energy consumption for a broader section of industry.

Benchmarking US refineries

As the TRI has evolved, an assortment of regulators, equipment vendors, financial analysts, researchers, activists for social justice and journalists have found novel ways to interpret and apply the information (Unison 1995). The Environmental Defense Fund (EDF), a US-based NGO, combined TRI data with other information to benchmark 166 oil refineries by their pollutant emissions and non-product output (Epstein *et al.* 1995).

This study normalised the environmental data by refinery capacity (e.g. emissions per barrel of oil) to adjust for size (see Box 10.7). EDF also ranked the 34 states with refineries, aggregating pollutant releases and waste generation across each state's refineries. EDF concluded that states' regulatory and enforcement efforts vary greatly in their attempts to reduce waste releases and off-site transfers. Those states with poorer refinery performance included West Virginia, Kansas and Texas. At the other end of the spectrum, New Jersey ranked among the best.

For US oil refineries, pollutant releases generally correlate with waste generation. Yet performance on both EPIs stretches over several orders of magnitude. Many factors lie behind the wide scatter in performance: the source of crude, product mix and refinery vintage can have a bearing on environmental performance. Beyond the specifics of petroleum refining, this external benchmarking of facility-level EPIs offers an important lesson. EPIs—at least those that are comparable and transparent—can provoke questions about the determinants of performance and opportunities for improvement.

Box 10.7 **Benchmarking US petroleum refineries**

IN A CREATIVE USE OF PUBLICLY REPORTED INFORMATION, THE ENVIRONMENTAL Defense Fund, a US-based NGO, ranked the environmental performance of 166 US oil refineries (Epstein *et al.* 1995). Detailed data on toxic waste generation and pollutant release were downloaded from the Right-to-Know Network (rtk.net) and normalised by refining capacity. The resulting EPIs reveal the relative performance across the sector, information that is valuable to communities, regulators and the industry itself.

These general patterns are borne out on closer inspection of these data. To illustrate, Table 10.1 presents results for the seven participating refineries in Illinois, along with their national ranking. The state, which stands fourth behind Texas, Louisiana and California in refinery capacity, is among the most efficient in controlling toxic pollutant releases. Yet, barrel for barrel, the Shell oil refinery at Wood River releases over 18 times more toxic chemicals to the environment than does Clark Oil's Hartford refinery. With regard to waste generation, the spread is even more dramatic, although this may reflect some lack of clarity in EPA's reporting instructions.

Comparing refinery performance in Illinois

COMPANY	LOCATION	CAPACITY barrels/day	POLLUTANT RELEASE EPI[†]	POLLUTANT RELEASE US Rank	WASTE GENERATION EPI[†]	WASTE GENERATION US Rank
Clark Oil and Ref.	Blue Island	65,000	0.67	24	0.64	15
Clark Oil and Ref.	Hartford	57,000	0.32	14	0.33	10
Indian Ref. L.P.	Lawrenceville	69,000	2.00	49	727	152
Marathon Oil Co.	Robinson	175,000	1.79	40	3.20	35
Mobil Oil Corp.	Joliet	180,000	0.58	22	55.2	107
Shell Oil Co.	Wood River	273,000	5.80	104	60.5	110
Uno-Ven Co.	Lemont	147,000	1.07	31	3,320	160

† Annual pollution, in pounds, divided by barrels of capacity per day
Note: EPI = environmental performance indicator

Table 10.1 **Benchmarking US petroleum refineries: results for the seven participating refineries in Illinois**

Source: Epstein *et al.* 1995

Performance reporting and alternative regulatory approaches

In some countries the demand for new strategies for environmental protection is coalescing around the idea of performance-based incentives with greater public involvement. In the United States, several regulatory initiatives, as well as the President's Council for Sustainable Development (PCSD), have sought novel ways to broker greater regulatory flexibility in exchange for better environmental performance (Aspen Institute 1996; PCSD 1996). As Leslie Carothers of UTC explains, 'Having meaningful, auditable, public information on business environmental performance can and should free government regulators to focus on non-compliers and lagging companies, where the need for traditional regulation and enforcement is the greatest.'[4] In essence, EPIs provide a means for demonstrating and taking credit for improved performance—a 'carrot and yardstick' approach to driving progress toward environmental goals.

In developing countries the challenges are very different but the importance of EPIs is still enormous. Rapidly growing economies, such as those in South-East Asia, are experiencing growth in their manufacturing sectors of about 10% a year, which means a doubling of their industrial base within a decade. Yet the regulatory institutions responsible

---

4 Personal communication with Leslie Carothers, Vice-President, Environment, Health and Safety, United Technologies Corporation, 1997.

for environmental protection typically lack the resources to implement, monitor and enforce compliance programmes effectively. For these countries new approaches to environmental management that go beyond traditional command and control are essential to steer clear of serious environmental costs as a side consequence of economic success.

Indonesia provides an interesting example of how to implement a practical information-based strategy to combat environmental pollution (see Box 10.8). The national environmental agency of Indonesia has launched an innovative programme known as PROPER (the pollution control, evaluation and rating programme) to rank the environmental performance of facilities in water pollution control. On the basis of self-reported data, site inspections and independent monitoring, 187 facilities were assigned to one of five colour-coded categories (see Chapter 7). The top gold ranking, which no facility has yet attained, requires cutting emissions to less than half of the regulatory limit, controlling air pollution and wastes, and 'extensive use of clean technology'. Although PROPER is far from perfect, it does show that EPIs are also applicable in developing countries and can be implemented in ways that suit local needs and institutional capacities. The World Bank is currently working in the Philippines, Colombia and Mexico to extend this approach.

International standards and environmental performance

From Oslo to Kuala Lumpur, attention is being lavished on the ISO 14000 series, the new series of international standards on environmental management. Although in some ways less demanding than Europe's EMAS, ISO 14001, the international environmental management system standard, is stimulating worldwide interest in corporate environmental management. Even firms that have chosen not to be registered are scrutinising the requirements and guidance developed over the past several years.

Box 10.8 **Harnessing the power of public opinion in Indonesia**

WEAK ENFORCEMENT OF EXISTING WATER REGULATIONS COMBINED WITH limited resources led Indonesia's National Pollution Control Agency, BAPEDAL, to try an alternative strategy for encouraging better environmental performance among its growing manufacturing sector. With the support of the World Bank, BAPEDAL launched a programme for rating and publicly disclosing the environmental performance of Indonesia's major water polluters (www.nipr.org). PROPER rates firms on their performance at preventing and controlling water pollution.

As demonstrated in Table 10.2, in less than two years the PROPER scheme has yielded significant improvements, especially among facilities previously out of compliance. Each of the 187 facilities selected was assigned to one of five colour-coded categories on the basis of performance. In the first three rounds of evaluation, no facility has yet achieved the highest gold rank, but there has been a large shift from red to blue, that is, into compliance. The number of laggards assigned to black has fallen from six to three to just one.

Rating	Environmental performance	Number of facilities June 1995	December 1995	September 1996
GOLD	World class, exceeds waste and air pollution standards	0	0	0
GREEN	Significantly exceeds water pollution standards	5	4	5
BLUE	Compliance with water pollution standards	61	72	94
RED	Below compliance, some pollution control	61	72	94
BLACK	Below compliance, no pollution control	6	3	1

Table 10.2 **Indonesia's environmental performance ratings**

Europe's EMAS distinguishes itself from ISO 14001 by virtue of its requirement for a site-specific public environmental statement. Using the model of financial auditing, EMAS provides for third-party verification of regular site audits and the public statement. The public environmental statement must provide information on use of raw materials, water and energy, on pollutant emissions, waste generation, noise and on any other significant environmental effects—data elements similar to the key EPIs proposed earlier in this chapter. Although EMAS was adopted in 1993 as a voluntary regulation, provision was made to allow it to be converted to a mandatory requirement for certain sectors and for installations above a certain size. It remains to be seen how firms seeking certification will choose between ISO 14001 and the more stringent requirements of EMAS. Meanwhile, the European Communities Commission has decided to consider ISO 14001 equivalent to EMAS, provided the facility publishes a public environmental statement.

ISO 14001, with associated guidance on life-cycle analysis, auditing, environmental performance evaluation and labelling, identifies the machinery needed to build an environmental management system in the widest sense. To be certified, firms must document a suite of policies and procedures for environmental management, including a commitment to continuous improvement (Tibor and Feldman 1996). Unfortunately, the standard is deliberately silent on what constitutes meaningful goals, leaving each firm to establish their own milestones and metrics. In fact, environmental performance is not a factor in certification, which has opened ISO 14001 to some serious criticism (BEC 1995).

In a sense, ISO 14001 is premised on the assumption that 'the system is the solution', that is, with better management comes better performance. Many NGOs, regulators and even some business managers are sceptical whether the emphasis on management systems will yield real performance improvements. ISO 14031, the guidance on environmental performance evaluation, currently in draft form, acknowledges the importance of measuring inputs and outputs, much like the EPIs presented earlier in this chapter. But the entire document, including a long list of illustrative EPIs, is optional (ISO 1996).

Although ISO 14001 is already a *fait accompli* it will take time before the value of certification is established. Firms can take advantage of its skeletal architecture, imbue it with concrete goals and EPIs and commit to forthright communication with their stakeholders. Companies that do this stand the best chance of earning greater respect from their employees, the communities in which they operate and perhaps their customers and investors. Those that aim for the bare minimum, who hedge their bets by concealing their goals and progress, may bear the costs of these international standards without realising the potential value. Meanwhile, outsiders in search of environmental performance will have to look behind ISO certification for concrete EPIs and evidence of the operational effectiveness of the management system.

EPIs and the financial community

Given that today's investment decisions shape future corporate development paths, how can the financial sector's pivotal role be harnessed to encourage firms to adopt cleaner, more efficient technology, products and processes? Although there has been much debate on the link between environmental and financial performance and a fresh batch of anecdotal evidence and research (Blumberg *et al.* 1997; Feldman *et al.* 1996; Schmidheiny and Zorraquín 1996), the financial community is dissatisfied with the current state of EPIs. When 85 financial analysts in London were asked about environmental disclosure and reporting, a mere third considered the information adequate for routine company assessments (EFBE 1994).

The attitude of the financial community may be changing. At the Third International Roundtable Meeting on Finance and the Environment, in May 1997, sponsored by the United Nations Environment Programme, over 300 participants, including representatives from business and the financial sector, met in New York to explore the connections between financial and environmental performance.

Halfway around the world from New York, the Jakarta Stock Exchange requires firms to have at least a 'blue' rating in Indonesia's PROPER scheme before they can go public—an illustration of how financial drivers can reinforce environmental performance and vice versa. A 1996 Natural Resources Defense Council report on the hidden environmental liabilities of power-plant ownership ranked electricity suppliers according to greenhouse gas emissions per dollar of operating revenue (Cavanagh *et al.* 1996). Control of greenhouse gas (e.g. carbon dioxide) emissions is the subject of intense international negotiation. From the least-carbon-intensive utility (Pacific Gas & Electric [PG&E]) to the most (PSI Energy Inc.) these indicators vary enormously. For each dollar of operating revenue, PSI-owned plants emitted 25 times more carbon dioxide than PG&E-owned plants.

Superior environmental performance can translate into reduced operating risk, lower costs and a competitive edge. But two prerequisites are necessary to make environmental considerations a routine part of investment and lending decisions. First, a clear quantifiable link must be made between environmental and financial performance. It is particularly important that leaders in the financial community work with businesses, academics, NGOs and others to expand the collective grasp of the 'dollars and sense' of EPIs. Second, information gaps on corporate environmental performance must be filled and a standardised framework constructed to govern reporting. Widespread corporate adoption and reporting of these EPIs are important steps toward closing the information loop between financial and environmental performance.

Lessons learned from external uses of corporate EPIs

Despite the lack of a universal framework for measuring and reporting environmental performance, these examples demonstrate a growing international transparency emerging in corporate environmental performance. Many countries now require firms to publicly report on some of the four key EPIs. Firms are also volunteering information on their environmental performance through corporate environmental reports, adherence to various codes of practice and participation in environmental management standards.

When in a transparent and comparable format, the very same EPIs that were useful to people inside the firms are also being used by people outside the firms. EPIs represent powerful tools for holding firms publicly accountable for the environmental impacts of their activities. Experience from North America, Europe and Asia suggests that the mere act of publicly reporting can drive firms to improve their environmental performance. Such information-based strategies can be implemented in ways that are compatible with the level of development and institutional capacity of the countries involved.

Corporate EPIs: the way forward

Important lessons

This broad sampling of internal and external uses of corporate EPIs holds some important lessons. First, there are a variety of users of EPIs inside companies and among their many stakeholders. This diversity of perspectives has fed the proliferation of different metrics. Different languages, regulatory cultures and priorities in countries compound the confusion. The four key EPIs proposed in this chapter focus managerial and public attention on a firm's basic material and energy inputs and outputs as products and wastes.

Second, EPI consumers share an interest in comparability, transparency and scope. This desire is reflected in the internal rationalisation of environmental information management systems in multinational corporations and in the push to reform reporting formats and requirements under existing regulatory programmes. These characteristics should serve as design principles for a common framework for tracking environmental performance.

Third, many opportunities exist for expanding the use of these key EPIs and weaving them into the overall fabric of environmental accountability. This final section highlights several of these opportunities and points the way toward progress on four broad fronts:

- ☐ Making environmental performance measurement standard business practice
- ☐ Recognising firms that demonstrate improved performance and greater disclosure
- ☐ Expanding public access to environmental performance information
- ☐ Facilitating standard formats for environmental performance reporting

For each challenge a set of recommendations is presented that specifies what needs to be done and some promising examples of steps in the right direction are pointed out.

Making environmental performance measurement standard business practice

Except for such mandatory reporting requirements as pollutant emissions, most firms are still in the dark about the environmental performance of their business activities. This makes the firms vulnerable to changing regulations, stakeholder expectations and customer demand. Excellence in environmental performance will become an integral part of business economic viability. The following recommendations are aimed at helping firms better measure, manage and improve environmental performance.

Firms should establish environmental goals corresponding to these EPIs. Indicators provide critical links in the chain of vision, strategy and implementation. With a strategy of resource productivity, pollution prevention and product stewardship, firms will realise economic and environmental benefits. A growing array of companies are using EPIs to quantitatively track progress toward corporate environmental performance goals. Judging by voluntary corporate environmental reports, companies such as Novo Nordisk, Kunert, Nortel and British Petroleum are moving in this direction. What is more, Dow Chemical recently announced ten-year environmental goals that include reducing chemical emissions and waste generation per pound of production by 50% and energy use per pound of production by 20% (Stavropoulos 1996).

Firms should use these EPIs to benchmark their performance internally and against other companies. Systematic comparison of environmental performance is valuable to provoke questions about performance and opportunities for improvement. Others outside companies are already beginning to do such benchmarking. With the expansion of publicly available datasets, such as national PRTRs, firms will benefit from greater access to comparable international information. Internally, firms should use these EPIs to measure the effectiveness of their environmental management systems.

Companies should revamp their information systems to integrate these EPIs into their internal management and reporting systems. Embedding quantitative measures of materials and energy flows in the corporate information architecture will help ensure that environmental performance is factored into business decisions throughout the firm. In

1992, for example, GM's Europe Technical Development Centre in Germany began developing an information system to support materials management as well as environmental reporting. The resulting system is now being piloted at other GM facilities in Europe and North America. Lessons learned in this experience will be incorporated into new plants in Thailand and Poland with the goal of creating a tool to support pollution prevention, life-cycle assessment and regulatory reporting on a global basis (GM 1996). Companies in the process of overhauling their internal systems for management information, accounting systems and technical communication have a golden opportunity to fold these EPIs into the design.

Firms should integrate these EPIs into managing their supply chains. Companies committed to improving environmental performance along the product chain should require standardised information on these EPIs from their tier-one suppliers. The resulting information can serve the downstream firm in two ways. First, it flags environmental issues that might otherwise pass unnoticed, issues that could interrupt supply or tarnish the reputation of their products. Second, it helps shift some of the responsibility for answering customer environmental queries to the suppliers. Certainly, environmental performance will not replace cost, quality and reliability as key criteria in choosing suppliers but it should be a factor in responsible management.

Firms should incorporate EPIs into internal incentive schemes. Since these EPIs are linked to resource productivity as well as environmental improvement, firms that factor them into internal incentives can drive the organisation in the desired direction. A few firms are beginning to tie salary bonuses and other rewards to such EPIs as pollutant emissions and non-product output, but the WRI–Tellus survey suggests such opportunities for aligning employee incentives with corporate goals remain largely untapped.

Recognising firms that demonstrate improved performance and greater disclosure

Experience has shown that mandatory disclosure of environmental performance is a powerful incentive for improvement. At the same time, many companies find that current key economic drivers—for example, tax policy, accounting systems, regulations and lending policies—often work in the opposite direction and fail to reward firms that invest in better environmental performance. These recommendations aim to better understand the interplay of environmental and economic performance and to identify policy changes to bring them into closer alignment.

NGOs and others should test these EPIs by benchmarking a variety of firms. Firms that excel in environmental performance cannot expect customers, communities or shareholders to be supportive if this information never sees the light of day, which is why external benchmarking of environmental performance is crucial to competitive advantage. By publicising EPIs, NGOs, advocacy groups, researchers and others raise the price of poor environmental performance. Comparative case studies of operations in different locations, analyses across whole industry sectors and other analyses will demonstrate that there is a demand for environmental performance. Practical research of this sort will also uncover where the most serious data gaps are and force greater comparability across political jurisdictions.

Further research is needed to make the link between these EPIs and financial performance. Research into the apparent relationship between environmental and financial performance is beginning to home in on specific economic benefits, such as the cost and availability of capital, insurance rates and, for publicly traded firms, the share price (Feldman *et al.* 1996). So far these analyses have relied on the conventional descriptions of environmental performance, including regulatory violations, spills and pollutant releases. These analyses should be extended to the measures of environmental performance proposed above, with a working hypothesis that firms with greater energy and material efficiency realise economic benefits simultaneously. Further research should examine the links and gaps between corporate environmental reporting and financial reporting.

EPIs need to be incorporated into alternative regulatory approaches. Calls for regulatory streamlining, reduced paperwork and other reforms in exchange for superior environmental performance require credible indicators to assure the effectiveness of the changes. As articulated in the Aspen principles, alternative paths for cleaner, cheaper environmental protection should be predicated on a set of clear and measurable environmental goals. Policy experiments such as US EPA's Project XL and the Common Sense Initiative, which are testing the feasibility of regulatory innovations, should be judged in part on their success in demonstrating improved environmental performance and in testing new ways of reporting this to other interested parties.

It should be required that these EPIs are reported in government procurement schemes. In many countries, government purchasing accounts for a large share of the marketplace. Consequently, public-sector procurement, which covers everything from office supplies to weapons systems, can considerably affect the economy. Procurement guidelines that require information on environmental performance will reward manufacturers and retailers committed to measuring and reporting their environmental performance. In the United States, Executive Order 12873 already calls on federal agencies to buy and use environmentally preferable products. Standardised reporting on EPIs provides common ground on which the environmental performance of products during manufacture, use and disposition can be evaluated.

Expanding public access to environmental performance information

An assortment of mandatory and voluntary initiatives currently governs the disclosure of corporate environmental performance information, which creates an incomplete patchwork. Not all companies report on environmental performance and those that do often focus on only selected aspects of performance. On the voluntary front it is typically the environmental leaders that report. Although these voluntary efforts are commendable, a mandatory approach is also needed to ensure information is publicly available on laggards as well as leaders. The following recommendations are targeted at a range of

US and international initiatives that has the potential to increase the information flow on environmental performance:

National governments should establish pollutant inventories and broaden them to include these EPIs. The growth of pollutant inventories is building a comparable international information base; however, most national inventories focus narrowly on large generators, the industrial sector and pollutant emissions alone. To increase their usefulness for policy-makers and the public these inventories must expand to include other crucial categories of environmental performance and sources other than large industries. An example of an expansion is US EPA's efforts to extend coverage of the TRI programme to other sectors, including mining and electric utilities and the reporting of chemical use. What is more, in many countries, small and medium-sized enterprises account for the majority of environmental effects, and non-industrial sectors, including transportation and municipalities, make sizeable contributions, too. Countries in the process of developing national pollutant inventories should incorporate indicators on non-product output, materials efficiency and energy consumption and apply them broadly across the economy.

Local governments should promote the disclosure of these EPIs where national efforts are lacking. Local governments have a natural interest in the performance of firms operating in their jurisdictions. In the absence of national inventories, state and local governments should establish their own inventories built on these EPIs. For example, the New Jersey and Massachusetts initiatives on environmental performance measurements have both reduced hazardous waste generation and saved local companies money. At an even more localised level, the City of Eugene, OR, has established a charter that requires all users of hazardous substances to file an annual public report listing inputs and outputs of all hazardous substances. Although these examples focus on toxics, other localities can tailor their efforts to match local concerns: for example, water usage in arid regions.

Corporate codes of practice should incorporate these EPIs to increase credibility and demonstrate progress. Voluntary codes, although hampered by their limited reach, are valuable in extending disclosure about environmental performance in locations where few formal requirements for public reporting are in place. They also represent an opportunity to introduce consistency and comparability in reporting approaches. Groups of business and organisations, such as the Public Environmental Reporting Initiative (PERI), CECODES and CERES, have made some progress on public reporting of corporate EPIs.

Facilitating standard formats for environmental performance reporting

The time has come to rationalise information and reporting strategies on environmental performance measurement. It is in the interests of firms, governments and others to converge on a standardised reporting system. Anyone who has tried to obtain information on corporate environmental performance will know only too well that public availability does not necessarily equate to accessibility. Facilities are identified and classified in conflicting ways, many government databases remain largely unlinked and specific pollutants are defined in multiple ways. Even where the information is publicly available, making practical use can require sifting through massive paper records, often inconveniently located. The current revolution in information technology provides the necessary tools to rationalise and integrate an assortment of disparate information schemes within a single framework.

National governments should develop and employ standardised data reporting schemes. A common-sense approach is to rationalise the conflicting definitions that have grown up separately under air, water and waste regulations. As a result, firms, regulators and others will be able to consolidate environmental performance information, including the four EPIs proposed in this chapter, lightening the reporting burden on the regulated community while improving the quality of the data collected. Admittedly, it can take years to upgrade a country's environmental information system, but, when a country undertakes this necessary task, agencies should organise their efforts around standard information elements and link these with environmental performance. As one example, US EPA's Facility Identification Initiative is working to standardise the definition of facilities subject to federal environmental reporting and permitting requirement.

Publicly available information on environmental performance should be made widely available on the Internet. The mushrooming of on-line access to public environmental information has boosted the utility of EPIs to companies, community groups, NGOs, researchers and others on every continent. As the experience in the United Kingdom demonstrates, the Internet represents a powerful tool for relaying environmental performance data worldwide. The World Wide Web might also serve as a supplemental repository for EPIs voluntarily reported by firms, governments, consortia and other providers.

National governments should link these EPIs at the facility, industry and sectoral level with national and global environmental goals. Merging EPIs at the micro and macro levels allows firms, officials and the public to judge for themselves the contribution of individuals toward overall environmental goals and the effectiveness of environmental policy. As environmental performance information expands and grows more consistent in quality and scope, national governments, industry associations, NGOs and others will be able to contrast the performance of specific facilities or companies to more aggregated trends. As exemplified by the North American initiative on pollutant inventories, consistent management of environmental information facilitates broad international comparisons.

Conclusions

Environmental performance indicators have a powerful appeal to business managers and outside parties, but, to be most useful, this information must adhere to basic guidelines of what gets measured and how. The four categories of environmental performance—materials use, energy consumption, non-product output and pollutant releases—help firms, regulators, communities and others to reach beyond the traditional focus on compliance, turning the spotlight instead on resource efficiency, pollution prevention and product stewardship. Business leaders committed to these goals can gauge their performance in these areas, establish goals and publicise their progress. At the same time, government reporting requirements and their information management needs major reorganisation: EPIs provide key elements for this overhaul. Communities, investors, NGOs, researchers and others with a stake in the environmental performance of business should press for credible, comparable environmental information and should reward those firms and sectors who rise to this challenge. From these collective efforts a new system will emerge of accountability for corporate environmental performance, and a robust framework for evaluating progress toward environmental objectives from the local to the global scale.

POSTSCRIPT

Melito Salazar and Warren Evans

The vision and determination of the architects of this book have resulted in an attractive design for achieving environmental objectives through policy reform—with the ideas and materials coming from government officials, politicians, industrialised and developing-country researchers, industrialists and bilateral and multilateral development specialists. The result is that we are one large step closer toward developing an agenda of win–win policies that can generate economic gains while achieving environmental objectives. But, in looking at the next steps, we must remember that any design is successful only if the blueprint is actually used—and Asia's developing countries have not demonstrated a propensity towards following environmentally sustainable designs up to now.

This book comes at a time when multilateral development banks such as the Asian Development Bank (ADB) recognise that policy reform is a key to sustainable development. Such organisations are increasing their emphasis on policy-based loans. These policy-based interventions in various sectors will provide a range of opportunities for policy integration between environment and development. This book also comes at a time when such institutions are focusing on performance-based lending, and the book's emphasis on environmental performance indicators is especially welcome. Also, most importantly, the emphasis on better policies for environment–development integration will be directly relevant to the growing consensus of international assistance agencies to give highest priority to programmes that directly tackle poverty, since the poor bear the brunt of failed development policies in urban and rural areas and suffer even more when these failures perpetuate the process of environmental degradation.

The quality and productivity of environment and natural resources in Asian developing countries is declining at an alarming rate. During the last three decades extensive efforts by bilateral and international assistance agencies have been made to assist developing countries to put in place effective environment management regimes and agencies. These countries have responded with the promulgation of numerous policies and laws intended to protect the environment. Significant progress has been made in strengthening environmental protection systems within many countries in East Asia.

There are also now a variety of examples of innovative policy approaches being adopted throughout the region. The same 30-year period is noteworthy in that several Asian developing countries experienced the highest rates of economic growth ever achieved. Despite these achievements, since the 1970s Asia has lost half its forest cover, as well as countless unique plant and animal species. A third of its agricultural land is degraded and being lost to production. Fish stocks have halved and coral reefs are under threat. The region has the largest number of heavily polluted cities in the world. Its rivers and lakes are also among the most polluted. This epidemic of environmental degradation is widespread, including in the rapidly growing East Asian sub-region, the transitional economies of Central Asia and South Asia.

Poverty remains crushing and endemic. The poor, particularly women and children, are the most victimised by environmental degradation. Environmental degradation exacerbates poverty, particularly among the rural poor, where such degradation impacts on soil fertility and the ability of fisheries and forests to meet subsistence needs. Owing to the dependence of poor rural societies on their natural resources, any degradation or loss of the resource base can result in the destitution of certain groups, even though, on average, the economy shows positive growth. Further, urban environmental degradation, through lack of or inappropriate waste treatment and sanitation, and industrial and transport-related pollution, adversely impacts on air and water quality, which impairs the health of the urban poor disproportionately. This, in turn, decreases opportunities for steady employment and schooling, diminishes productivity and perpetuates poverty.

The blame for the current state of environmental degradation in the region has been placed at many doorsteps. *Asia's Clean Revolution* provides a succinct analysis of the origins of the region's environmental problems. A study by the ADB (1997) identified policy and market failures as the root causes. Policy-makers have too often failed to recognise the environmental effects of rapid economic growth, and governments have too often failed to adopt environmental and development policies that are effectively designed and adequately implemented. We all understand that economic growth is essential if adequate investments are to be made in infrastructure and the human and institutional resources that are needed for arresting environmental degradation. For example, protection of the population against water-borne diseases in a number of countries has become possible only with economic growth, which provided the investment needed for sanitation, waste treatment and protected water supply.

Unfortunately, demonstrations of real commitments to take the necessary actions for achieving sustainability are rare. It is noteworthy that during the 'growth spurt' in East Asia the increase in government spending on the environment was not substantial compared with increases in other forms of public and private investment. This fact raises serious doubts as to the political commitment to improve the environment. A reasonable surcharge on new luxury cars and the petrol consumed would have financed a sizeable urban air-quality management programme. Reasonable taxes on property booms would have financed a major urban environmental infrastructure programme. A serious attempt to recover costs of water supplied to, and solid wastes taken away from, industries would have encouraged greater efficiency in industrial production and reduced waste loads and resource depletion rates and would have facilitated private-sector participation in the delivery of environmental services. All of these ideas were known, but perhaps not well understood, by decision-makers and politicians during the boom times. A real opportunity was lost. Now the boom and, presumably, the bust are over and the regions' developing countries are now in recovery and growth. Convincing decision-makers to invest in environment will be more difficult than ever, but more important than ever.

The challenge today—and *Asia's Clean Revolution* is a powerful start—is to stimulate developing-country governments and the private sector to remedy the policy deficiencies that have undermined sustainable development in the past. Environmental protection and management must become an internalised cost and an expected benefit arising from development efforts aimed at economic growth and poverty alleviation in the Asian region. The coming decades are expected to bring ever-increasing environmental risks as a rapidly developing Asia expands both in terms of population and in terms of economic development. Given existing trends, Asia is expected to emerge from the current economic turbulence to embark further on what may become the largest urban and industrial expansion ever witnessed. Asia's population is set to grow by 50% within the space of the next generation. Most importantly, there is expected to be a tidal shift from rural to urban areas and from agriculture to industry. The urban population in the newly industrialised countries of East Asia has been projected by the World Resources Institute to grow from 550 million in 995 to almost 1.2 billion in 2025 (WRI 1998). Another study predicts that more than half of Asia's population will reside in urban centres by 2020 (UNDP 1999).

These and other projections of our world in the first quarter of the 21st century present a daunting future, implying a need for a fundamental shift in development patterns and lifestyles. This in turn requires an equally fundamental shift towards the integration between development and environment policy to ensure sustainability. The premise of this book is that tinkering around the edges will not work and that only a change in approach will achieve the level of success required. As one example, we have not yet made the essential shift from end-of-pipe mitigation of pollution, which, given expected astronomical growth in population and development, cannot be sustainable, to the upfront influencing of the basic processes of investment, technology change and market development to ensure cleaner production.

Three major trends in Asia present exciting opportunities for the necessary integration of development and environment policies on which Asia's environmental future will be built. These trends are the increasing emphasis on better governance, the accelerating rate of technology development and information exchange and the burgeoning role of private-sector financing and investment. A few examples will illustrate what these trends might mean.

Take governance as an example. When the region's major national environmental policies and regulations were prepared, largely in the mid-1980s and early 1990s, there were expectations that governments were coming to grips with the major environmental issues. Unfortunately, implementation of the policies and regulations was fragmented and, perhaps more importantly, the policies and regulations were not synchronised with those already in place in the major development sectors. It is clear that the new environmental agencies are unable to compete with the older and more powerful sectoral agencies for scarce funds and qualified personnel, causing them to be marginalised in the overall national decision-making structure.

One of the main failures—a failure of governance—has been in the implementation of those environmental rules and policies that do exist. The state of urban environmental compliance in most developing countries provides perhaps the best illustration of the failure to enforce environmental policies and regulations. Although countries have applicable regulations and policies, cease-and-desist orders are used more frequently than the imposition of fines, and fines are rarely collected. Fines stipulated in the law have lost their deterrence value as inflation makes the level of fines meaningless or nearly so. Performance bonds either are not sought or are not forfeited. There is an unwillingness to impose criminal sanctions in cases where the law permits this and monitoring is weak or is not strongly linked to environmental management needs. The growing emphasis on good governance could have a profound influence on basic compliance and the rule of law, opening new opportunities for market-based instruments and other flexible and economically efficient approaches to environmental management.

The second trend relates to access to appropriate technology for development. This is crucial for improving environmental conditions in the region. Although new technologies are still being developed mainly in industrialised countries, most developing countries still lag behind in obtaining information on these technologies. Nonetheless, the availability of information is growing rapidly and, more importantly, that raw information is being transformed into knowledge about alternative industrial processes. As this book points out, it is now, or soon will be, technically and economically possible for manufacturers in Asia to import, adapt and innovate on industrial processes and equipment developed in countries in the Organisation for Economic Co-operation and Development (OECD). In many cases, these technologies will tend to be cleaner simply because they are newer. An exciting opportunity exists because the vast majority of the capital stock of factories, machinery and infrastructure that will be in place over the next century has not yet been built. The challenge is to create the policy and regulatory environment that will provide incentives for the use of cleaner technology in the next phase of urban and industrial investment.

A further consideration related to the technology and information trend is the growing awareness among the general population that they deserve an everyday environment that is cleaner and healthier. The combination of two relatively new and growing phenomena—the increasing influence of non-governmental organisations (NGOs) and other organised groups of civil society, and the growing access to the Internet and other sources of global information and knowledge—presents a potentially powerful influence on governments and industries to act on environmental imperatives.

The third trend is the growing role of the private sector in general but also specifically in the environmental arena. Globalisation, increased trade and resource flows, deregulation and privatisation provide new opportunities for environmental investment by the private sector. The capacity of governments to take on the problems of environmental degradation is limited, particularly in developing countries. Policies are required that will attract the interest of the private sector in investing in urgently needed environmental infrastructure. Although this is normally thought as applying only to urban settings, it is equally applicable to the conservation of forests and biodiversity in rural areas. However, to date no Asian developing country has implemented a comprehensive policy designed to harness private-sector support and foster effective partnerships between the public and private sectors to finance environmental activities and reduce the cost that is borne by governments. In addition, enabling policies are needed that involve such crucial industries as banking and insurance, given that these are often among the strongest and most politically influential in much of Asia.

Against this backdrop, it is appropriate to state that the policies to be advocated and pursued may, for the most part, *not* be environmental policies *per se*. Although this last statement may seem paradoxical, if we are to make a meaningful change in our approach to environmental management and improvements then the policy innovations required will be largely those in the realms of trade, investment, industry and agriculture. The critical challenge is to reshape the policies of these and other real development sectors so that environmental protection, improvement and management become accepted and attractive dimensions of these sectors rather than vaguely understood impositions from outside. Asian countries have the opportunity to harness and use the enabling trends of good governance, technology development and private investment for a cleaner and healthier environment or pay the consequences of 'business as usual'.

One more critical parameter in the environment–development policy integration formula must be recognised: development assistance from multilateral and bilateral assistance agencies, NGOs and academic research institutions will continue to be needed to help Asian and Pacific developing countries meet the environment challenge. In general, such assistance has been disjointed and, as a result, numerous opportunities for synergism have been lost. The process initiated by the US–Asia Environmental Partnership in securing collaboration of institutions such as the ADB and the Greening of Industry Network Asia was clearly synergistic. The concept of environment policy integration emerged from consultations around this work, becoming the primary focus of a new large-scale analysis by the ADB of how policy dialogue between multilateral development banks and borrower governments can shift to better address borrowers' environmental objectives. Such collaboration must be seen as mandatory. As funding for assistance becomes increasingly difficult to obtain, it is essential that organisations such as the ADB and the US–Asia Environmental Partnership work together, and work with other private and public organisations, to maximise development and environmental benefit.

BIBLIOGRAPHY

ADB (Asian Development Bank) (1995) *Water Utilities Data Book: Asian and Pacific Region* (Manila: ADB).
ADB (Asian Development Bank) (1997) *Emerging Asia: Changes and Challenges* (Manila: ADB).
ADB (Asian Development Bank) (1999) *Asian Development Outlook 1999* (Hong Kong: Oxford University Press).
Aden, J., and M.T. Rock (1999) 'Initiating Environmental Behavior in Manufacturing Plants in Indonesia', *Journal of Environment and Development* 8.4: 357-75.
AET (Asia Environment Technology) (1999) www.asianenviro.com.
Afsah, S. (1998) *Impact of Financial Crisis on Industrial Growth and Environmental Performance in Indonesia* (Washington, DC: US–Asia Environmental Partnership).
Afsah, S., B. Laplante and N. Makarim (1995) 'Program-Based Pollution Control Management: The Indonesian PROKASIH Program' (Working paper 1602, Washington, DC: World Bank; www.nipr.org/work_paper/1602/index.htm).
Afsah, S., B. Laplante and D. Wheeler (n.d.) *Controlling Industrial Pollution: A New Paradigm* (Washington, DC: World Bank).
Agence France-Presse (1998) 'China admits logging bans will more than double timber imports', 19 December 1998.
Aiken, R.S., C.H. Leigh, T.R. Leinbach and M.R. Moss (1982) *Development and Environment in Peninsular Malaysia* (Singapore: McGraw-Hill).
AJEM (*Asia Journal of Environmental Management*) (1994) Special Issue on Community-Based Environmental Management in Asia, *Asia Journal of Environmental Management* 2.1 (ed. Y.S.F. Lee and K. Lowry).
Amsden, A. (1979) 'Taiwan's Economic History: A Case of *Etatisme* and a Challenge to Dependency', *Modern China* 5.3: 341-80.
Anderson, K. (1995) 'Social Policy Dimensions of Economic Integration: Environmental and Labor Standards' (Paper presented at the East Asia Seminar on Regional versus Multilateral Trade Arrangements, sponsored by the National Bureau of Economic Research, Seoul, 15–17 June 1995).
Anex, R.P. (1999) *Stimulating Innovation in Green Technology: Policy Alternatives and Opportunities* (Norman, OK: Science and Public Policy Program, University of Oklahoma).
Angel, D.P., M. Rock and T. Feridhanusetyawan (1999) 'Towards Clean Shared Growth in Asia' (Working paper; Washington, DC: US–Asia Environmental Partnership, July 1999).
APEC (Asia–Pacific Economic Co-operation) (1995) *Feasibility of Improving Market Information on Seafood Trade in the APEC Region* (Singapore: APEC Secretariat).
APEC (Asia–Pacific Economic Co-operation) (1998) *Asia–Pacific Profiles* (Hong Kong: APEC).
Ard-am, O., and K. Soonthorndhada (1994) 'Household Economy and Environmental Management in Bangkok: The Cases of Wat Chonglom and Yen-ar-kard', *Asian Journal of Environmental Management* 2.1 (May 1994): 22-30.
Arora, S., and T. Cason (1995) 'An Experiment in Voluntary Environmental Regulation: Participation in EPA's 33/50 Program', *Journal of Environmental Economics and Management* 28: 217-86.
Aspen Institute (1996) *The Alternative Path: A Cleaner, Cheaper Way to Protect and Enhance the Environment* (Washington, DC: Aspen Institute).
Athukorala, P. (1989) 'Export Performance of "New Exporting Countries": How Valid the Optimism?', *Development and Change* 20.1: 89-120.
Atkinson, G., and K. Hamilton (1996) 'Accounting for Progress: Indicators for Sustainable Development', *Environment*, September 1996: 16-24.

Aucott, M., D. Wachspress and J. Herb (1996) 'Industrial Pollution Prevention Trends in New Jersey' (Trenton, NJ: New Jersey Department of Environmental Protection, December 1996).

Ausubel, J.H. (1996) 'The Liberation of the Environment', *Daedalus* 125.3: 1-18.

Baldwin, R. (1992) 'Measurable Dynamic Gains from Trade', *Journal of Political Economy* 100.1: 162-74.

Barber, B. (1992) 'Jihad vs. McWorld', *The Atlantic* 269: 53-65.

Barber, C. (1997) *Project on Environmental Scarcities, State Capacity and Civil Violence: The Case of Indonesia* (Cambridge, MA: American Academy of Arts and Sciences).

Barber, C. (1998) 'Forest Resource Scarcity and Social Conflict in Indonesia', *Environment* 40: 4-9.

Barron, W., and J. Cottrell (1996) 'Making Environmental Law in Asia More Active' (Hong Kong: Centre for Urban Planning and Environmental Management).

Bartone, C., J. Bernstein, J. Leitmann and J. Eigen (1994) 'Toward Environmental Strategies for Cities: Policy Considerations for Urban Environmental Management in Developing Nations' (Urban Management Programme Policy Paper 18; Washington, DC: World Bank).

Battat, J., I. Frank and X. Shen (1996) 'Suppliers to MNCs: Linkage Programs to Strengthen Local Companies in Developing Countries' (Occasional Paper 6; Washington, DC: Foreign Investment Advisory Service, World Bank).

Bebbington, A., and J. Farrington (1993) 'Governments, NGOs, and Agricultural Development: Perspectives on Changing Inter-Organisational Relationships', *Journal of Development Studies* 29.2: 199-219.

BEC (Benchmark Environmental Consulting) (1995) *ISO 14000: An Uncommon Perspective* (Brussels: The European Environmental Bureau).

Becker, M., and K. Geiser (1997) *Evaluating Progress: A Report on the Findings of the Massachusetts Toxics Use Reduction Program Evaluation* (Boston, MA: Toxics Use Reduction Programme).

Bell, D., and J. Bauer (1999) *The East Asian Challenge for Human Rights* (Cambridge, UK: Cambridge University Press).

Bell, M., and K. Pavitt (1992) 'Accumulating Technological Capability in Developing Countries', in *Proceedings of the World Bank Annual Conference on Development Economics* (Washington, DC: World Bank): 257-81.

Bellamy, C. (1999) 'Public, Private and Civil Society', Address to Harvard International Development Conference, Cambridge, MA, 16 April 1999 (available from www.unicef.org).

Bello, W., and S. Rosenfeld (1990) *Dragons in Distress: Asia's Miracle Economies in Crisis* (San Francisco: Institute for Food Policy and Development).

Bergsten, C.F. (1994) 'APEC and World Trade: A Force for Worldwide Liberalisation', *Foreign Affairs* 73.3: 20-27.

Bernadini, O., and R. Galli (1993) 'Dematerialisation: Long Term Trends in the Intensity of Use of Materials and Energy', *Futures*, May 1993: 431-48.

Birdsall, N., and D. Wheeler (1992) 'Trade Policy and Industrial Pollution in Latin America: Where are the Pollution Havens?', in P. Low (ed.), *International Trade and the Environment* (Washington, DC: World Bank): 159-68.

Blumberg, J., Å. Korsvold and G. Blum (1997) *Environmental Performance and Shareholder Value* (Geneva: World Business Council for Sustainable Development).

Boyer, R., and D. Drache (1996) *States against Markets: The Limits of Globalization* (London: Routledge).

Boyle, A., and M. Anderson (1996) *Human Rights Approaches to Environmental Protection* (Oxford: Clarendon Press).

BPPT (Badan Pengkajiandan Penerapan Teknologi) (1993) *Science and Technology Indicators of Indonesia* (Jakarta: BPPT).

Brack, D. (1996) *International Trade and the Montreal Protocol* (London: Royal Institute of International Affairs).

Brandon, C. (1994) 'Reversing Pollution Trends in Asia', *Finance and Development* 31.2 (June 1994): 21-23.

Brandon, C., and R. Ramankutty (1993) 'Toward an Environmental Strategy for Asia' (Discussion Paper 224; Washington, DC: World Bank).

Branscomb, L., and J.H. Keller (eds.) (1997) *Investing in Innovation* (Cambridge, MA: MIT Press).

Broad, R., and J. Cavanagh (1993) *Plundering Paradise* (Berkeley, CA: University of California Press).

Broadcast Engineering (1998) 'CNN: Charting a New Frontier, 66 M for News Gathering', *Broadcast Engineering*, 30 November 1998.

Brown, H.S., J.J. Himmelberger and A.L. White (1993) 'Development–Environment Interactions in the Export of Hazardous Technologies', *Technological Forecasting and Social Change* 43: 125-55.

Brown, L., et al. (1998) *State of the World* (New York: Worldwatch Institute).

Brown Weiss, E., P. Szasz and D. Magraw (eds.) (1992) *International Environmental Law: Basic Instruments and References* (Dobbs Ferry, NY: Traditional Publishers).

Bruun, O., and A. Kalland (1995) *Asian Perceptions of Nature* (Richmond, UK: Curzon).

Carmichael, G., and F.S. Rowland (1998) 'Development of Asian Mega Cities: Environmental, Economic, Social and Health Implications' (Seminar Series; Washington, DC: US Global Change Research Programme, June 1998).

Carson, R. (1962) *Silent Spring* (Boston, MA: Houghton Mifflin).

Cassel, D. (1996) 'Corporate Initiatives: A Second Human Rights Revolution', *Fordham International Law Journal* 19: 1963-84.

Cavanagh, R., D. Lashof and S. Schwab (1996) *Risky Business: Hidden Environmental Liabilities of Power Plant Ownership* (New York: Natural Resources Defense Council).

CEC (Commission for Environmental Co-operation) (1996) *Putting the Pieces Together: The Status of Pollutant Release and Transfer Registers in North America* (Montreal: CEC).

CEC (Commission for Environmental Co-operation) (1997) *Taking Stock: North American Pollutant Releases and Transfers 1994* (Montreal: CEC).

CGCAP (California Global Corporate Accountability Project) (1999) 'Hard Issues, Innovative Approaches: Defining the Scope and Exploring the Mechanisms of Corporate Social Accountability' (Natural Heritage Institute, Human Rights Advocates, and the Nautilus Institute for Security and Sustainable Development; www.nautilus.org/cap).

Charnovitz, S. (1997) 'Two Centuries of Participation: NGOs and International Governance', *Michigan Journal of International Law* 18.2: 183-286.

Chia, S.Y., and T.Y. Lee (1993) 'Subregional Economic Zones: A New Motive Force', in C.F. Bergsten and M. Noland (eds.), *Asia Pacific Development, Pacific Dynamism and the International Economic System* (Washington, DC: Institute for International Economics).

Chooi, C.F. (1984) 'Ponding System for Palm Oil Mill Effluent Treatment', in *Proceedings of the Workshop on Review of Palm Oil Mill Effluent Technology vis-à-vis Department of Environment Standard* (Palm Oil Research Institute of Malaysia [PORIM] workshop proceedings 9; Bandar Baru Bangi, Malaysia: PORIM).

Christensen, K., B. Nielson, P. Doelman and R. Schellman (1995) 'Cleaner Technologies in Europe', *Journal of Cleaner Production* 3: 67-70

Chun-Chieh, C. (1994) 'Growth with Pollution: Unsustainable Development in Taiwan and its Consequences', *Studies in Comparative International Development* 29.2: 23-47.

Clifford, M. (1988) 'Too Many People Looking for Too Few Houses', *Far Eastern Economic Review*, 8 September 1988: 84-85.

Cohen, J., and A. Arato (1997) *Civil Society and Political Theory* (Cambridge, MA: MIT Press).

Conference Board (1996) *Corporate Environment, Health and Safety Reward Programs: A Research Report* (New York: Conference Board).

Cook, E. (ed.) (1996) *Ozone Protection in the United States: Elements of Success* (Washington, DC: World Resources Institute).

Cooper, R.N. (1994) *Environment and Resource Policies for the World Economy* (Washington, DC: The Brookings Institution).

Cordella, T. (1998) 'Can short-term capital controls promote capital inflows?' (Working Paper 98/131; New York: International Monetary Fund).

Council on Competitiveness (1988) *Picking up the Pace: The Commercial Challenge to American Innovation* (Washington, DC: Council on Competitiveness).

Dasgupta, S., A. Mody, S. Roy and D. Wheeler (1997) 'Environmental Regulation and Development: A Cross-Country Empirical Analysis', *New Issues in Pollution Regulation* (Washington, DC: World Bank; www.worldbank.org/nipr).

Davies, T., and J. Mazurek (1996) *Industry Incentives for Environmental Improvement: Evaluation of US Federal Initiatives* (Washington, DC: Global Environmental Management Initiative).

Dean, J.M. (1992) 'Trade and Environment: A Survey of the Literature', in P. Low (ed.), *International Trade and the Environment* (World Bank Discussion Paper 159; Washington, DC: World Bank): 15-28.

De Tocqueville, A. (n.d.) *Democracy in America* (available at xroads.virginia.edu/~hyper/detoc/).

Dharmapatni, I.A.I., and H. Prabatmodjo (1994) 'Community-Based Urban Environmental Management: A Bandung Case Study', *Asian Journal of Environmental Management* 2.1 (May 1994): 31-42.

Dicken, P. (1998) *Global Shift* (London: Paul Chapman, 3rd edn).

Digregorio, M. (1993) 'Labor Systems and Management of the Urban Environment: An Analysis of Waste Recovery in Hanoi, Vietnam' (MA thesis, Urban and Regional Planning, University of Hawaii, Honolulu, HI).

Dillon, P.S. (1996) *Extended Product Responsibility in the Electronics Industry* (Medford, MA: Tufts University Press).
Ditz, D., and J. Ranganathan (1997) *Measuring Up: A Common Framework for Tracking Corporate Environmental Performance* (Washington, DC: World Resources Institute).
Ditz, D., J. Ranganathan and R.D. Banks (eds.) (1995) *Green Ledgers: Case Studies of Corporate Environmental Accounting* (Washington, DC: World Resources Institute).
DOE (Department of Environment, Malaysia) (1985) *Environmental Quality Report 1981–84* (Kuala Lumpur: DOE).
DOE (Department of Environment, Malaysia) (1992) *Environmental Quality Report 1991* (Kuala Lumpur: DOE).
Dorman, P. (2000) 'Actually Existing Globalization', in P.S. Aulakh and M.G. Schecter (eds.), *Rethinking Globalization: From Corporate Transnationalism to Local Interventions* (New York: St Martin's Press).
Douglass, M. (1991) 'Planning for Environmental Sustainability in the Extended Jakarta Metropolitan Region', in N. Ginsburg, B. Koppel and T.G. McGee (eds.), *The Extended Metropolis: Settlement Transition in Asia* (Honolulu, HI: University of Hawaii Press): 239-73.
Douglass, M. (1992) 'The Political Economy of Urban Poverty and Environmental Management in Asia: Access, Empowerment and Community-Based Alternatives', *Environment and Urbanization* 4.2: 9-32.
Douglass, M. (1993) 'Urban Poverty and Policy Alternative in Asia', in *State of Urbanisation in Asia* (Bangkok: United Nations Economic and Social Commission for Asia and the Pacific).
Douglass, M. (1997) 'Urban Poverty and Environmental Management: A Comparative Analysis of Community Activation in Asian Cities', in P. Hills and C. Chan (eds.), *Community Mobilization and the Environment in Hong Kong* (Hong Kong: Centre of Urban Planning and Environmental Management, Hong Kong University): 53-95.
Douglass, M. (1998a) 'World City Formation on the Asia Pacific Rim: Poverty, "Everyday" Forms of Civil Society and Environmental Management', in M. Douglass and J. Friedmann (eds.), *Cities for Citizens: Planning and the Rise of Civil Society in a Global Age* (Chichester, UK: John Wiley): 107-37.
Douglass, M. (1998b) 'East Asian Urbanization: Patterns, Problems and Prospects' (Shorenstein Distinguished Lecture, Institute for International Studies, Stanford University, Stanford, CA).
Douglass, M. (1998c) 'Sustainability and Strategic Planning for Jabotabek: Overview and Summary of the Seminar Proceedings' (International Seminar on Strategies for a Sustainable Greater Jabotabek, Jakarta, 8–10 July 1996).
Douglass, M., and Y.F. Lee (1996) 'Urban Priorities for Action', in *World Resources 1996–97* (Washington, DC: World Resources Institute): 103-24.
Douglass, M., and M. Zoghlin (1994) 'Sustainable Cities from the Grassroots: Livelihood, Habitat and Social Networks in Suan Phlu, Bangkok', *Third World Planning Review* 16.2: 171-200.
Douglass, M., with O. Ard-Am and I.K. Kim (2000) 'Urban Poverty and the Environment: Social Capital and State–Community Synergy in Seoul and Bangkok', in P. Evans (ed.), *Livable Cities? The Politics of Urban Livelihood and Sustainability* (Berkeley, CA: University of California Press).
Dua, A., and D.C. Esty (1997) *Sustaining the Asia Pacific Miracle: Environmental Protection and Economic Integration* (Washington, DC: Institute for International Economics).
Dybrig, P., and C. Spatt (1983) 'Adoption Externalities as Public Goods', *Journal of Public Economy* 20.23: 1.
Earth Council (1999) 'The Earth Charter Campaign', www.earthcharter.org.
Economist (1993) 'Pay Now, Save Later: Pollution in Asia', *The Economist*, 11 December 1993: 36-37.
Eder, N. (1996) *Poisoned Prosperity: Development, Modernization and the Environment in South Korea* (Armonk, NY: M.E. Sharpe).
Edwards, M., and D. Hulme (1992) *Making a Difference: NGOs and Development in a Changing World* (London: Earthscan).
EFBE (Extel Financial and Business in the Environment) (1994) *City Analysts and the Environment* (London: Business in the Environment).
EIA (Environment Investigation Agency) (1999) *International Energy Outlook 1999* (Washington, DC: US Department of Energy)
Elkington, J. (1998) *Cannibals with Forks, The Triple Bottom Line of 21st Century Business* (Oxford, UK: Capstone Publishing).
Elkington, J., and J. Hailes (1988) *The Green Consumer Guide. From Shampoo to Champagne: High Street Shopping for a Better Environment* (London: Victor Gollancz).
Ellickson, R.C. (1979) 'Public Property Rights: A Government's Rights and Duties when its Landowners Come into Conflict with Outsiders', *Southern California Law Review* 52.6: 1627.

EPA (US Environmental Protection Agency) (1995) *A Conceptual Framework to Support Development and Use of Environmental Information in Decision-Making* (Washington, DC: Environmental Statistics and Information Division, Office of Policy, Planning and Evaluation, US EPA, April 1995).

EPA (US Environmental Protection Agency) (1997a) *Environmentally Preferable Purchasing Program: Cleaning Products Pilot Project* (Washington, DC: EPA).

EPA (US Environmental Protection Agency) (1997b) *1995 Toxics Release Inventory Public Data Release* (Washington, DC: EPA).

EPA (US Environmental Protection Agency) (1997c) 'Addition of Facilities in Certain Industry Sectors, Final Rule', *Federal Register* 62.84: 23833-92.

Epstein, L.N., S. Greetham and A. Karuba (eds.) (1995) *Ranking Refineries: What do we know about oil refinery pollution from right-to-know data?* (Washington, DC: Environmental Defense Fund).

Epstein, M. (1996) *Measuring Corporate Environmental Performance: Best Practices for Costing and Managing an Effective Environmental Strategy* (Chicago: Irwin Professional Publishing).

Eskeland, G.S., and A.E. Harrison (1997) 'Moving to Greener Pastures? Multinationals and the Pollution Haven Hypothesis' (Working Paper Series No. 1744; Washington, DC: World Bank).

Esty, D.C. (1994a) *Greening the GATT: Trade, Environment, and the Future* (Washington, DC: Institute for International Economics).

Esty, D.C. (1994b) 'The Case for a Global Environmental Organisation', in P. Kenen (ed.), *Managing the World Economy: Fifty Years after Bretton Woods* (Washington, DC: Institute for International Economics): 287-309.

Esty, D.C. (1994c) 'Making Trade and Environmental Policy Work Together: Lessons from NAFTA', *The Swiss Review of International Economic Relations (Aussenwirtschaft)* 49: 59-79.

Esty, D.C. (1996) 'Revitalizing Environmental Federalism', *Michigan Law Review* 95.3: 570.

Esty, D.C. (1997) 'Stepping up to the Global Environmental Challenge', *Fordham Environmental Law Journal* 8.1: 103.

Esty, D.C. (1998a) 'Non-Governmental Organizations at the World Trade Organization: Co-operation, Competition, or Exclusion', *Journal of International Economic Law* 1: 123.

Esty, D.C. (1998b) 'Linkages and Governance: NGOs at the World Trade Organization', *University of Pennsylvania Journal of International Economic Law* 19: 709.

Esty, D.C., and B.S. Gentry (1997) 'Foreign Investment, Globalization, and Environment', in T. Jones (ed.), *Globalization and the Environment* (Paris: Organisation for Economic Co-operation and Development).

Esty, D.C., and D. Geradin (1997) 'Market Access, Competitiveness, and Harmonization: Environmental Protection in Regional Trade Agreements', *Harvard Environmental Law Review* 21.2: 265.

Esty, D.C., and R. Mendelsohn (1995) *Powering China* (New Haven, CT: Yale Centre for Environmental Law and Policy).

Esty, D.C., and R. Mendelsohn (1998) 'Moving From National to International Environmental Policy', *Policy Sciences* 31.3: 225.

Esty, D.C., and M. Porter (1998) 'Industrial Ecology and Competitiveness', *Journal of Industrial Ecology* 2.1: 35.

Etzioni, A. (ed.) (1995) *New Communitarian Thinking* (Charlottesville, VA: University Press of Virginia).

Evans, P. (1997) 'Introduction: Development Strategies across the Public–Private Divide', in P. Evans (ed.), *State–Society Synergy: Government and Social Capital in Development* (Berkeley, CA: International and Area Studies, University of California at Berkeley): 1-11.

Feldman, I.R., and M.C. Shiavo (1995) 'A Private Road to Environmental Excellence: The Working of Green Track', *Corporate Environmental Strategy* 3.1 (Summer 1995): 13-18.

Feldman, S.J., P.A. Soyka, and P. Ameer (1996) 'Does improving a firm's environmental management system and environmental performance result in higher stock price?' (Fairfax, VA: ICF Kaiser International Inc., November 1996).

Fischel, W.A. (1975) 'Fiscal and Environment Considerations in the Location of Firms in Suburban Communities', in E.S. Mills and W.E. Oates (eds.), *Fiscal Zoning and Land Use Controls: The Economic Issues* (Lexington, MA: Lexington Books).

Fischer, K., and J. Schot (1993) *Environmental Strategies for Industry* (Washington, DC: Island Press).

Flavin, C., and O. Tunali (1996) *Climate of Hope: New Strategies for Stabilizing the World's Atmosphere* (Washington, DC: Worldwatch Institute).

Freeman, H., T. Harten, J. Springer, P. Randall, M.J. Curran and K. Stone (1995) 'Industrial Pollution Prevention: A Critical Review' (Paper prepared for the 85th annual Meeting of the Air and Waste Management Association, Kansas City, MO, 21-26 June 1995).

Friedmann, J. (1988) *Life Space and Economic Space* (New York: Transaction Books).

Friedmann, J. (1998) 'The New Political Economy of Planning: The Rise of Civil Society', in M. Douglass and J. Friedmann (eds.), *Cities for Citizens: Planning and the Rise of Civil Society in a Global Age* (Chichester, UK: John Wiley): 19-38.

Funabashi, Y. (1995) *Asia–Pacific Fusion: Japan's Role in APEC* (Washington, DC: Institute for International Economics).

Gentry, B.D. (ed.) (1998) *Private Capital Flows and the Environment: Lessons from Latin America* (Cheltenham, UK: Edgar Elgar).

Gibney, M., and R.D. Emerick (1996) 'The Extraterritorial Application of US Law and the Protection of Human Rights', *Temple International and Comparative Law Journal* 10: 123-45.

Ginsburg, N., B. Koppel and T.G. McGee (eds.) (1991) *The Extended Metropolis: Settlement Transition in Asia* (Honolulu, HI: University of Hawaii Press).

GM (General Motors) (1996) *General Motors Corporation Environmental Health and Safety Report* (Detroit: GM).

Graedel, T.E., and B.R. Allenby (1995) *Industrial Ecology* (Upper Saddle River, NJ: Prentice–Hall)

Graham, C. (1994) *Safety Nets, Politics and the Poor: Transitions to Market Economies* (Washington, DC: The Brookings Institution).

Gray, C.B. (1983) 'Regulation and Federalism', *Yale Journal on Regulation* 1: 93.

Greider, W. (1997) *One World, Ready or Not: The Manic Logic of Global Capitalism* (New York: Simon & Schuster).

Grossman, G., and A. Krueger (1994) 'Economic Growth and the Environment' (Working paper W4634; Boston, MA: National Bureau of Economic Research, February 1994).

Haas, P.M., R.O. Keohane and M.A. Levy (eds.) (1993) *Institutions for the Earth: Sources of Effective International Environmental Protection* (Cambridge, MA: MIT Press).

Haggard, S. (1990) *Pathways from the Periphery* (Ithaca, NY: Cornell University Press).

Hahn, R.W. (1989) 'Economic Prescriptions for Environmental Problems: How the Patient Followed the Doctor's Orders', *Journal of Economic Perspectives* 3.2: 95-114.

Hammer, J.S., and S. Shetty (1995) 'East Asia's Environment: Principles and Priorities for Action' (Discussion paper 287; Washington, DC: World Bank).

Hanley, N., S.S. Shogren and B. White (1997) *Environmental Economics in Theory and Practice* (New York: Oxford University Press).

Hann, C., and E. Dunn (eds.) (1996) *Civil Society: Challenging Western Models* (London: Routledge).

Hardin, G. (1968) 'The Tragedy of the Commons', *Science* 162: 1243.

Harpham, T., P. Garner and C. Surjadi (1990) 'Planning for Child Health in a Poor Urban Environment: The Case of Jakarta, Indonesia', *Environment and Urbanization* 1.2: 63-72.

Harrison, A. (1996) 'Determinants and Effects of Direct Foreign Investment in Côte d'Ivoire, Morocco, and Venezuela', in M. Roberts and J. Tybout (eds.), *Industrial Evolution in Developing Countries* (New York: Oxford University Press): 163-86.

Hausman, J.A. (1978) 'Specification Tests in Econometrics', *Econometrica* 46: 1251-72.

Havel, V. (1988) 'Anti-political Politics', in J. Keane (ed.), *Civil Society and the State* (London: Verso).

Hayes, D.J. (1996) 'The Business Risk Audit', *The Environmental Forum*, November/December 1996: 19-23.

Healey, P. (1997) 'Collaborative Approaches to Urban Planning and their Contribution to Institutional Capacity-Building in Urban Regions', *International Journal of Urban Science* 1.2: 167-83.

Heaton, Jr, G.R. (1990) *Regulation and Technological Change* (Washington, DC: World Resources Institute).

Heaton, Jr, G.R. (1997a) *High Technology Programs in the US, Japan and Europe* (Paris: Directorate of Science, Technology and Industry, OECD).

Heaton, Jr, G.R. (1997b) *Regulation and Innovation: A Scoping Paper* (Paris: Directorate of Science, Technology and Industry, OECD).

Heaton, Jr, G.R., and R.D. Banks (1997) 'A New Generation of Environmental Technology', *Journal of Industrial Ecology* 1.2: 23-32.

Heaton, Jr, G.R., R.D. Banks and D. Ditz (1994) *Missing Links: Technology and Environmental Improvement in the Industrializing World* (Washington, DC: World Resources Institute).

Heaton, Jr, G.R., R. Repetto and R. Sobin (1991) *Transforming Technology: An Agenda for Sustainable Growth in the 21st Century* (Washington, DC: World Resources Institute).

Heaton, Jr, G.R., R. Repetto and R. Sobin (1992) *Back to the Future: US Government Policy toward Environmentally Critical Technology* (Washington, DC: World Resources Institute).

Held, D., A.G. McGrew, D. Goldblatt and J. Perraton (1999) *Global Transformations: Politics, Economics and Culture* (Stanford, CA: Stanford University Press).

Henderson, H. (1997) 'Social Capital and Economic Development' (Paper presented at the International Conference on Governance for Sustainable Growth and Equity, 28 July 1997; New York: United Nations).

Hettige, H., M. Mani and D. Wheeler (1997) *Industrial Pollution in Economic Development: Kuznets Revisited* (Washington, DC: Development Research Group, World Bank).

Hettige, H., P. Martin, M. Singh, and D. Wheeler (1995) *The Industrial Pollution Projection System* (Washington, DC: World Bank).

Hewison, K. (1996) 'The Ebb and Flow of Civil Society and the Decline of the Left in Southeast Asia', in G. Rodan (ed.), *Political Opposition in Industrializing Asia* (London: Routledge): 40-71.

Hill, H. (1996) 'Indonesia's Industrial Policy and Performance: Orthodoxy Vindicated', *Economic Development and Culture Change* 45 (October 1996): 146-74.

Hirsch, P., and C. Warren (1998) *The Politics of Environment in South East Asia* (London: Routledge).

Hirst, P., and G. Thompson (1996) *Globalization in Question: The International Economy and the Possibilities of Governance* (Cambridge, MA: Polity Press).

Ho, K.C. (1997) 'From Port City to City-State: Forces Shaping Singapore's Built Environment', in W.B. Kim, M. Douglass, S.C. Choe and K.C. Ho (eds.), *Culture and the City in East Asia* (Oxford: Oxford University Press): 212-33.

Homer Dixon, T. (1999) *Environment, Scarcity and Violence* (Princeton, NJ: Princeton University Press).

House, G. (1995) 'Raising a Green Standard', *Industry Week* 244.14 (17 July 1995): 73-74.

Hsiao, H.H.M. (1995) 'Assessing Taiwan's Environmental Movement' (Paper presented at the 1st Workshop on Asia's Environmental Movements in Comparative Perspective', Honolulu, HI, 29 November-1 December, 1995; Honolulu, HI: East-West Center Program on Environment).

Hurrell, A., and B. Kingsbury (1992) *The International Politics of the Environment: Actors, Interests, and Institutions* (Oxford: Clarendon Press).

IDB (Industrial Development Bureau) (1995) *Development of Industries in Taiwan Republic of China* (Taipei: Ministry of Economic Affairs).

IDB (Industrial Development Bureau) (n.d.a) *Industrial Pollution Control in Taiwan ROC* (Taipei: Ministry of Economic Affairs).

IDB (Industrial Development Bureau) (n.d.b) *Strategy of Industrial Pollution Control in Taiwan ROC* (Taipei: Ministry of Economic Affairs).

IDB (Industrial Development Bureau) (n.d.c) *Promotion and Accomplishment of Industrial Pollution Prevention and Control in the Republic of China* (Taipei: Ministry of Economic Affairs).

IHDP (International Human Dimensions Programme) (1999) 'Industrial Transformation: Science Plan' (Report 12; Bonn: IHDP).

IIED (International Institute for Environment and Development) (1994) *Environmental Synopsis of Indonesia* (London: IIED).

IISD (International Institute for Sustainable Development) (1999) 'Report on the WTO's High-Level Symposium on Trade and Environment', www.wto.org/hlms/sumhlenv.htm.

IMF (International Monetary Fund) (1991) *International Financial Statistics 1990* (Washington, DC: IMF).

IMF (International Monetary Fund) (1997a) *Balance of Payments Statistics Yearbook* (Washington, DC: IMF).

IMF (International Monetary Fund) (1997b) *International Financial Statistics Yearbook* (Washington, DC: IMF).

Intal, P.S. (1996) 'Perspectives from the Philippines and ASEAN', in S. Tay and D. Esty (eds.), *Asian Dragons and Green Trade: Environment, Economics and International Law* (Singapore: Times Academic Press).

IPCC (Intergovernmental Panel on Climate Change) (1995) *Climate Change 1995: The IPCC Second Assessment Report* (New York: Cambridge University Press).

IRRC (Investor Responsibility Research Center) (1996) *Corporate Environmental Profiles Directory 1996: Executive Summary* (Washington, DC: IRRC).

Irwin, F., T. Natan, W.R. Muir, E.S. Howard, L. Lobo and S. Martin (1995) *A Benchmark for Reporting on Chemicals at Industrial Facilities* (Washington, DC: World Wide Fund for Nature).

ISO (International Organization for Standardization) (1996) *Environmental Performance Evaluation* (Committee draft 14031, ISO/TC 207/SC 4 N 207; Geneva: ISO, 5 December 1996).

Jaffe, A.B., S.R. Peterson, P.R. Portney and R.N. Stavins (1995) 'Environmental Regulation and the Competitiveness of US Manufacturing: What does the evidence tell us?', *Journal of Economic Literature* 33.1: 132-63.

Japan Economic Newswire (1998) 'ASEAN heads to "spare no effort for economic recovery"', *Japan Economic Newswire*, 16 December 1998.

JCIE (Japan Center for International Exchange) (1998) *Globalization, Governance and Civil Society* (Tokyo: JCIE).
Johnson, C. (1987) 'Political Institutions and Economic Performance: The Government–Business Relationship in Japan, South Korea, and Taiwan', in F.C. Deyo (ed.), *The Political Economy of New Asian Industrialism* (Ithaca, NY: Cornell University Press): 136-64.
Johnson, J.M. (1997) *The Science and Technology Resources of Japan: A Comparison with the United States* (Washington, DC: National Science Foundation).
Johnson, J.M. (1998) *Human Resources for Science and Technology: The Asian Region* (Washington, DC: National Science Foundation).
Johnston, B.F., and P. Kilby (1975) *Agriculture and Structural Transformation* (New York: Oxford University Press).
Jomo, K.S. (1998) 'Financial Liberalization, Crises and Malaysian Policy Responses', *World Development* 26.8: 1563.
Jones, L., and I. Sakong (1980) *Government, Business, and Entrepreneurship: The Korean Case* (Cambridge, MA: Harvard University Press).
Jorgensen, E.V., Inc. (1977) 'Palm Oil Sludge: A Profitable Investment' (Mimeograph; Kuala Lumpur: Department of Environment).
Jorgensen, H.K. (1982) 'The UP Decanter–Drier System for Reduction of Palm Oil Effluent', in *Proceedings of Regional Workshop on Palm Oil Mill Technology and Effluent Treatment* (Palm Oil Research Institute of Malaysia [PORIM] Workshop Proceedings No. 4; Bandar Baru Bangi, Malaysia: PORIM).
Judge, G.G., W.E. Griffiths, R.C. Hill, H. Lutkepohl and T.C. Lee (1985) *The Theory and Practice of Econometrics* (New York: John Wiley, 2nd edn).
Kalland, A., and G. Persoon (eds.) (1998) *Environmental Movements in Asia* (London: Curzon Press).
Katz, M.L., and C. Shapiro (1985) 'Network Externalities, Competition and Compatibility', *American Economic Review* 75: 424.
Katzen, S. (1997) Administrator of the Office of Information and Regulatory Affairs, Office of Management and Budget, quoted in 'On Site: Statistical Yardsticks Taking the Nation's Measure', *The Washington Post*, 23 May 1997: A27.
Keane, J. (1988) 'Despotism and Democracy: The Origins and Development of the Distinction between Civil Society and the State 1750–1850', in J. Keane (ed.), *Civil Society and the State* (London: Verso Press): 35-72.
Keesing, D. (1988) 'The Four Exceptions' (Occasional Paper 2; New York: Trade Expansion Programme, United Nations Development Programme).
Kellert, S.R. (1996) *The Value of Life: Biological Diversity and Human Society* (Washington, DC: Island Press).
Kelly, H. (1990) *Energy and Economic Growth Revisited* (Washington, DC: World Resources Institute).
Khalid Abdul Rahim (1991) 'Internalisation of Externalities: Who bears the cost of pollution control?', *The Environmentalist* 11.1: 19-25.
Khalid Abdul Rahim and Wan Mustafa Wan Ali (1992) 'External Benefits of Environmental Regulation', *The Environmentalist* 12.4: 277-85.
Khalid Abdul Rahim and J.B. Braden (1993) 'Welfare Effects of Environmental Regulation in an Open Economy: The Case of Malaysian Palm Oil', *Journal of Agricultural Economics* 44.1: 25-37.
Khera, H.S. (1976) *The Oil Palm Industry of Malaysia* (Kuala Lumpur: Penerbit Universiti Malaya).
Kiesling, F. (1994) *Minnesota P2 Planning Survey: Results and Technical Report* (Minneapolis, MN: Minnesota Survey Research Center, University of Minnesota).
Kim, I.K. (1994) 'The Environmental Problems in Urban Communities and the Protection of the Environment in Korea', *Korea Journal of Population and Development* 23.1: 63-76.
Kim, L.K. (1997) *Imitation to Innovation: The Dynamics of Korea's Technological Learning* (Boston, MA: Harvard Business School Press).
KNCFH (Korean NGOs and CBOs Forum for Habitat II) (1996) 'Voices of the Korean NGOs and CBOs to Habitat II Istanbul', 30 May 1996.
Krier, J.E., and M. Brownstein (1992) 'On Integrated Pollution Control', *Environmental Law* 22: 119.
Krugger Consultants/NESDB (National Economic and Social Development Board) (1996) *Urban Environmental Management in Thailand: A Strategic Planning Process* (Bangkok: Krugger Consultants).
Lall, S. (1992) 'Technological Capabilities and Industrialization', *World Development* 20.2: 165-82.
Laughlin, J., and L. Corson (1995) 'A Market-Based Approach to Fostering P2', *Pollution Prevention Review* 5.1: 11-16.
Lawrence, R.Z. (1996) *Regionalism, Multilateralism, and Deeper Integration* (Washington, DC: The Brookings Institution).

Lee, S.H. (1995) 'Environmental Movements in South Korea' (Paper presented at the 1st Workshop on Asia's Environmental Movements in Comparative Perspective, Honolulu, HI, 29 November–1 December 1995; Honolulu, HI: East–West Center Program on Environment).
Lee, Y.S.F. (1998) 'Intermediary Institutions, Community Organisations, and Urban Environmental Management: The Case of Three Bangkok Slums', *World Development* 26.6: 993-1011.
Lee, Y.S.F., and A. So (1999) *Asia's Environmental Movements: Comparative Perspectives* (Armonk, NY: M.E. Sharpe).
Lim, K.H., S.K. Quah, D. Gillies and B.J. Wood (1984) 'Palm Oil Effluent Treatment and Utilization in Sime Darby Plantations: The Current Position', in *Proceedings of Regional Workshop on Palm Oil Mill Technology and Effluent Treatment* (Palm Oil Research Institute of Malaysia [PORIM] Workshop Proceedings No. 9; Bandar Baru Bangi, Malaysia: PORIM).
Lim, K.H. (1984) 'Problems of Implementation by MOPGC Members', in *Proceedings of Regional Workshop on Palm Oil Mill Technology and Effluent Treatment* (Palm Oil Research Institute of Malaysia [PORIM] Workshop Proceedings No. 9; Bandar Baru Bangi, Malaysia: PORIM).
Lim, L. (1998) 'Whose model failed? Implications of the Asian Crisis', *Washington Quarterly* 21: 25-36
Lohani, B. (1998) *Environmental Challenges in Asia in the 21st Century* (Manila: Asian Development Bank).
Lohmann, L. (1995) 'No Rules of Engagement: Interest Groups, Centralization and the Creative Politics of the Environment in Thailand', in J. Rigg (ed.), *Counting the Costs: Economic Growth and Environmental Change in Thailand* (Singapore: Institute for South East Asian Studies): 211-34.
Low, P., and A. Yeats (1992) 'Do dirty industries migrate?', in P. Low (ed.), *International Trade and the Environment* (World Bank Discussion Paper 159; Washington, DC: World Bank): 89-104.
Ma, A.N., C.S. Chow, C.K. John, A. Ibrahim and Z. Isa (1982) 'Palm Oil Mill Effluent Treatment: A Survey', in *Proceedings of Regional Workshop on Palm Oil Mill Technology and Effluent Treatment* (Palm Oil Research Institute of Malaysia [PORIM] Workshop Proceedings No. 4; Bandar Baru Bangi, Malaysia: PORIM).
Ma, A.N., Yusof Basiron and Mohd Nasir Amiruddin (1980) 'The Interdependence of Economic Development and Environmental Quality in South East Asia: Malaysia as a Case Study' (Unpublished manuscript; Bandar Baru Bangi, Malaysia: Palm Oil Research Institute of Malaysia).
McAllister, B. (1993) 'The United Nations Conference on Environment and Development: An Opportunity to Forge a New Unity in the World Bank on Human Rights, the Environment, and Sustainable Development', *Hastings International and Comparative Law Review* 16: 689.
McGee, T.G., and I. Robinson (eds.) (1995) *The New Southeast Asia: Managing the Mega-Urban Regions* (Vancouver: University of British Columbia Press).
McVey, R. (ed.) (1992) *Southeast Asian Capitalists* (Ithaca, NY: Cornell University Press).
Maheswaran, A. (1984) 'Legislative Measures in the Control of Palm Oil Mill Effluent Discharge' (Mimeograph; Malaysia Department of Environment).
Maheswaran, A., and G. Singam (1977) 'Pollution Control in the Palm Oil Industry: Promulgation of Regulations', *Planter* 53: 470-76.
Maheswaran, A., Abu Bakan Jaafar and G. Singh (1980) 'Water Quality Management in Malaysia' (Paper presented at The Interdependence of Economic Development and Environmental Quality in South-East Asia Symposium, Miami University, Oxford, OH, 5–7 August 1980; Kuala Lumpur: Departments of Environment, Malaysia).
Mallet, V. (1999) *The Trouble with Tigers: The Rise and Fall of Southeast Asia* (London: HarperCollins).
Mapes, J. (1994) 'Clean Industry Technologies to Asia: The Window of Opportunity', *Far Eastern Economic Review* 28 (10 November 1994): 10.
Marcus, P.A., and J.T. Willing (eds.) (1997) *Moving Ahead with ISO 14000: Improving Environmental Management and Achieving Sustainable Development* (Toronto: John Wiley).
Mardon, R. (1990) 'The State and the Effective Control of Foreign Capital', *World Politics* 43 (October 1990): 111-38.
Matthews, J.T. (1997) 'Power Shift', *Foreign Affairs* 76: 50-66.
Mazurek, J. (1999) *Making Microchips: Policy, Globalization, and Economic Restructuring in the Semiconductor Industry* (Cambridge, MA: MIT Press).
Meadows, D. (1972) *The Limits to Growth* (New York: Universe Books).
Mekani, K., and H. Stengel (1995) 'The Role of NGOs and Near NGOs', in O.G. Ling (ed.), *Environment and the City: Sharing Singapore's Experience and Future Challenges* (Singapore: Institute for Policy Studies/Times Academic Press).
Metzembaum, S. (1998) *Making Measurement Matter: The Challenge and Promise of Building a Performance-Focused Environmental Protection System* (Washington, DC: The Brookings Institution).

Murphy, R. (1996) *A Dependent Private Sector: No Prospects for Civil Society in China* (Murdoch, Western Australia: Asia Research Centre on Social Political and Economic Change, Murdoch University).
NAPA (National Academy of Public Administration) (1995) *Setting Priorities, Getting Results: A New Direction for the Environmental Protection Agency* (Washington DC: NAPA).
Nelson, K. (1994) 'Funding and Implementing Projects that Reduce Waste', in R. Socolow, C. Andrews, E. Berkhout and V. Thomas (eds.), *Industrial Ecology and Global Change* (Cambridge, UK: Cambridge University Press): 371-82.
New Straits Times (1998) *New Straits Times (Malaysia)*, 23 June 1998.
New York Times (1999) 'Asia Pollution Reaches West Coast', *New York Times*, 4 March 1999.
Nichols, P.M. (1996) 'Extension of Standing in the World Trade Organization Disputes to Non Governmental Entities', *University of Pennsylvania Journal of International Law* 17: 295.
NJDEP (New Jersey Department of Environmental Protection) (1996) *Evaluation of the Effectiveness of Pollution Planning in New Jersey: A Program-Based Evaluation* (Trenton, NJ: NJDEP, May 1996).
NIPR (New Issues in Pollution Reduction) (1999) (Washington, DC: World Bank; www.worldbank.org/nipr).
NISSD (Nautilus Institute for Security and Sustainable Development) (1999) 'Proposal for a Global Environment Facility (GEF) Technology Risk Guarantee Mechanism' (www.nautilus.org/papers/energy/index.html#fcc).
Noble, L.G. (1998) 'Environmental Activism', in G.S. Silliman and L.G. Noble (eds.), *Organizing for Democracy: NGOs, Civil Society and the Philippine State* (Honolulu, HI: University of Hawaii Press).
Norberg-Bohm, V. (1997) *Stimulating 'Green' Technological Innovation: An Analysis of Alternative Policy Mechanisms* (Cambridge, MA: Environmental Technology and Public Policy Programme, Massachusetts Institute of Technology).
Nortel (1997) *Fulfilling Our Commitments: A Progress Report on Environment, Health and Safety* (Brampton, ON: Nortel).
NRC (National Research Council, Board on Sustainable Development) (1999) *Our Common Journey: A Transition toward Sustainability* (Washington, DC: National Academy Press).
NRIISAS (Nomura Research Institute and Institute of Southeast Asian Studies) (1995) *The New Wave of Foreign Direct Investment in Asian* (Tokyo: NRIISAS).
NSF (National Science Foundation) (1993) *Human Resources for Science and Technology: The Asian Region* (Washington, DC: National Science Board).
NSF (National Science Foundation) (1996) *Asia's New High-Tech Competitors* (Washington, DC: National Science Board).
NSF (National Science Foundation) (1996) *Science and Engineering Indicators, 1996* (Washington, DC: National Science Board).
NSF (National Science Foundation) (1997) *Science and Technology Resources of Japan* (Washington, DC: National Science Board).
NSF (National Science Foundation) (1998) *Science and Engineering Indicators, 1998* (Washington, DC: National Science Board).
NSTC (National Science and Technology Council) (1994) *Technology for a Sustainable Future* (Washington, DC: NSTC).
Oates, W. (1993) 'Fiscal Decentralization and Economic Development', *National Tax Journal* 46.2 (June 1993): 237-43.
Oates, W.E., and R.M. Schwab (1988) 'Economic Competition among Jurisdictions: Efficiency Enhancing or Distortion Inducing?', *Journal of Public Economics* 5.3: 333.
O'Connor, D. (1994) *Managing the Environment with Rapid Industrialization: Lessons from the Asia Pacific Experience* (Paris: Organisation for Economic Co-operation and Development).
OECD (Organisation for Economic Co-operation and Development) (1992) *The OECD Environment Industry: Situation, Prospect, and Government Policies* (Paris: OECD).
OECD (Organisation for Economic Co-operation and Development) (1996a) *Promoting Cleaner Production in Developing Countries: The Role of Development Co-operation* (Paris: OECD).
OECD (Organisation for Economic Co-operation and Development) (1996b) *Pollutant Release and Transfer Registers: A Tool for Environmental Policy and Sustainable Development. Guidance Manual for Governments* (Paris: OECD).
OECD (Organisation for Economic Co-operation and Development) (1997) *Reforming Industrial Regulation in OECD Countries* (Paris: OECD)
Oldeman, L.R., R.T.A. Hakkeling and W.G. Sombrock (1990) *World Map of the Status of Human-Induced Soil Degradation* (Wageningen, Netherlands: International Soil Reference and Information Centre/ United Nations Environment Programme).

O'Neil, W. (1980) 'Pollution Permits and Markets for Water Quality' (PhD dissertation, University of Wisconsin, Madison, WI).
O'Neil, W. (1983) 'The Regulation of Water Pollution Permit Trading under Conditions of Varying Streamflow and Temperature', in E. Joeres and M. David (eds.), *Buying a Better Environment: Cost-Effective Regulation through Permit Trading* (Madison, WI: University of Wisconsin Press).
Ong, A.S.H., A. Maheswaran and A.N. Ma (1987) 'Malaysia', in L.S. Chia (ed.), *Environmental Management in Southeast Asia* (Singapore: National University of Singapore Press).
Ooi, G.L. (1998) *Environment and the City: Sharing Singapore's Experience and Future Challenges* (Singapore: Times Academic Press, 3rd edn).
Ooi, G.L., and G. Koh (1999) 'Achieving State–Society Synergies in Singapore: New Stakes, New Partnership' (Draft paper; Singapore: Institute for Policy Studies).
OSTA (Office of Science and Technology Advisors) (1995) *A Cleaner Home and a Better Image Abroad: Taiwan's Environmental Efforts* (Taipei: Taiwan Environmental Protection Agency).
Ostro, B. (1992) 'Estimating the Health and Economic Effects of Particulate Matter in Jakarta: A Preliminary Assessment' (Paper presented at the 4th Annual Meeting of the International Society for Environmental Epidemiology, 26–29 August 1992, Cuernavaca, Mexico/Policy Research Working Paper 1301; Washington, DC: World Bank).
Panayotou, T., and C. Zinnes (1994) 'Free Lunch Economics for Industrial Ecologists', in R. Scolow, C. Andrews, F. Berkhout and V. Thomas (eds.), *Industrial Ecology and Global Change* (Cambridge, UK: Cambridge University Press): 383-97.
Pangestu, M., and K. Roesad (1996) 'Experiences from Indonesia and other ASEAN Countries', in S. Tay and D. Esty (eds.), *Asian Dragons and Green Trade: Environment, Economics and International Law* (Singapore: Times Academic Press).
Pargal, S., and D. Wheeler (1995) 'Informal Regulation of Industrial Pollution in Developing Countries: Evidence from Indonesia' (Policy Research Working Paper 1416; Washington, DC: World Bank; www.nipr.org/work_paper/1416/index.htm).
PCSD (President's Council for Sustainable Development) (1996) *Sustainable America: A New Consensus* (Washington, DC: PCSD).
Pelczynski, Z.A. (1988) 'Solidarity and the Rebirth of Civil Society in Poland, 1976–81', in J. Keane (ed.), *Civil Society and the State* (London: Verso).
Pesapane, A. (1998) 'ISO 14000 and Environmental Cost Accounting: The Gateway to the Global Market', *Law and Policy in International Business* 29.50: 1.
Petri, P. (1995) 'The Interdependence of Trade and Investment in the Pacific', in E.K.Y. Chen and P. Drysdale (eds.), *Corporate Links and Foreign Direct Investment in Asia and the Pacific* (Sydney: Harper Educational).
Poffenberger, M. (1990) *Keepers of the Forest: Land Management Alternatives in Southeast Asia* (West Hartford, CT: Kumarian Press).
PORIM (Palm Oil Research Institute of Malaysia) (1986) 'Environmental Quality and Standard' (Paper prepared for 8th Palm Oil Mill Engineer/Executives Training Course, 25 April 1986; Bandar Baru Bangi, Malaysia: PORIM).
Porter, M. (1991) 'America's Green Strategy', *Scientific American* 264: 168.
Porter, M., and C. van der Linde (1995a) 'Toward a New Conception of the Environment–Competitiveness Relationship', *Journal of Economic Perspectives* 9.4: 97-118.
Porter, M.E., and C. van der Linde (1995b) 'Green and Competitive: Ending the Stalemate', *Harvard Business Review* 73.5 (September/October 1995): 120-34.
Princen, T., and M. Finger (1994) *Environmental NGOs in World Politics: Linking the Local and the Global* (London: Routledge).
Quah, S.K., K.H. Lim, D. Gillies, B.J. Wood and J. Kanagaratnam (1982) 'Sime Darby POME Treatment and Land Application Systems', in *Proceedings of Regional Workshop on Palm Oil Mill Technology and Effluent Treatment* (Palm Oil Research Institute of Malaysia [PORIM] Workshop Proceedings No. 4; Bandar Baru Bangi, Malaysia: PORIM).
Radelet, S., and J. Sachs (1997) 'Asia's Reemergence', *Foreign Affairs* 76: 44-59
Rausch, L.M. (1995) *Asia's New High Tech Competitors* (Washington, DC: National Science Foundation).
Reed, D. (1993) 'The Global Environment Facility and Non-Governmental Organizations', *American University Journal of International Law and Policy* 9.1: 19.
Reichert, W.M. (1996) 'Note. Resolving the Trade and Environment Conflict: The WTO and NGO Consultative Relations', *Minnesota Journal of Global Trade* 5: 219.
Reid, A. (1988) *South East Asia in the Age of Commerce 1450–1680* (New Haven, CT: Yale University Press).
Reinecke W. (2000) 'The Other World Wide Web: Global Public Policy Networks', *Foreign Policy*, Winter 1999/2000: 44-57.

Reiterer, M. (1996) 'The WTO's Committee on Trade and the Environment', in S. Tay and D. Esty (eds.), *Asian Dragons and Green Trade: Environment, Economics and International Law* (Singapore: Times Academic Press).
Resosudarmo, A.P., B.P. Resosudarmo and B. Isham (1997) 'The Indonesian Clean River Program (Prokasih) as Perceived by the People Residing along the River in Jakarta', *Indonesian Journal of Geography* 29.74: 47-64.
Resosudarmo, B.P. (1996) 'The Impact of Environmental Policies on a Developing Economy: Application to Indonesia' (Unpublished PhD dissertation; Ithaca, NY: Cornell University).
Revesz. R.L. (1992) 'Rehabilitating Interstate Competition: Rethinking the "Race-to-the-Bottom": Rationale for Federal Environmental Regulation', *New York University Law Review* 67: 1210.
Rhee, Y.H., G. Pursell and B. Ross-Larson (1984) *Korea's Competitive Edge: Managing Entry into World Markets* (Baltimore, MD: Johns Hopkins University Press).
Rigg, J. (1997) *Development in South East Asia* (London: Routledge).
Riker, J. (1995) 'Reflections on Government–NGO Relations', in N. Heyzer, J. Riker and A. Quizon (eds.), *Government–NGO Relations in Asia* (London: Macmillan): 194-96.
Rikhardsson, P. (1996) 'Developments in Danish Environmental Reporting', *Business Strategy and the Environment* 5.4 (December 1996): 269-72.
Rikhardsson, P. (1999) 'Statutory Environmental Reporting in Denmark: Status and Challenges', in M. Bennett and P. James (eds.), *Sustainable Measures: Evaluation and Reporting of Environmental and Social Performance* (Sheffield, UK: Greenleaf Publishing): 344-52.
Roberts, M., and J. Tybout (eds.) (1996) *Industrial Evolution in Developing Countries* (New York: Oxford University Press).
Rock, M.T. (1995) 'Thai Industrial Policy: How irrelevant was it to export success?', *Journal of International Development* 7.5: 745-57.
Rock, M.T. (1996a) 'Pollution Intensity of GDP and Trade Policy: Can the World Bank be wrong?', *World Development* 24: 471-79
Rock, M.T. (1996b) 'Toward More Sustainable Development: The Environment and Industrial Policy in Taiwan', *Development Policy Review* 14.3: 255-72.
Rock, M.T. (1999) 'Reassessing the Effectiveness of Industrial Policy in Indonesia: Can the neo-liberals be wrong?', *World Development* 27.4: 691-704.
Rock, M.T., D.P. Angel and T. Feridhanusetyawan (1999a) 'Industrial Ecology and Clean Development in Asia', *Journal of Industrial Ecology* 3: 29-42.
Rock, M.T., Y. Fei, and C. Zhang (1999b) China's Environmental Examination System: Has it made an environmental difference?' (Mimeograph; Arlington, VA: Winrock International).
Rodrik, D. (1997) *Has globalization gone too far?* (Washington, DC: Institute for International Economics).
Rodrik, D. (1999) 'The New Global Economy and Developing Countries: Making Openness Work' (Policy essay 24; Washington, DC: Overseas Development Council).
Roht-Arriaza, N. (1995) 'Shifting the Point of Regulation: The International Organization for Standardization and Global Lawmaking on Trade and the Environment', *Ecology Law Quarterly* 22: 479-539.
Roome, N. (1998) *Sustainable Strategies for Industry* (Washington, DC: Island Press)
Roos, D., F. Field and J. Neely (1997) 'Industry Consortia', in L. Branscombe and J.H. Keller (eds.), *Investing in Innovation* (Cambridge, MA: MIT Press).
Rose, C.M. (1994) *Property and Persuasion: Essay on the History, Theory, and Rhetoric of Ownership* (Boulder, CO: Westview Press).
Ruckelshaus, W.D. (1998) *The Environmental Protection System in Transition: Toward a More Desirable Future* (Final Report of the Enterprise for the Environment Project; Washington DC: Centre for Strategic and International Studies, January 1998).
Russell, M. (1990) 'Monitoring and Enforcement', in P.R. Portney (ed.), *Public Policies for Environmental Protection* (Washington, DC: Resources for the Future): 243-74
Sachs, A. (1995) *Eco-Justice: Linking Human Rights and the Environment* (Worldwatch Paper 127; Washington, DC: Worldwatch Institute).
Sachs, J., and A. Warner (1995) 'Economic Reform and the Process of Global Integration', in W. Brainard and G. Perry (eds.), *Brookings Papers on Economic Activity* (Washington, DC: The Brookings Institute): 1.
Salzhauer, A.L. (1991) 'Obstacles and Opportunities for a Consumer Eco-label', *Environment* 33.9: 11.
Salzman, J. (1997) 'Informing the Green Consumer: The Debate over the Use and Abuse of Environmental Labels', *Journal of Industrial Ecology* 1.2: 11.
Schmidheiny, S., and F. Zorraquín (1996) *Financing Change: The Financial Community, Eco-efficiency and Sustainable Development* (Cambridge, MA: MIT Press).

Schwartz, P., and B. Gibb (1999) *When Good Companies Do Bad Things* (New York: John Wiley).
SCOPE (Studies and Competence Centre for Organisational and Policy Research in European Business) (1999) *International Codes of Conduct and Corporate Social Responsibility* (Rotterdam: SCOPE, Erasmus University).
Seligman, A. (1992) *The Idea of Civil Society* (New York: Free Press).
Silliman, G.S., and L.G. Noble (1998) *Organizing for Democracy: NGOs, Civil Society and the Philippine State* (Honolulu, HI: University of Hawaii Press).
Sinnatamby, G. (1990) 'Low Cost Sanitation', in J.E. Hardoy, S. Cairncross and D. Satterthwaite (eds.), *The Poor Die Young* (London: Earthscan): 127-57.
Smart, B. (1992) *Beyond Compliance* (Washington, DC: World Resources Institute).
Smith, A, (1776) *An Inquiry into the Nature and Causes of the Wealth of Nations* (Edinburgh: W. Creech).
Socolow, R., C. Andrews, F. Berkhout and V. Thomas (1994) *Industrial Ecology and Global Change* (Cambridge, UK: Cambridge University Press).
Solidarity (1989) 'Our Threatened Heritage', in C. Fay (ed.), *Proceedings of the Solidarity Seminar* (Manila: The Philippines Solidarity Foundation).
Sonnenfeld, D.A. (1998) 'Social Movements, Environment, and Technology in Indonesia's Pulp and Paper Industry', *Asia Pacific Viewpoint* 39.1 (April 1998): 36-41.
Stavropoulos, W. (1996) 'Environmentalism's Third Wave: Managing for Global Competitiveness' (Address to the National Press Club, Washington, DC, 26 April 1996).
Steering Committee (1989) *Taiwan 2000: Balancing Economic Growth and Environmental Protection* (Taipei: Republic of China).
Steger, U. (1996) 'Managerial Issues in Closing the Loop', *Business Strategy and the Environment* 5.4 (December 1996): 252-68.
Steinzor, R.I. (1998) 'Reinventing Environmental Regulation', *The Harvard Environmental Law Review* 22: 103-202.
Stewart, R.B. (1977) 'Pyramids of Sacrifice? Problems of Federalism in Mandating State Implementation of National Environmental Policy', *Yale Law Journal* 86: 1196.
Stewart, R.B. (1992) 'International Trade and Environment: Lessons for the Federal Experience', *Lee Law Review* 49: 1315.
Stiglitz, J. (1998) 'Boats, Planes and Capital Flows', *Financial Times*, 25 March 1998: 32.
Straits Times (1997) *Straits Times*, 28 June 1997.
Sucharipa-Behrmann, L. (1994) 'Eco-labels for Tropical Timber: The Austrian Experience', in *Lifecycle Management and Trade* (Paris: Organisation for Economic Co-operation and Development).
Sunderlin, W.D. (1998) *Between Damage and Opportunity: Indonesia's Forests in an Era of Economic Crisis and Political Change* (Bogor, Indonesia: Centre for International Forestry Research).
Susskind, L.E. (1994) *Environmental Diplomacy: Negotiating More Effective Global Agreements* (New York: Oxford University Press).
Suryodipuro, L. (1995) 'Towards an Environmentally Desirable Urban Form: The Case of Jabotabek' (Honolulu, HI: Department of Urban and Regional Planning, University of Hawaii).
Svendson, A. (1998) *The Stakeholder Strategy* (San Francisco: Barrett-Koehler).
SWAPO (South West Organising Project) (1995) *Intel inside New Mexico: A Case Study of Environmental and Economic Injustice* (Albuquerque, NM: SWAPO).
Tam, T.K., H.K. Yeow and Y.C. Poon (1982) 'Land Application of Palm Oil Mill Effluent (POME): H&C Experience', in *Proceedings of Regional Workshop on Palm Oil Mill Technology and Effluent Treatment* (Palm Oil Research Institute of Malaysia [PORIM] Workshop Proceedings No. 4; Bandar Baru Bangi, Malaysia: PORIM).
Tan, H.A., and G. Batra (1995) 'Enterprise Training in Developing Countries: Incidence, Productivity Effects, and Policy Implications' (Washington, DC: Private Sector Development Department, World Bank).
Tay, S. (1997a) 'Trade and the Environment: Perspectives from the Asia–Pacific', *World Bulletin* 13.1-2: 1-22.
Tay, S. (1997b) 'Human Rights, Culture and the Singapore Example', *McGill Law Journal* 41.4: 743-80.
Tay, S. (1998a) 'Asia's Economic Crisis: Impact and Opportunities for Sustainable Development' (Special report to the 4th Asia–Pacific NGOs Environmental Conference, 26-27 November 1998; Singapore: Singapore Institute of International Affairs).
Tay, S. (1998b) 'Towards a Singaporean Civil Society', in D. Da Cunha and J. Furmston (eds.), *South East Asian Affairs* (Singapore: Institute for South East Asian Studies): 244-64.
Tay, S. (1999a) 'Globalization and Civil Society in the Asia–Pacific' (Paper presented at the Conference on Globalization and Security in the Asia–Pacific, Honolulu, HI, 22-25 February 1999; Singapore: Singapore Institute of International Affairs).

Tay S. (1999b) 'South East Asian Fires: The Challenge for International Law and Sustainable Development', *The Georgetown International Environmental Law Review* 11.2: 241-305.

Tay, S., and D. Esty (eds.) (1996) *Asian Dragons and Green Trade: Environment, Economics and International Law* (Singapore: Times Academic Press).

Tay, S., with G.C. Yen (1999) *The Asian Economic Crisis, Asian Values and Human Rights* (Report to the Canadian International Development Agency; Singapore: Singapore Institute of International Affairs).

TEI (Thailand Environment Institute) (1996) *Environmental Non-Governmental Organizations in Thailand* (Bangkok: Thailand Environment Institute).

TEPA (Taiwan Environmental Protection Agency) (1993) *State of the Environment* (Taipei: TEPA).

Thillaimuthu, J. (1978) 'The Environment and the Palm Oil Industry. A New Solution: The Incineration of Sludge', *Planter* 54: 228-36.

Tibor, T., with I. Feldman (1996) *ISO 14000: A Guide to the New Environmental Management Standards* (Burr Ridge, IL: Irwin).

Tiebout, C. (1956) 'A Pure Theory of Local Expenditures', *Journal of Political Economy* 64.5: 416.

Tietenberg, T.H. (1988) *Environmental and Natural Resource Economics* (Glenview, IL: Scott, Foresman & Co., 2nd edn).

Tietenberg, T.H. (1990) 'Economic Instruments for Environmental Regulation', *Oxford Review of Economic Policy* 6.1: 17-33.

Tismaneanu V. (1992) *Reinventing Politics: Eastern Europe after Communism* (New York: Free Press).

Tsong-Juh, C. (1994) 'A Review of Present Status of Laws and Regulations on Environmental Protection in the Republic of China', in *Taiwan 2000: Balancing Economic Growth with Environmental Protection* (Taipei: Taiwan Environmental Protection Agency): 438-41.

TURI (Toxics Use Reduction Institute) (1997) *Evaluating Progress: A Report on the Findings of the Massachusetts Toxics Use Reduction Program Evaluation* (Lowell, MA: TURI).

UNCSD (United Nations Commission on Sustainable Development) (1995) *Work Programme on Indicators of Sustainable Development* (New York: Division for Sustainable Development, UNCSD, November 1995).

UNCTAD (United Nations Conference on Trade and Development) (1995) *World Investment Report 1995* (Geneva: UNCTAD).

UNCTAD (United Nations Conference on Trade and Development) (1998) *World Investment Report* (Washington, DC: UNCTAD).

UNCTAD (United Nations Conference on Trade and Development) (1999) *World Investment Report 1999: Foreign Direct Investment and the Challenge of Development* (New York: UNCTAD).

UNDP (United Nations Development Programme) (1999) *Human Development Report 1998* (New York: Oxford University Press).

UNEP (United Nations Environment Programme) (1992) *Environmental Data Report* (New York: Blackwell).

UNEP (United Nations Environment Programme) (1994) *Environmental Data Report* (New York: Blackwell).

UNEP (United Nations Environment Programme) (1997) *Global Environment Outlook* (New York: Oxford University Press).

UNEP/GEMS (United Nations Environment Programme/Global Environment Monitoring Service) (1993) *Global Biodiversity* (Nairobi: UNEP/GEMS Environment Library II).

UNEP (United Nations Environment Programme)/SustainAbility (1996) *Engaging Stakeholders: The Second International Progress Report on Company Environmental Reporting* (London: UNEP/SustainAbility).

UNESCAP (United Nations Economic and Social Commission for Asia and the Pacific) (1993) *State of Urbanization in Asia* (Bangkok: UNESCAP).

UNESCAP (United Nations Economic and Social Commission for Asia and the Pacific) (1995) *State of the Environment in Asia and the Pacific* (Bangkok: UNESCAP).

UNESCO (United Nations Economic, Social and Cultural Organisation) (1995) *Statistical Yearbook, 1995* (Paris: UNESCO Publishing).

Unison (1995) *The Right Stuff: Using the Toxics Release Inventory* (Washington, DC: The Unison Institute).

United Nations (1992) *World Investment Directory 1992: Foreign Direct Investment, Legal Framework and Corporate Data. I. Asia and the Pacific* (New York: United Nations).

United Nations (1994) *World Population Prospectus: The 1994 Revision* (New York: United Nations).

USAEP (US–Asia Environmental Partnership) (1997) *Country Assessments* (Washington, DC: USAEP).

USDC (US Department of Commerce) (1997a) *International Plans, Policies, and Investments in Science and Technology* (Washington, DC: Office of Technology Policy, USDC).

USDC (US Department of Commerce) (1997b) *Patent and Trademark Office Review: Fiscal Year 1997* (Washington, DC: Patent and Trademark Office, USDC).

USDC (US Department of Commerce) (1997c) *International Science and Technology: Emerging Trends in Government Policies and Expenditures* (Washington, DC: Office of Technology Policy, USDC).

USDC (US Department of Commerce) (1998) *The US Environmental Industry* (Washington, DC: Office of Technology Policy, USDC).

USOTA (US Congress Office of Technology Assessment) (1994) *Industry, Technology, and the Environment: Competitive Challenges and Business Opportunities* (Washington, DC: USOTA).

USOTA (US Congress Office of Technology Assessment) (1995) *Environmental Policy Tools* (Washington, DC: USOTA).

USPIRGEF (US Public Interest Research Group Education Fund) (1997) *Outlook for 1997: A Full Year* (Working Notes on Community Right-To-Know; Boston, MA: USPIRGEF, January/February 1997).

Utterback, J. (1994) *Mastering the Dynamics of Innovation* (Boston, MA: Harvard Business School Press).

Victor, D.G., K. Raustiala and E.B. Skolnikoff (1997) *The Implementation and Effectiveness of International Environmental Commitments* (Cambridge, MA: MIT Press).

Vicusi, W.K. (1995) 'Equivalent Frames of Reference for Judging Risk Regulation Policies', *New York University Environmental Law Journal* 4.1: 437.

Vincent, J. (1993) *Reducing Effluent while Raising Affluence: Water Pollution Abatement in Malaysia* (Cambridge, MA: Harvard Institute for International Development).

Vincent, J.R., and Rozali Mohamed Ali (1997) *Environment and Development in a Resource-Rich Economy* (Cambridge, MA: Harvard University Press).

Vogler, J.R. (1995) *The Global Commons: A Regime Analysis* (Chichester, UK: John Wiley).

Vossenaar, R., and V. Jha (1996) 'Asian Perspectives on Competitiveness', in S. Tay and D. Esty (eds.), *Asian Dragons and Green Trade: Environment, Economics and International Trade* (Singapore: Times Academic Press).

VROM (Netherlands Ministry of Housing, Spatial Planning, and the Environment) (1997) *Environmental Policy of the Netherlands: An Introduction* (The Hague: VROM, 17 January 1997).

Wackernagel, M., and W. Rees (1996) *Our Ecological Footprint: Reducing Human Impact on the Earth* (Gabriola Island, BC: New Society Publishers).

Wade, R. (1990) *Governing the Market* (Princeton, NJ: Princeton University Press).

Wallace, G.J. (1997) 'Linked to Slavery', *Legal Times*, 1–8 December 1997: 25-26.

Walley, N., and B. Whitehead (1994) 'It's Not Easy Being Green', *Harvard Business Review* 72.3: 46.

Walzer, M. (ed.) (1995) *Toward a Global Civil Society* (Providence, RI: Berghahn Books).

Warnick, I., R. Herman, S. Govind and J. Ausubel (1996) 'Materialization and De-materialization: Measures and Trends', *Daedalus* 125.3: 171-98.

Watabe, A., and K. Yamaguchi (1996) 'Asian Structural Interdependency and the Environment', in P.R. Kleindorfer, H.C. Kunreuther and D.S. Hong (eds.), *Energy, Environment, and the Economy: Asian Perspectives* (Cheltenham, UK: Edward Elgar).

WCED (World Commission on Environment and Development) (1987) *Our Common Future* ('The Brundtland Report'; New York: Oxford University Press).

Webster, D. (1995) 'The Urban Environment in Southeast Asia: Challenges and Opportunities', in *Southeast Asian Affairs* (Singapore: Institute for South East Asian Studies).

Weiss, L. (1998) *State Capacity: Governing the Economy in a Global Era* (Cambridge, MA: Polity Press).

Wheeler, D., and S. Afsah (1996) 'Going Public on Polluters in Indonesia: BAPEDAL's PROPER PROKASIH Program' (East Asian Executive Reports, May 1996; www.nipr.org/work_paper/proper/index.htm).

Wheeler, D., and P. Martin (1992) 'Prices, Policies and the International Diffusion of Clean Technology: The Case of Wood Pulp Production', in P. Low (ed.), *International Trade and the Environment* (Washington, DC: World Bank): 197-224.

White, A., and D. Zinkl (1997) *Corporate Environmental Performance Indicators: A Benchmark Survey of Business Decision Makers* (Boston, MA: Tellus Institute).

Whitehead, L. (1997) 'Bowling in the Bronx: The Uncivil Interstices between Civil and Political Society', in R. Fine and S. Rai (eds.), *Civil Society: Democratic Perspectives* (London: F. Cass).

WHO (World Health Organisation) (1992) 'Our Planet, Our Health', *Environment and Urbanization* 4.1: 65-76.

WHO/UNEP (World Health Organisation/United Nations Environment Programme) (1992) *Urban Air Pollution in Megacities of the World* (Cambridge, MA: Basil Blackwell).

Williams, L. (1994) 'Asia's Urban Meltdown', *World Press Review* 41.2: 46-47.

Wood, B.J. (1977) 'A Review of Current Methods for Dealing with Palm Oil Effluents', *Planter* 53: 477-95.

Woolcock, M. (1997) 'Social Capital and Economic Development: Towards a Theoretical Synthesis and Policy Framework', *Theory and Society* 27.1: 1-57.
World Bank (1992) *World Development Report* (Washington, DC: World Bank).
World Bank (1993) *The East Asian Miracle: Economic Growth and Public Policy* (New York: Oxford University Press).
World Bank (1994a) *Indonesia: Environment and Development* (Washington, DC: World Bank).
World Bank (1994b) *World Development Report* (Washington, DC: World Bank).
World Bank (1997a) *World Development Indicators* (Washington, DC: World Bank).
World Bank (1997b) *1997 Country Reports: Indonesia* (Washington, DC: World Bank, May 1997).
World Bank (1997c) *Malaysia: Enterprise Training, Technology, and Productivity* (Washington, DC: World Bank).
World Bank (1997d) *Can the environment wait? Priorities for East Asia* (Washington, DC: World Bank).
World Bank (1998a) *East Asia: The Road to Recovery* (Washington, DC: World Bank).
World Bank (1998b) 'Environmental Implications of the Economic Crisis and Adjustment in East-Asia' (Memorandum; Washington, DC: World Bank, 1 July 1998).
World Bank (1998c) *Global Economic Prospects* (Washington, DC: World Bank).
World Bank (1998d) *World Development Indicators* (Washington, DC: World Bank)
World Bank (1999) *Greening Industry: New Roles for Communities, Markets and Governments* (New York: Oxford University Press)
WRI (World Resources Institute) (1994) *World Resources, 1994-95* (New York: Oxford University Press).
WRI (World Resources Institute) (1996) *World Resources, 1996-97* (New York: Oxford University Press).
WRI (World Resources Institute) (1997) *World Resources, 1997-98* (New York: Oxford University Press).
WRI (World Resources Institute) (1998) *World Resources, 1998-99* (New York: Oxford University Press).
Yamamoto, T. (1995) *Emerging Civil Society in the Asia Pacific Community* (Singapore: Institute of South East Asian Studies).
Yamamoto, T. (1999) *Deciding the Public Good: Governance and Civil Society in Japan* (Tokyo: Japan Centre for International Exchange).
Yusof Basiron and A.N. Ma (1992) 'Current Status of Environmental Management and Regulations in the Palm Oil Industry in Malaysia' (Unpublished manuscript; Bandar Baru Bangi, Malaysia: Palm Oil Research Institute of Malaysia).
Zarsky, L. (1997) 'Stuck in the Mud? Nation-States, Globalization and the Environment', in Organisation for Economic Co-operation and Development (OECD), *Globalization and the Environment, Preliminary Perspectives* (Paris: OECD): 27-52.
Zarsky, L. (1998) 'APEC, Globalization, and the "Sustainable Development" Agenda', *Asian Perspective* 22.2: 133-68.
Zarsky, L. (1999a) 'Havens, Halos and Spaghetti: Untangling the Evidence about Foreign Direct Investment and the Environment', in Organisation for Economic Co-operation and Development (OECD), *OECD, FDI and the Environment* (Paris: OECD).
Zarsky, L. (1999b) 'International Investment Rules and the Environment: Stuck in the Mud?', *Foreign Policy in Focus* 4.22 (August 1999).
Zarsky, L. (2000) 'Reflections on Seattle. US Arrogance and Incompetence: A Lethal Mix' (Nautilus Institute for Security and Sustainable Development; www.nautilus.org).
Zarsky, L., and J. Hunter (1999) 'Communities, Markets, and City Government: Innovative Roles for Coastal Cities in Reducing Marine Pollution in the Asia–Pacific Region', in T. Inoguchi, E. Newman and G. Paolettl (eds.), *Cities and the Environment: New Approaches to Eco-Societies* (Tokyo: United Nations University Press): 216-29.

ABBREVIATIONS

ADB	Asian Development Bank
AFTA	ASEAN Free Trade Agreement
AJEM	*Asia Journal of Environmental Management*
AMDAL	Analisis Mengenai Dampak Lingkungan, Indonesia
APEC	Asia–Pacific Economic Co-operation
APKINDO	Indonesian Plywood Association
ASEAN	Association of South-East Asian Nations
ASEAN-ISIS	ASEAN Institutes of Strategic and International Studies
BAPEDAL	Badau Pengendalian Dampak Lingkungan, Indonesia
BEC	Benchmark Environmental Consulting
BOD	biochemical oxygen demand
BPPT	Badan Pengkajiandan Penerapan Teknologi, Indonesia
BTU	British thermal unit
CAPD	Council for Agricultural Planning and Development, Taiwan
CEC	Commission for Environmental Co-operation, North America
CECODES	Consejo Empresarial Colombiano para el Desarrollo Sostenible
CEPD	Council for Economic Planning and Development, Taiwan
CERES	Coalition for Environmentally Responsible Economies
CFC	chlorofluorocarbon
CGCAP	California Global Corporate Accountability Project
CITES	Convention on International Trade in Endangered Species
CMA	Chemical Manufacturers' Association
CO	carbon monoxide
CO_2	carbon dioxide
COD	chemical oxygen demand
CPO	crude palm oil
$CuCl_2$	copper(II)chloride
DME	developing market economy
DOE	Department of the Environment, Malaysia
EDF	Environmental Defense Fund, USA
EFBE	Extel Financial and Business in the Environment
EFSG	Economic and Financial Special Group, Taiwan
EIA	Environment Investigation Agency, USA
EMAS	EU Eco-Management and Audit Scheme
EPA	US Environmental Protection Agency
EPI	environmental performance indicator
EPIndex	environmental performance index
EQA	Environmental Quality Act, Malaysia
ETI	Environmental Technology Initiative, USA
EU	European Union

FASB	Federal Accounting Standards Board, USA
FDI	foreign direct investment
FELDA	Federal Land Development Authority, Malaysia
FSC	Forest Stewardship Council
GAAP	generally accepted accounting principles
GATT	General Agreement on Tariffs and Trade
GDP	gross domestic product
GEMI	Global Environmental Management Initiative
GIS	geographical information system
GM	General Motors
GNP	gross national product
HIID	Harvard Institute for International Development
HPE	high-performing economy
ICC	International Chamber of Commerce
IDB	Industrial Development Bureau, Taiwan
IHDP	International Human Dimensions Programme
i.i.d.	independently and identically distributed
IISD	International Institute for Sustainable Development
IMF	International Monetary Fund
IRRC	Investor Responsibility Research Center, USA
ISO	International Organization for Standardization
ITRI	Industrial Technology Research Institute, Taiwan
KIP	Kampung Improvement Program, Indonesia
KMT	Kuomintang, Taiwan
KNCFH	Korean NGOs and CBOs Forum for Habitat II
MAI	Multilateral Agreement on Investment
MCA	marginal cost of abatement
MCCP	marginal cost of cleaner production
MEA	multilateral environmental agreement
MFN	most-favoured nation
MITI	Ministry of International Trade and Industry, Japan
MNC	multinational corporation
MOEA	Ministry of Economic Affairs, Taiwan
MRA	mutual recognition arrangement
MSC	Marine Stewardship Council
NAFTA	North American Free Trade Agreement
NAPA	National Academy of Public Administration, USA
NEP	New Economic Policy, Malaysia
NFI	National Federation of Industries, Taiwan
NGO	non-governmental organisation
NIC	newly industrialised country
NIE	newly industrialising economy
NIMBY	'not in my back yard'
NIPR	New Ideas in Pollution Reduction
NISSD	Nautilus Institute for Security and Sustainable Development
NJDEP	New Jersey Department of Environmental Protection
NO_2	nitrogen dioxide
NPO	non-product output
NPRI	National Pollutant Release Inventory, Canada
NRC	National Research Council, Board on Sustainable Development, USA
NRIISAS	Nomura Research Institute and Institute of Southeast Asian Studies, Japan
NSF	National Science Foundation, USA
NSTC	National Science and Technology Council, USA
O_3	Ozone
OECD	Organisation for Economic Co-operation and Development

ABBREVIATIONS

OLS	ordinary least squares
OSTA	Office of Science and Technology Advisors, Taiwan
PCB	polychlorinated biphenyl
PCP	pentachlorophenol
PCSD	President's Council for Sustainable Development, USA
PERI	Public Environmental Reporting Initiative
PG&E	Pacific Gas & Electric
PM10	particulate matter 10 microns in diameter and smaller
POME	palm oil mill effluent
PORIM	Palm Oil Research Institute of Malaysia
ppb	parts per billion
ppm	parts per million
PPM	production process or method
PRDEI	Policy Research Department, Environment and Infrastructure Division, World Bank
PROKASIH	Program Kali Bersih, Indonesia
PROPER	Program Penilaian Peringkat Kinerja Perusahaan, Indonesia
PRTR	pollutant release and transfer register
PSI	pollution standards index, Taiwan
R&D	research and development
RITE	Research Institute for Innovative Technology for the Earth, Japan
RRIM	Rubber Research Institute of Malaysia
SCOPE	Studies and Competence Centre for Organisational and Policy Research in European Business, Netherlands
SEC	Securities and Exchange Commission, USA
SME	small or medium-sized enterprise
SO_2	sulphur dioxide
SO_x	sulphur oxides
TEPA	Taiwan Environmental Protection Agency
TNC	transnational corporation
TRI	Toxic Release Inventory of US EPA
TSP	total suspended particulates
TURI	Toxics Use Reduction Institute, USA
UNCSD	United Nations Commission on Sustainable Development
UNCTAD	United Nations Conference on Trade and Development
UNDP	United Nations Development Programme
UNEP	United Nations Environment Programme
UNICEF	United Nations Children's Fund
USAEP	US–Asia Environmental Partnership
USAID	US Agency for International Development
USDC	US Department of Commerce
USPIRGEF	US Public Interest Research Group Education Fund
UTC	United Technologies Corporation
VROM	Netherlands Ministry of Housing, Spatial Planning, and the Environment
WBCSD	World Business Council for Sustainable Development
WCED	World Commission on Environment and Development
WHO	World Health Organisation
WRI	World Resources Institute
WTO	Word Trade Organisation
WWF	World Wide Fund for Nature

AUTHORS' BIOGRAPHIES

Shakeb Afsah is senior manager at the International Resources Group in Washington, DC. Mr Afsah specialises in designing and implementing environmental benchmarking and performance analysis systems for corporations and industrial enterprises. He also provides consulting services to corporations on environmental reputation assurance and optimal disclosure management. Before joining IRG, Mr Afsah worked at the World Bank for six years and served as resident advisor to the ministers of environment in Indonesia and the Philippines and was responsible for implementing public disclosure programmes in these countries.

David P. Angel holds the Leo L. and Joan Kraft Laskoff professorship in economics, technology and the environment at Clark University, Worcester, MA, USA, where he is also the Dean of Graduate Studies and Research. His background and training are in economic geography, focusing on issues of industrial and technological change. Dr Angel is currently involved in several research and policy projects concerning globalisation, industrial change and the environment in Asia, North America and Eastern Europe.

Owen Cylke is co-ordinator of the policy group for the US–Asia Environmental Partnership and senior advisor at Winrock International. Mr Cylke is a graduate of Yale University and Yale Law School. Earlier, Mr Cylke was a member of the US Senior Foreign Service at the Agency for International Development, 1966–89, serving, among other assignments, as director of the US Economic Assistance Mission to India and retiring with the rank of career minister. From 1990–93 he served as president of the Association of Big Eight Universities, and from 1993–97 he served as senior fellow at the Tata Energy & Resources Institute (New Delhi and Washington, DC). Mr Cylke's area of professional interest is international development, specifically the interface between economic growth and the environment.

Daryl Ditz is currently Director of Environmental Management Programs at the Environmental Law Institute, a non-profit centre for research, publication and training in Washington, DC. Previously, he worked with Janet Ranganathan on issues involving corporate management and public policy in the Technology Program at the World Resources Institute.

Mike Douglass is Professor of Urban and Regional Planning at the University of Hawaii. He has been engaged in research and practice on urban and regional development issues in Asia for many years. His recent books include *Cities for Citizens? Planning and the Rise of Civil Society in a Global Age* (John Wiley, 1999).

Daniel C. Esty is Professor of Environmental Law and Policy at the Yale Law School and the Yale School of Forestry and Environmental Studies. He is also Director of the Yale Center for Environmental Law and Policy and Associate Dean of the Yale School of Forestry and Environmental Studies. Dan is the author or editor of four books and numerous articles on environmental policy issues and the relationships between the environment and trade, security, competitiveness, international institutions and development. His recent books include: *Thinking Ecologically: The Next Generation of Environmental Policy* (ed. with M. Chertow; Yale University Press, 1997) and *Sustaining the Asia Pacific Miracle: Environmental Protection and Economic Integration* (with A. Dua; Institute for International Economics, 1997).

J. Warren Evans is Manager of the Environment Division at the Asian Development Bank in Manila, Philippines. Established in 1966, the Asian Development Bank is a multilateral development finance institution dedicated to reducing poverty in Asia and the Pacific.

Tubagus Feridhanusetyawan is a Senior Economist at the Centre for Strategic and International Studies, Jakarta. He is also a lecturer at the graduate programme, Faculty of Economics, University of Indonesia, Jakarta. His fields of research are labour economics, international economics, econometrics, general equilibrium modelling and applied macroeconomics. He has served as an economic consultant for various international organisations, and has published in several journals, reports and other internationally circulated publications.

George R. Heaton, Jr teaches at the Worcester Polytechnic Institute in Massachusetts, and consults widely on technology and environmental policy. Trained as a lawyer, Mr Heaton's career in university and government service has combined teaching and research on the relationship between public policy and technological change in industry. His strong interest in Asia has led to university and government appointments in Japan, as well as assignments for the World Bank and the USAEP in China and elsewhere.

Somporn Kamolsiripichaiporn is Deputy Director of the Environmental Research Institute, Chulalongkorn University, Thailand, and is a member of the Biochemistry Department, Faculty of Science. Since 1996 she has been involved in establishing an industry–university co-operative research centre for hazardous waste management, which is a collaboration of five universities in Thailand and includes the development of an international course on environmental management. She is the co-ordinator of Greening of Industry Network-Asia (GIN-Asia).

Khalid Abdul Rahim is Associate Professor in the Faculty of Economics and Management at Universiti Putra Malaysia. His area of expertise is environmental economics.

Victor J. Kimm, a former Deputy Assistant Administrator at the US Environmental Protection Agency, serves as Distinguished Practitioner in Residence on the faculty of the University of Southern California's Washington-based Graduate School of Policy, Planning and Development. Since retirement from the federal service in the summer of 1998, he has begun a consulting practice dealing with complex environment management problems, largely in developing countries.

Ooi Giok Ling is Senior Research Fellow at the Institute of Policy Studies and, concurrently, Associate Professor (Adjunct) at the National University of Singapore. She was Director of Research in the Ministry of Home Affairs in Singapore. Giok Ling is a member of Singapore's Environment Council. She has been consultant to the United Nations and other international organisations. Her research and publications focus on the environment, housing and urban studies, including local governance issues and ethnicity and healthcare in third-world development.

Mari Elka Pangestu is a member of the board of directors of the Centre for Strategic and International Studies, Jakarta. She is also a lecturer at various universities in Indonesia, serves on various international boards, and is a consultant to international organisations including the World Bank and the Asian Development Bank. Dr Pangestu has published widely in Indonesian and international journals and media. She received her PhD from the Department of Economics, University of California-Davis in 1986.

Janet Ranganathan is a Senior Associate in the Management Institute for Environment and Business at the World Resources Institute in Washington, DC, where she works on the development and implementation of corporate accountability tools for measuring and driving progress toward more sustainable business activities. She is co-manager of a major collaborative effort to develop an international standard for measuring and reporting business greenhouse gas emissions. In addition, she works extensively with business and other stakeholder groups on the development of business sustainability metrics.

Budy P. Resosudarmo received a Sarjana (bachelor's) degree in Electrical Engineering from the Bandung Institute of Technology, a master's degree in Operations Research from the University of Delaware, and a doctoral degree in Agricultural Economics from Cornell University. Currently, he is a researcher at the Indonesian Government Agency for the Assessment and Application of Technology and at the Inter-University Center, Economics, University of Indonesia. His research focuses on the impact of development policies on people and the environment.

Michael T. Rock is Professor of Economics and Chair of the Department of Economics and Management at Hood College in Frederick, MD, USA. Prior to joining Hood College, Dr Rock was the senior economist of the Winrock International Institute for Agricultural Development, a development NGO established by the Rockefeller family. Rock's published research focuses on trade and the environment, deforestation in poor countries, the environmental behaviour of manufacturing plants in the first- and second-tier newly industrialised countries (NICs) of East Asia, water policy and development and the role of the state in the development of the South-East Asian second-tier NICs.

Rozali Mohamed Ali is Executive Director of Commerce, Asset-Holding Berhad, Malaysia. He was previously head of the Bureau of Science, Technology, Energy, Natural Resources and the Environment and of the Center for Environmental Studies at the Institute of Strategic and International Studies in Malaysia.

Melito S. Salazar, Jr has held various positions in government and industry in the Philippines, including Governor of the Board of Investments and Undersecretary of the Department of Trade and Industry. Since July 1999 he has been Director of the Monetary Board of the Bangko Sentral ng Pilipinas. Mr Salazar is the only Filipino to receive the World Association of Small and Medium Enterprise Special Honor Award. Presently he is also on the Board of Advisers of the Greening of Industry Network, Adviser to the Chamber of Commerce of the Philippines Foundation and a Trustee of the Finex Foundation and the Luen Thai Philippines Foundation.

Hadi Soesastro is the Executive Director of the Centre for Strategic and International Studies, Jakarta; and a member of the National Economic Council, President Abdurrahman Wahid's advisory board on economic matters. Dr Soesastro is also a lecturer at various universities in Indonesia and an adjunct professor at The Australian National University, Canberra. His research interest is on issues of globalisation, the links between economics and security, regional economic co-operation, energy as well as national development.

Simon S.C. Tay LLB Hons (National University of Singapore), LLM (Harvard) teaches international and constitutional law at the Faculty of Law, National University of Singapore. He is a publicly nominated Member of the Singapore Parliament and also Chairman of the Singapore Institute of International Affairs. His work focuses on the environment, human rights and civil society in Asia. A Fulbright scholar, he won the Laylin Prize at the Harvard Law School for the best thesis in international law

Jeffrey R. Vincent is a Fellow at the Harvard Institute for International Development. He is an environmental economist with research interests related to tropical forestry policy, national income accounts and the environment, and economics-based approaches to pollution management in developing countries.

Lyuba Zarsky is Program Director of the Nautilus Institute for Security and Sustainable Development, an innovative research and advocacy group based in Berkeley, CA. The Institute works to build global institutions and norms that enhance accountability and promote human rights, peace and sustainable development. An economist by academic training, she has written widely on international trade, investment and sustainable development. She is currently co-directing the California Global Corporate Accountability Project (www.nautilus.org).

INDEX

Air pollution 158
  in Indonesia 25
  in Taiwan 199, 205
Air quality 13, 106
  standards 91
Asian financial crisis 8, 22-23, 25, 66, 80, 113, 128-35, 157
  environmental impacts of 131-35
Asia–Pacific Economic Co-operation (APEC) 15, 22, 32, 40, 66-68, 74, 85-86, 134
Asia–Pacific Environmental Summit 120
Asian Development Bank (ADB) 11, 13, 132, 246, 250
Asian economic miracle 8, 50, 129, 132, 134, 157, 247
Association of South-East Asian Nations (ASEAN) 15, 22, 29, 66, 70, 86, 100, 131, 134, 146
  AFTA 66, 134
  Haze Action Plan 70, 71
  ISIS 146

B&Q 223
Bandung 116, 118
Bangladesh 53, 103
BAPEDAL 102, 156, 159-72, 236
Basic needs 104
Best available technology 92
Biochemical oxygen demand (BOD) 11, 25, 158-60, 178-91
Body Shop 223
Brundtland Report 9
Brunei 71, 131
Burma
  see Myanmar

Cambodia 88, 96, 100, 103, 131
Capacity-building
  localisation of 109-11
CFCs (chlorofluorocarbons) 44, 78-79, 205
Chaebols 98, 110

Chemical Manufacturers' Association (CMA) 32
  Responsible Care 19, 32, 102
China 13, 18, 20-21, 23, 27, 29, 36, 53, 65, 68, 70, 77, 80-81, 88, 96-97, 100, 139-40, 142, 146, 148, 206, 208
China Petroleum Company 200, 204
China Steel 200, 204
Chung Hwa Pulp Corporation 200
Civil society 22, 111, 123, 128-54
  concepts of 135-38
Cleaner production 89, 94, 97
  see also Pollution prevention
Cleaner technology 20, 26
Climate change 27
Coalition for Environmentally Responsible Economies (CERES) 32, 217, 224, 243
Collaborative governance 33, 111-20, 124
Colombian Business Council for Sustainable Development (CECODES) 225, 243
Commission for Environmental Co-operation (CEC) 232
Common Sense Initiative 242
Conditional adjustment programmes 79-80
Côte d'Ivoire 151

De Tocqueville, A. 136
Developing market economies (DMEs) 88, 99
Developmental state 108
Discharge limits 92

Earth Summit 162
Eco-labels 30, 32, 82-83, 103, 152, 207, 225

Eco-Management and Audit Scheme (EMAS) 82, 211, 217, 236-37
Ecological footprint 43, 213
Economic crisis
  see Asian financial crisis
Economic miracle
  see Asian economic miracle
Electroplating 96, 103, 204-206
End-of-pipe technology 19-20, 26, 28, 35-36, 39, 50, 57, 62, 89-101, 115, 159, 215, 248
Energy intensity 15, 19, 23-34
Energy and materials efficiency 27, 225
Energy prices 58
Environmental performance 29, 95, 121, 209-45
  information on 15, 30, 59, 84, 125-27, 225, 231
  comparability of 222
  disclosure 152, 162-63, 210
  indicators (EPIs) 209-45
  in the Netherlands 229-30
  PROPER 157-72
  non-regulatory drivers 36
Environmental policy 10, 12, 28-31, 40
  in Indonesia 157-72
  integration 29
  in Malaysia 173-93
  non-regulatory drivers 80-85
  in OECD countries 17, 57, 62, 91
  in Taiwan 194-208
Environmental regulations
  codes of conduct 91, 103, 150-52, 243
  command and control 57, 92-95, 100, 112, 143, 148, 158, 189
  effluent charges 180
  in Malaysia 177-87

INDEX

market-based instruments 12, 35-36, 93-95, 100, 152, 249
marginal abatement costs 89, 185-87, 189-93
polices 28-31, 91-95
private-law model 32
self-regulation 149-52
systems 12, 24
in Taiwan 199
Environmental Defense Fund 234
Environmental industry 46
Environmental infrastructure 113
Environmental Protection Agency (EPA), USA 74, 198, 202, 225, 229-31, 234, 242-44
33/50 programme 159, 229-30
Project XL 242
Environmental security 130
Environmental standards 72-74, 82
harmonisation of 73-74
Environmental technology 42, 46-50, 54-56
in Japan 49, 61
in OECD countries 47
R&D 47
in the US 49
see also Cleaner technology
Environmental Technology Initiative (ETI) 49
EPIs
see Environmental performance

Federal Accounting Standards Board (FASB), USA 84
Federal Land Development Authority (FELDA), Malaysia 175, 183
Foreign aid 80
Foreign direct investment (FDI) 14, 54-55, 64-65, 75-78, 133-34, 138, 140, 151, 197
Forest Stewardship Council (FSC) 82
Framework Convention on Climate Change 32, 132

General Agreement on Tariffs and Trade (GATT) 32, 40, 65, 83, 162
Uruguay Round 65-66, 162
General Motors (GM) 217, 222-24, 241
Generally accepted accounting principles (GAAP) 211, 215

Geographical information system (GIS) 124
Global Environment Facility 148
Global Environmental Management Initiative (GEMI) 32
Global Reporting Initiative (GRI) 217
Globalisation 9, 13-16, 20-22, 35, 63-87, 140, 249
and civil society 140-43
and FDI 64-65
Governance 22, 128-54, 249-50
and civil society 34
public–private partnerships 148-49
stakeholders 147-49
strategic engagement 152-53
Green design 58, 60-61
Green labelling
see Eco-labels
Greenhouse gas emissions 12-13, 16, 24, 33, 68, 215-18, 220, 223-24, 238
Greening of Industry Network 8-9, 250
Greenpeace 141

Hewlett-Packard 150, 223
High-performing economies (HPEs) 194, 197
Hong Kong 36, 58, 67, 88, 100, 103, 110, 129, 194
Human rights 129, 137-42, 150

ICI 221-22
India 27, 53, 103, 145
Indonesia 13, 17-29, 34, 36, 51-55, 59, 67-68, 70-71, 80-82, 88, 96-102, 105, 111, 117, 129, 131, 138, 144, 146, 156, 157-72, 194, 197, 236-38
Industrial Development Bureau (IDB), Taiwan 101, 195, 196, 199, 201-208
Industrial-led development 17
Industrial policy 16, 30-31, 41-62, 96-99
in Taiwan 194-208
Industrial Technology Research Institute (ITRI), Taiwan 101, 196, 202-203, 207
Industrial transformation 9
Integrated fertiliser/bio-gas recovery system 182-83
Intel 150
Inter-city networks 109, 120-22

International Chamber of Commerce (ICC) 32
International Monetary Fund (IMF) 14, 34, 80, 81, 128, 131, 133
International Organization for Standardization (ISO) 82, 149-50
ISO 14000 series 19, 30-32, 35, 77, 82, 99, 102-103, 198, 202, 207-208, 211-13, 236
ISO 14001 149-50, 152, 236-38
ISO 14031 213, 238
in Taiwan 207
Investor Responsibility Research Center (IRRC) 219

JAGANUSA 160, 163, 165-66
Japan 23, 47-57, 60-61, 64-65, 67-68, 105, 108, 113, 137, 139, 145, 194-95, 197, 204, 231
Jati Dua 118-20

Kampung Improvement Program (KIP), Indonesia 117
Korea 18, 21, 30-31, 52-55, 61, 67, 80, 88, 96, 98, 100-103, 105, 108, 110-11, 113, 123, 129, 194-95
Kuomintang (KMT) 195

Laos People's Democratic Republic 88, 96, 100, 103, 131
Learning-by-doing 98
Levi-Strauss 150
Life-cycle analysis 30, 32, 82-83, 202, 215, 237, 241

Malaysia 13, 17-18, 23-24, 27, 36, 51, 53-55, 66, 71, 88, 96, 98, 100-102, 129, 131-32, 138, 146, 156, 173-93, 194
Environmental Quality Act 1974 177-78
Marine Stewardship Council (MSC) 82
Market-based instruments
see Environmental regulations
Materials intensity 15, 19, 23-34
McWorld 140
Mega urban regions 105, 112, 120
Mexico 151
Ministry of International Trade and Industry (MITI), Japan 60

Montreal Protocol   32, 79, 132
Morocco   151
Most-favoured-nation (MFN)   66
Multilateral Agreement on Investment (MAI)   78
Multilateral environmental agreements (MEAs)   78-79
Multinational corporations (MNCs)   31-32, 59-60, 75, 84, 99, 128, 147
Myanmar (Burma)   88, 113, 131, 140, 150

Newly industrialised countries (NICs)   13, 17-31, 36, 55
NIMBY ('not in my back yard')   115, 206
Non-governmental organisations (NGOs)   34, 70, 84, 101, 111, 120, 128, 135, 138-41, 144-46, 153, 162, 165, 249
Asian   135, 147
Nortel   220-21, 229
North American Free Trade Agreement (NAFTA)   14, 66, 74, 84

OECD (Organisation for Economic Co-operation and Development)   8-9, 11-31, 39-42, 46-50, 54-62, 78, 85, 91, 93, 96, 98, 100, 102-103, 149-150, 151-53, 197, 212-13, 249

Pacific Gas & Electric (PG&E)   238
Palm oil   173-93
Palm oil mill effluent (POME)   173, 176-77
Palm Oil Research Institute of Malaysia (PORIM)   182
Patents   53-54
Philippines   13, 17, 18, 21, 23-24, 27, 36, 54-55, 64, 88, 96-97, 100-101, 111, 113, 129, 131, 145, 151, 153, 236
Pollutant release and transfer registers (PRTRs)   231, 233
Pollution intensity   15, 19, 23, 25-26, 42-43, 89-91, 94, 197
in Taiwan   204
see also Energy intensity/Materials intensity
Pollution control   195
Pollution havens   151
see also Race to the bottom
Pollution loads   173
Pollution prevention   93-94, 225
see also Cleaner production

Polychlorinated biphenyls (PCBs)   43
Poverty   13, 39, 63, 79-80, 85, 103-104, 111, 131-33, 137, 145, 157, 175, 177, 246-48
President's Council for Sustainable Development (PCSD)   235
Printed circuit board industry   204, 215
Private capital   19, 32, 248
PROKASIH   158-60, 163, 165-72
PROPER   36, 102, 156, 157-72, 169, 236, 238
Public policy   88-103
environmental policy   35

R&D   44, 46, 52-53, 227
education   52
in Malaysia   178
Race to the bottom   72, 151
see also Pollution havens
Race to the top   151
Reebok   150
Regional co-operation   67, 70
Research Institute for Innovative Technology for the Earth (RITE)   49, 61
Resource depletion   27
Resource pricing   58
Responsible Care
see Chemical Manufacturers' Association
Right-to-Know Network   234
Rubber Research Institute of Malaysia (RRIM)   182

Semiconductors   215
Silent Spring   9
Singapore   18, 21, 23, 36, 51-54, 61, 67, 71, 88, 96, 100-103, 129, 131-32, 146, 194
Small and medium-sized enterprises (SMEs)   31, 35, 99, 103, 133, 147, 171, 202
South Korea   23, 36-37, 64, 68, 129, 145, 194
Sri Lanka   103
Students   52-53
Sulphur dioxide   68, 93, 158, 205
Sun Microsystems   223
Supply chains   19, 23, 30, 35, 99, 102-103, 215, 222-25, 241
Sustainable development/Sustainability   9-10, 12, 15-16, 19, 39-40, 41-62, 63, 68, 74-76, 80, 86, 88, 130, 134, 147, 153, 246-48
Tacit knowledge   98
Taipower   200, 204

Taiwan   23, 30-31, 36-37, 51-55, 67, 88, 96, 98, 100-105, 111, 113, 129, 132, 145, 194-208
Taiwan Environmental Protection Agency (TEPA)   198-200, 202, 204, 207
Taiwan VCM Corporation   200
Technology   41-62
catch-up   17
diffusion   45-46
change   43
innovation   44, 58
OECD   50
policy   41-62, 96-99
technological learning   100
transfer   45
transformation   57-61
Thailand   13, 17, 18, 20-21, 23-24, 27, 36, 51, 53-55, 68, 80, 88, 96, 98, 100-101, 111, 113, 129, 131, 137-38, 144-45, 194, 241
Toxic Release Inventory (TRI), USA   59, 84, 217, 226, 228-34, 243
Trade
agreements   72-85
APEC   74-75
and environmental standards   72-74
and investment policy   31-33
free-riders   78
intra-Asian   64-65
policy   63-87
Transboundary environmental problems   68-70, 146
greenhouse gas emissions   68
sulphur dioxide   68

United Nations Children's Fund (UNICEF)   138
United Nations Development Programme (UNDP)   142
United Nations Environment Programme (UNEP)   85
United Technologies Corporation (UTC)   222
Urban collaborative planning   117-18
Urban–industrial investment   19, 23
Urban infrastructure   112
Urban land use   117
Urban policy   33-34, 104-27
urban environmental management   122
Urban population   13, 20, 104-105, 112, 248
Urban solid waste   116, 118-20
Urban transition   105-107
US–Asia Environmental Partnership   8, 77, 250

Values, Asian   139
Venezuela   151
Vietnam   27, 88, 96, 100, 103, 118, 129, 131, 206

Waste collection   116
Waste intensity   15, 19, 23
Wat Chonglom   117
Water pollution   11, 26, 106
  in Indonesia   158-60
  in Malaysia   173-93
  in Taiwan   206

Water quality standards   91
Win–win opportunities   19, 94, 96
World Bank   13, 19, 23, 25-27, 34, 80, 96, 129, 133, 151, 158-60, 163, 165, 171-72, 174, 208, 236
World Business Council for Sustainable Development (WBCSD)   32, 150
World Health Organisation (WHO)   88, 158

World Trade Organisation (WTO)   15, 29, 32, 67, 74-75, 78, 84-86, 134, 141-42
  Seattle meeting   74, 142
World Wide Fund for Nature (WWF)   141, 144